T0140836

Julian Berberich

Stability and robustness in data-driven predictive control

Logos Verlag Berlin

λογος

Bibliographic information published by the Deutsche Nationalbibliothek

The Deutsche Nationalbibliothek lists this publication in the Deutsche Nationalbibliografie; detailed bibliographic data are available on the Internet at http://dnb.d-nb.de

D93

ISBN 978-3-8325-5531-3

Logos Verlag Berlin GmbH
Georg-Knorr-Str. 4, Geb. 10,
D-12681 Berlin
Germany

Tel.: +49 (0)30 / 42 85 10 90
Fax: +49 (0)30 / 42 85 10 92
http://www.logos-verlag.de

Stability and robustness in data-driven predictive control

Von der Fakultät Konstruktions-, Produktions- und Fahrzeugtechnik der Universität Stuttgart zur Erlangung der Würde eines Doktor-Ingenieurs (Dr.-Ing.) genehmigte Abhandlung

Vorgelegt von

Julian Berberich

aus Würzburg

Hauptberichter: Prof. Dr.-Ing. Frank Allgöwer
Mitberichter: Prof. Dr. Florian Dörfler
Prof. Giancarlo Ferrari Trecate, PhD

Tag der mündlichen Prüfung: 02.06.2022

Institut für Systemtheorie und Regelungstechnik

Universität Stuttgart

2022

Acknowledgements

I would like to thank many people who have supported me during the last years, both professionally and personally. Research is a team effort and I would not have gotten to this point without them.

First and foremost, my sincerest gratitude belongs to my PhD advisor Prof. Frank Allgöwer. At the Institute for Systems Theory and Automatic Control (IST) of the University of Stuttgart, he has created an extraordinary research environment, in which I had the privilege to work during the last four years. I am grateful to Frank for the invaluable freedom, for the insightful discussions, for allowing me to attend many international conferences, for the excellent professional and personal advice, and for being a true role model.

I would like to thank Prof. Florian Dörfler and Prof. Giancarlo Ferrari Trecate for their interest in my work, for their valuable comments, and for being members of my examination committee. A special thanks goes to Florian for hosting me in Zürich for a 3-month research stay, and to all IfA members for their great hospitality and the fruitful interactions.

I am very grateful to the *MPC dream team*, Johannes Köhler and Prof. Matthias Müller, for the long-term collaboration which greatly contributed to the results in this thesis. I would like to thank Matthias for his ongoing support and guidance, starting with my first research project as a student up to the collaboration on data-driven MPC. I would like to thank Johannes for teaching me how to develop an intuition about MPC and for always knowing the right hammer when I was stuck with a problem.

I would also like to thank Prof. Carsten Scherer for the fruitful collaboration and for teaching me a lot, not only about robust control theory but also mathematics in general and how to write concise papers.

I would like to thank my colleagues at the IST for making my time there not only productive but also enjoyable, and for being not only colleagues but also friends.

Special thanks go to the following colleagues for the excellent collaboration: Dennis Gramlich, Michael Hertneck, Anne Koch, Matthias Köhler, Patricia Pauli, Lukas Schwenkel, Robin Strässer, Janani Venkatasubramanian, Stefan Wildhagen. I would also like to thank the administrative staff at the IST for their support.

I am thankful to my external collaborators for the very interesting and productive collaboration, in particular: Mohammad Alsalti, Prof. Jie Chen, Christian Klöppelt, Victor Lopez, Prof. Jian Sun, Prof. Mario Sznaier, Prof. Roland Tóth, Chris Verhoek, Prof. Gang Wang, Xin Wang. Moreover, I would like to thank my former students for the valuable interactions: Joscha Bongard, Fabian Jakob, Said Jamal Mohamad, Yujia Wang, Nils Wieler, Zhibo Zhou. Finally, I would like to thank the International Max Planck Research School for Intelligent Systems (IMPRS-IS) for their support.

I would like to thank my friends and especially my band *Aalbachtal Express* for making my life beyond research fun.

I am most thankful to my family for their endless support: to my brothers Roman and Benni and their families for showing me that being an uncle can be more fun than doing research, and to my parents for enabling me to pursue this path and for always being there for me.

Finally, I will be eternally grateful to you, Alisa, for your patience, for your big heart, and for always being my best adviser.

Kirchheim am Neckar, June 2022
Julian Berberich

Table of Contents

Notation

In the following, we list the main symbols and acronyms used in this thesis. Additional notation is defined in the corresponding sections.

Abbreviations and acronyms

CSTR	continuous stirred tank reactor
DeePC	data-enabled predictive control
IOSS	input-output-to-state stability
LICQ	linear independence constraint qualification
LMI	linear matrix inequality
LP	linear program
LTI	linear time-invariant
LTV	linear time-varying
MPC	model predictive control
QP	quadratic program

Integers, real numbers, and sets

$\mathbb{I}_{\geq 0}$	set of nonnegative integers
$\mathbb{I}_{>0}$	set of positive integers
$\mathbb{I}_{\geq n}$	set of integers larger than or equal to $n \in \mathbb{I}_{\geq 0}$
$\mathbb{I}_{[a,b]}$	set of integers in the interval $[a, b]$
$\mathbb{R}$	set of real numbers
$\mathbb{R}_{\geq 0}$	set of non-negative real numbers
$\mathcal{K}_\infty$	class of functions $\alpha : \mathbb{R}_{\geq 0} \to \mathbb{R}_{\geq 0}$ which are continuous, strictly increasing, unbounded, and satisfy $\alpha(0) = 0$
$\text{int}(X)$	interior of a set $X \subseteq \mathbb{R}^n$

Sequences and corresponding vector and matrix structures

$\{x_k\}_{k=0}^{N-1}$	sequence of length $N \in \mathbb{I}_{>0}$ with values $x_k \in \mathbb{R}^n$, $k \in \mathbb{I}_{[0,N-1]}$
$x_{[a,b]}$	column vector containing the entries of the sequence $\{x_k\}_{k=0}^{N-1}$ with indices $\mathbb{I}_{[a,b]}$ for $a \leq b$, i.e., $x_{[a,b]} := \begin{bmatrix} x_a^\top & x_{a+1}^\top & \dots & x_b^\top \end{bmatrix}^\top$
x	column vector containing all entries of $\{x_k\}_{k=0}^{N-1}$, i.e., $x := x_{[0,N-1]}$
$H_L(x)$	Hankel matrix containing shifted versions of $\{x_k\}_{k=0}^{N-1}$, i.e.,

$$H_L(x) := \begin{bmatrix} x_0 & x_1 & \dots & x_{N-L} \\ x_1 & x_2 & \dots & x_{N-L+1} \\ \vdots & \vdots & \ddots & \vdots \\ x_{L-1} & x_L & \dots & x_{N-1} \end{bmatrix}$$

Vectors, matrices, and norms

I_n	identity matrix of dimension $n \times n$
$0_{n\times m}$	$n \times m$ matrix of zeros
	(indices of I_n and $0_{n\times m}$ are omitted when the dimensions are obvious)
$\mathbb{1}_n$	n-dimensional column vector with all entries equal to 1
$\star$	matrix entries which can be inferred from symmetry
$A^\dagger$	Moore-Penrose inverse of a matrix $A \in \mathbb{R}^{n\times m}$
$A^\top$	transpose of a matrix $A \in \mathbb{R}^{n\times m}$
$A \otimes B$	Kronecker product of matrices $A \in \mathbb{R}^{n_1\times n_2}$, $B \in \mathbb{R}^{n_3\times n_4}$
$P \succ 0$ ($P \succeq 0$)	the symmetric matrix $P = P^\top \in \mathbb{R}^{n\times n}$ is positive (semi-)definite
$P \prec 0$ ($P \preceq 0$)	the symmetric matrix $P = P^\top \in \mathbb{R}^{n\times n}$ is negative (semi-)definite
$\lambda_{\min}(P)$	minimum eigenvalue of a symmetric matrix $P = P^\top \in \mathbb{R}^{n\times n}$
$\lambda_{\max}(P)$	maximum eigenvalue of a symmetric matrix $P = P^\top \in \mathbb{R}^{n\times n}$
$\lambda_{\min}(P_1, P_2)$	minimum value among all eigenvalues of two symmetric matrices $P_1 = P_1^\top \in \mathbb{R}^{n_1\times n_1}$, $P_2 = P_2^\top \in \mathbb{R}^{n_2\times n_2}$, i.e., $\lambda_{\min}(P_1, P_2) := \min\{\lambda_{\min}(P_1), \lambda_{\min}(P_2)\}$
$\lambda_{\max}(P_1, P_2)$	maximum value among all eigenvalues of two symmetric matrices $P_1 = P_1^\top \in \mathbb{R}^{n_1\times n_1}$, $P_2 = P_2^\top \in \mathbb{R}^{n_2\times n_2}$, i.e., $\lambda_{\max}(P_1, P_2) := \max\{\lambda_{\max}(P_1), \lambda_{\max}(P_2)\}$

$\|x\|_p$	p-norm of a vector $x \in \mathbb{R}^n$ with $p \in \mathbb{I}_{>0} \cup \{\infty\}$
$\|x\|_Q$	weighted 2-(semi-)norm of a vector $x \in \mathbb{R}^n$, i.e., $\|x\|_Q := \sqrt{x^\top Q x}$ with $Q \succeq 0$
$\|x\|_{\mathcal{Z}}$	point-to-set distance of a vector $x \in \mathbb{R}^n$ w.r.t. a set $\mathcal{Z}$, i.e., $\|x\|_{\mathcal{Z}} := \inf_{x' \in \mathcal{Z}} \|x' - x\|_2$
$\|A\|_p$	induced p-norm of a matrix $A \in \mathbb{R}^{n \times m}$ with $p \in \mathbb{I}_{>0} \cup \{\infty\}$

Abstract

We present a framework for data-driven model predictive control (MPC) with theoretical guarantees on closed-loop stability and robustness. The proposed approach relies on Willems' Fundamental Lemma which parametrizes all trajectories of an unknown linear system based on one measured input-output trajectory. This result can be used to design MPC schemes simply from measured trajectories of the system rather than from its model, which need not be known. However, when applying such a scheme in closed loop, stability is not necessarily guaranteed. To close this gap, we develop a framework for designing and analyzing MPC schemes, which are only based on measured data and do not require explicit model knowledge, and come with rigorous closed-loop guarantees. To this end, we consider three distinct problem setups of increasing difficulty: data-driven MPC for linear systems with noise-free data, for linear systems with noisy data, and for nonlinear systems, respectively.

Data-driven MPC for linear systems with noise-free data

Chapter 3 addresses MPC for linear time-invariant (LTI) systems based on one noise-free input-output trajectory. More specifically, three data-driven MPC schemes are investigated for this scenario. First, we propose to use terminal equality constraints which enforce the input and output to be equal to the setpoint for several time steps at the end of the prediction horizon. Next, we develop a data-driven MPC scheme that contains general terminal ingredients, i.e., a terminal cost and a terminal constraint, and we provide a procedure to design such terminal ingredients based on input-output data. Finally, we suggest a tracking MPC formulation containing an artificial setpoint which is optimized online. For each of these schemes, we prove recursive feasibility, satisfaction of input and output constraints, as well as stability of the closed loop.

Data-driven MPC for linear systems with noisy data

In Chapter 4, we consider MPC for LTI systems based on one input-output trajectory which is affected by measurement noise. We prove that, by adding a slack variable and suitable regularization terms, the data-driven MPC with terminal equality constraints practically exponentially stabilizes the closed loop. Additionally, we derive a constraint tightening to ensure that the output robustly satisfies the constraints. Finally, we present a robust data-driven tracking MPC scheme which combines robustifying ingredients such as a slack variable and regularization with an artificial equilibrium. Based on a novel separation argument of nominal and robust data-driven MPC, we prove that this scheme practically exponentially stabilizes the optimal reachable equilibrium for the given setpoint.

Data-driven MPC for nonlinear systems

In Chapter 5, we address MPC for unknown nonlinear systems based only on input-output data. In contrast to the approaches for linear systems, we update the data online at each step, thereby implicitly exploiting local linear approximations of the underlying nonlinear dynamics. We prove that, if a certain cost parameter is sufficiently small and the initial state is close to the steady-state manifold, then the proposed data-driven MPC scheme practically exponentially stabilizes the optimal reachable equilibrium for the unknown nonlinear system. As an intermediate result of independent interest, we also develop and analyze a model-based MPC scheme to control nonlinear systems by using the linearized dynamics at the current state as the prediction model.

In summary, the main goal of this thesis is to develop a framework for MPC based only on measured data with stability and robustness guarantees for the closed loop. To this end, we merge existing model-based MPC results with Willems' Fundamental Lemma, and we provide a detailed analysis of the effect of noise or nonlinear dynamics on the closed-loop behavior. Moreover, we demonstrate with numerical as well as experimental examples that the proposed data-driven MPC framework not only admits strong theoretical guarantees, but is also simple to apply and provides high performance for challenging control problems.

Deutsche Kurzfassung

Diese Doktorarbeit befasst sich mit der Entwicklung und Analyse von Verfahren zur prädiktiven Regelung (Englisch: *model predictive control*, MPC) basierend auf gemessenen Daten mit theoretischen Garantien für Stabilität und Robustheit. Der vorgestellte Ansatz beruht auf dem *Fundamental Lemma* von Willems et al., welches sämtliche Trajektorien eines unbekannten linearen Systems basierend auf einer gemessenen Eingangs-Ausgangs-Trajektorie parametrisiert. Mit Hilfe dieses Resultats können MPC-Algorithmen entworfen werden, indem das übliche Zustandsraummodell durch datenabhängige Hankel-Matrizen ersetzt wird. Im Allgemeinen ist nicht garantiert, dass der aus der Anwendung dieses Schemas resultierende geschlossene Kreis stabil ist. Zur Lösung dieses Problems entwickeln wir verschiedene Ansätze zum Entwurf und zur Analyse von MPC-Algorithmen, welche ausschließlich auf gemessenen Daten basieren und kein explizites Modellwissen benötigen, jedoch rigorose Garantien für den geschlossenen Regelkreis aufweisen. Wir betrachten drei verschiedene Problemstellungen mit zunehmendem Schwierigkeitsgrad: Datenbasierte MPC-Methoden für lineare Systeme mit rauschfreien Daten, für lineare Systeme mit verrauschten Daten und für nichtlineare Systeme.

Datenbasierte MPC-Methoden für lineare Systeme mit rauschfreien Daten

In Kapitel 3 behandeln wir MPC für lineare, zeitinvariante Systeme basierend auf einer rauschfreien Eingangs-Ausgangs-Trajektorie. Im Speziellen entwickeln wir drei MPC-Algorithmen für dieses Szenario. Zunächst verwenden wir Gleichheitsbeschränkungen, welche erzwingen, dass die prädizierten Signale gegen Ende des Prädiktionshorizonts gleich dem Sollwert sind. Anschließend stellen wir einen datenbasierten MPC-Algorithmus vor, welcher allgemeine Endbedingungen enthält, d.h. eine Endkostenfunktion sowie eine Zielmenge. Ferner entwickeln wir eine Methode zur Auslegung solcher Endbedingungen basierend auf Eingangs-Ausgangs-Daten. Schließlich präsentieren wir eine MPC-Formulierung basierend auf einer künstlichen Referenz, welche während des Betriebs optimiert wird. Für jeden der Algorithmen beweisen wir rekursive Lösbarkeit, die Erfüllung von Eingangs- und Ausgangs-Beschränkungen sowie Stabilität des geschlossenen Kreises.

Datenbasierte MPC-Methoden für lineare Systeme mit verrauschten Daten

Kapitel 4 beschreibt MPC-Methoden für lineare, zeitinvariante Systeme basierend auf einer Eingangs-Ausgangs-Trajektorie, welche von Messrauschen beeinflusst ist. Wir beweisen, dass der datenbasierte MPC-Algorithmus mit Gleichheitsbeschränkungen durch Hinzufügen einer Schlupfvariable sowie geeigneter Regularisierungsterme zu praktischer exponentieller Stabilität führt. Ebenso leiten wir striktere Beschränkungen her, welche sicherstellen, dass der Ausgang die originalen Beschränkungen trotz der verrauschten Daten robust einhält. Schließlich entwickeln wir einen robusten datenbasierten MPC-Algorithmus, welcher robustheitserzeugende Komponenten (Schlupfvariable und Regularisierung) mit einer künstlichen Referenz verknüpft. Mit Hilfe eines neuartigen Separationsarguments für nominelle und robuste datenbasierte MPC-Algorithmen beweisen wir, dass dieser Algorithmus das bestmögliche Gleichgewicht für die gegebene Referenz praktisch exponentiell stabilisiert.

Datenbasierte MPC-Methoden für nichtlineare Systeme

Kapitel 5 adressiert MPC-Methoden für unbekannte nichtlineare Systeme basierend auf Eingangs-Ausgangs-Daten. Im Gegensatz zu den Ansätzen für lineare Systeme werden die Daten in jedem Zeitschritt durch neue Messungen ersetzt, wodurch implizit lokale lineare Approximationen des zugrundeliegenden nichtlinearen Systems ausgenutzt werden. Unter den Annahmen, dass eine bestimmte Kostenmatrix klein genug gewählt wird und der Anfangszustand sich nahe an der Gleichgewichtsmannigfaltigkeit befindet, beweisen wir, dass der vorgestellte MPC-Algorithmus das bestmögliche Gleichgewicht für das unbekannte nichtlineare System praktisch exponentiell stabilisiert. Als Zwischenresultat von unabhängigem Interesse entwickeln und analysieren wir einen modellbasierten MPC-Algorithmus für nichtlineare Systeme, welcher zur Prädiktion nur die linearisierte Systemdynamik um den aktuellen Zustand verwendet.

Das Hauptziel dieser Doktorarbeit ist die Entwicklung von Entwurfs- und Analyseverfahren für MPC-Algorithmen basierend auf gemessenen Daten mit Stabilitäts- und Robustheitsgarantien. Hierzu verbinden wir existierende modellbasierte MPC-Resultate mit dem *Fundamental Lemma* von Willems et al. und wir analysieren den Einfluss von Messrauschen oder Nichtlinearitäten auf das geregelte System. Außerdem demonstrieren wir anhand von numerischen sowie experimentellen Ergebnissen, dass der vorgestellte datenbasierte MPC-Ansatz nicht nur wünschenswerte theoretische Garantien aufweist, sondern einfach anzuwenden ist und eine hohe Regelgüte für anspruchsvolle praktische Regelungsprobleme aufweist.

Chapter 1

Introduction

1.1 Motivation

Classical controller design requires model knowledge of the plant. For example, if a controller is to be designed based on methods from nonlinear [67], robust [160], or model predictive [118] control, then typically a state-space model of the underlying plant must be available. However, determining an accurate model can be cumbersome and is in many cases the most time-consuming task in the controller design. On the other hand, it is often easily possible and cheap to gather data by exciting the system in an experiment. With this motivation, the field of Reinforcement Learning [130] has received huge interest in recent years and often exhibits remarkable performance in practical applications. Yet, despite their empirical success, many of these approaches do not admit strong theoretical guarantees such that methods can behave unpredictably or fail, even in very simple applications, see, e.g., [120].

Therefore, there has been an increasing effort in employing control-theoretic tools to use data for control purposes. Arguably, the most intuitive approach to design controllers based on data consists of a two-step procedure: 1) estimating a model from the available data and 2) using the identified model to design a controller via model-based techniques. The field of system identification, which deals with the estimation of models from data, is well-established and contains a wide range of different approaches, see, e.g., [85]. When aiming at theoretical guarantees for a controller designed via an identified model, then two ingredients are required: 1) a model estimate and 2) a bound on the estimation error. In particular, if the model does not exactly represent the underlying system and no bound on the model error is known, then it cannot be guaranteed that the resulting controller will satisfy control goals, e.g., stabilize the underlying system. However, determining the above two ingredients based on data is challenging and an active field of research even for linear time-invariant (LTI) systems, in particular in the inevitable scenario where the data are affected by noise, see, e.g., [32, 100, 101] for stochastic noise. If the noise admits a deterministic description, set membership estimation can be used to obtain a model, where obtaining computationally tractable models with tight error bounds is a key challenge [12, 106]. For nonlinear systems, obtaining accurate models with corresponding error bounds is even more involved: Only few approaches admit strong guarantees and they often rely on Lipschitz continuity-like properties [21, 105], which may lead to complex models, or they require appropriate choices of basis functions [128].

As an alternative to the above *indirect* approach, consisting of sequential system identification and model-based controller design, there have been many works on *direct* data-driven control which avoid the intermediate identification step. This includes techniques such as virtual reference feedback tuning [22], iterative feedback tuning [58], or unfalsification-based approaches [73], see the survey [59] for details and further approaches. As for the indirect approach, however, providing rigorous guarantees based on a finite set of noisy data points is challenging, not only for nonlinear systems but also in the LTI case. To summarize, while the established literature contains different indirect and direct approaches for using data in control, typically no strong theoretical guarantees can be given, in particular in the presence of noise or for nonlinear systems.

The *Fundamental Lemma* proposed by Willems et al. [144] has shown to be a promising basis for data-driven control. The result allows to directly parametrize all trajectories of an unknown LTI system based on one input-output trajectory. Although the paper [144] was published in 2005, the Fundamental Lemma has only received increasing attention in recent years, with many different applications ranging from signal processing and system analysis to controller design, as surveyed in [92]. One of the most prominent and powerful applications of the Fundamental Lemma is the design of data-driven model predictive control (MPC) approaches, as first recognized by [27, 150]. Based on these early works, different extensions and modifications have been developed to cope with noisy data and nonlinear systems, e.g., via regularization [92, Section 5.2.2]. These approaches not only admit rigorous guarantees but also have remarkable performance when used in challenging real-world applications [92, Section 5.2.4]. However, the mentioned theoretical results mainly focus on certifying robustness and performance of the underlying *open-loop* optimal control problem. On the contrary, if we want to implement a feedback controller, then these optimal control problems need to be solved repeatedly in a receding-horizon fashion as in standard (model-based) MPC [118]. For this scenario, the above approaches provide no theoretical guarantees, e.g., on stability or robustness, neither for linear nor nonlinear systems and neither for noise-free nor noisy data.

This motivates the purpose of the present thesis: We develop a framework for designing and analyzing data-driven MPC schemes based on the Fundamental Lemma using only measured input-output data and no explicit model knowledge. Motivated by the above discussion, we derive theoretical guarantees on *closed-loop* stability and robustness, not only for the case of linear systems with noise-free data, but also for the challenging scenarios of noisy data and nonlinear systems. In the following, we discuss literature related to this thesis in more detail, followed by a description of the contributions and outline of the thesis.

1.2 Related work

In this section, we briefly survey selected topics on MPC and data-driven control which are relevant for this thesis. Further details on the relation of our results to existing works will be discussed in the respective sections later in the thesis.

Model predictive control

MPC is a well-established control technique which can handle hard constraints, performance criteria, and nonlinear multi-input multi-output systems. It is based on repeatedly solving an open-loop optimal control problem at each time instant and only applying the first part of the optimal input trajectory in closed loop, see, e.g., [118] for an introduction. To predict future trajectories in this optimal control problem, a *model* of the plant is required. Hence, standard MPC approaches are model-based, although model inaccuracies can also be handled via robust MPC techniques [71, 103, 117]. Further, MPC typically relies on state measurements, whereas output-feedback MPC schemes require an additional state estimation step, see [118, Section 5] and references therein.

Due to the easier availability of data, recent research in MPC has also addressed possibilities to exploit data. In adaptive MPC approaches [2, 11, 53, 87, 132], online data measurements are used to refine an initial model estimate and thus improve performance and reduce conservatism in comparison to robust MPC. Further, learning-based MPC approaches [14, 56, 57, 158] merge MPC with machine learning techniques, which are used to identify the system or directly tune the MPC scheme. A different approach which predicts future trajectories based on past data is proposed by [124], however, no closed-loop guarantees are provided.

In summary, the existing MPC literature focuses on model-based state-feedback MPC. Approaches handling data are typically based on an additional system identification step such that similar restrictions as described in Section 1.1 apply. On the contrary, the MPC schemes presented in this thesis are based directly on input-output data without any intermediate estimation step, they admit closed-loop stability guarantees for noisy data and nonlinear systems, and they are inherently output-feedback MPC schemes.

Willems' Fundamental Lemma

The Fundamental Lemma proven by Willems et al. [144] shows that data-dependent Hankel matrices span all trajectories of an LTI system, assuming that the system is controllable and the input generating the data is persistently exciting. This result, which was originally formulated in the behavioral approach to control [113, 143], provides a powerful basis for data-driven control since it leads to a data-driven system representation without any intermediate model estimation step. Contributions on data-driven optimal control [95] and data-driven simulation [94] were among the first works exploiting the Fundamental Lemma. After relatively small interest in the result over the first decade after its publication, the Fundamental Lemma has been revived in recent years, leading to a surge of research on data-driven control. We refer to [92] for an excellent survey summarizing and discussing many recent contributions in this field. In the following, we mention selected works which are of particular importance to this thesis.

First, the Fundamental Lemma can be used to *analyze* unknown systems based on measured data. In [JB21, 102, JB27], it is employed to check whether the system is dissipative or satisfies an integral quadratic constraint over a finite time-horizon. Moreover, [JB19, JB20, 138] verify dissipativity for all systems that are consistent with the available data and a given noise bound, thus ensuring that the underlying true system is also dissipative. The data-driven inference of dissipativity and more general properties (e.g., the nonlinearity measure) for nonlinear systems (e.g., polynomial systems) is investigated in [96, 97, 98]. One key motivation for all these results is that the availability of input-output properties, such as dissipativity, allows for designing controllers with rigorous mathematical guarantees, e.g., via feedback theorems [34]. In Section 4.2, we propose data-driven procedures to estimate observability and controllability constants, which are required to construct a constraint tightening for robust data-driven MPC and which can also be interpreted as a type of data-driven system analysis.

Moreover, the Fundamental Lemma has found use in designing explicit feedback controllers, e.g., state-feedback gains, with closed-loop guarantees. This idea has first been suggested by [30] with many subsequent improvements and extensions, e.g., on relaxing persistence of excitation requirements [140], handling noise via robust control [JB4], bridging direct and indirect data-driven state-feedback design [37], and many more, compare [92, Section 5.3] and references therein. A closely related approach has recently been developed by [137], which computes a bound on the unknown system matrices from noisy data which is then employed for control via robust control arguments, conceptually similar to set membership estimation [105, 106]. Using related ideas, [JB14] provides a framework for combining data with prior knowledge for the purpose of robust controller design which can handle, e.g., different forms of prior knowledge on the system structure or parameters, general noise descriptions, and nonlinear uncertainties in the loop. More general noise descriptions (i.e., pointwise-in-time bounds) have also been addressed by [17, 98]. In Section 3.2, we propose a procedure to construct terminal ingredients for data-driven MPC based on data-driven output-feedback design from [JB14].

Data-driven optimal control

Finally and most importantly for this thesis, the Fundamental Lemma can be used to set up data-driven optimal control algorithms, simply by optimizing over all vectors in the image of a data-dependent Hankel matrix instead of using a state-space model. After the initial inception of this idea [27, 150], many follow-up works have been developed, with a strong focus on improving robustness, e.g., via regularization, and deriving theoretical guarantees for the corresponding open-loop optimal control problem.

A data-enabled predictive control (DeePC) method which handles stochastic noise via tools from distributionally robust optimization can be found in [28]. This work contains a tractable reformulation of the underlying distributionally robust optimal control problem, including a relaxation of chance constraints on the output, as well as bounds on the achieved performance. Performance guarantees of a min-max formulation of DeePC with noisy data can be found in [63] along with different tractable reformulations of the min-max optimization problem. While the analyses in both [28] and [63] rely on a robustness-enhancing regularization term in the cost, the paper [36] has shown that this regularization can also be interpreted as relaxing the corresponding indirect (identification-based) data-driven control problem. An alternative approach based on a maximum-likelihood framework can be found in [152], which has advantages regarding data compression but relies on approximate solutions of a nonlinear optimization problem via sequential quadratic programming. Nonetheless, this approach has shown good practical performance for data-driven optimal control [153] and can also be used to derive confidence bounds in data-driven simulation [151].

A data-driven optimal control algorithm which combines the Fundamental Lemma with the system level synthesis framework [8] can be found in [149] along with results on the influence of the noise on the suboptimality gap. While this work only addresses state-feedback, a data-driven output-feedback approach has been suggested in [43, 44] based on the input-output parametrization [45]. These results also provide suboptimality bounds for the realized performance depending on the noise level and, additionally, a constraint tightening to ensure robust constraint satisfaction. Moreover, [147, 148] consider robust data-driven optimal control based on noisy input-output data, relying on the solution of a semidefinite program, and they also allow for input-output constraints. Further results on data-driven optimal control address Kalman filtering for data compression [7] or extensions to nonlinear systems [77, 79] and disturbance feedback [134]. Finally, data-driven optimal control has shown remarkable performance in many challenging control applications such as power systems [60, 61, 62], quadcopters [27, 28, 29, 39], synchronous motor drives [24, 25], and building control [78].

To summarize, the existing literature on data-driven optimal control focuses on *open-loop* guarantees, e.g., performance bounds or constraint satisfaction in finite-horizon optimal control. While these approaches also show remarkable performance in practical applications of a *closed-loop* receding-horizon implementation, they do not provide any guarantees on the closed loop. In this thesis, we present a framework for data-driven optimal control with guarantees on closed-loop stability and robustness when applied in a receding-horizon fashion. To emphasize this difference to data-driven (open-loop) optimal control and the similarity to MPC, we refer to the proposed approach as "data-driven MPC". Additionally, some of our results also provide novel insights into open-loop optimal control: In Section 4.3.4, we show that our results in Chapter 4 yield performance bounds and suboptimality estimates similar to the references discussed above for both the finite-horizon open-loop as well as the infinite-horizon closed-loop performance. Further, while most works mentioned above address linear systems, we provide a first theoretical analysis of data-driven MPC when controlling unknown *nonlinear* systems. Finally, we not only validate our framework with multiple simulation examples but we also confirm the empirical success of data-driven optimal control in Section 5.3.4 with an experimental application to a nonlinear four-tank system.

1.3 Contributions and outline of the thesis

In the following, we provide an outline of the thesis and summarize the contributions of each chapter.

Chapter 2: Background

In Chapter 2, we provide the background required for the main results in the thesis. After describing selected concepts from discrete-time LTI systems in Section 2.1, we introduce the Fundamental Lemma, which forms the backbone for the proposed data-driven MPC framework, in Section 2.2. Further, in Section 2.3 we provide an introduction to model-based linear MPC.

Chapter 3: Data-driven MPC for linear systems with noise-free data

In Chapter 3, we present a framework for data-driven MPC for linear systems based on noise-free data. We develop three different MPC formulations, each containing suitable refinements in comparison to the original formulation from [27, 150] in order to ensure closed-loop stability in a receding-horizon implementation.

In Section 3.1, we include terminal equality constraints which ensure that the predicted input-output trajectory is equal to the setpoint for several time steps at the end of the prediction horizon. We prove that the resulting scheme exponentially stabilizes the desired setpoint while satisfying input-output constraints in closed loop.

In Section 3.2, we extend this idea by considering more general terminal ingredients, consisting of a terminal set constraint and a terminal cost, and we show closed-loop stability and constraint satisfaction. Since the design of terminal ingredients may be challenging in the absence of model knowledge, we also provide a linear matrix inequality (LMI) based procedure to construct a suitable terminal cost and terminal set ensuring closed-loop stability directly from data.

In Section 3.3, we propose a data-driven tracking MPC scheme based on an artificial setpoint which is optimized online. In contrast to the previous sections, we consider the more general class of systems with affine dynamics since the tracking objective necessitates an explicit treatment of non-zero setpoints. Thus, we first present an extension of the Fundamental Lemma from [144] to such affine systems. Based on this result, we then prove that the data-driven tracking MPC scheme exponentially stabilizes the setpoint or the optimal reachable equilibrium, in case this setpoint is not feasible for the constraints or the system dynamics.

In Section 3.4, we apply the proposed data-driven MPC schemes to a linearized four-tank system and discuss their respective advantages and drawbacks. The results in Sections 3.1–3.3 provide a framework for designing and analyzing MPC schemes which rely only on measured input-output data but admit the same theoretical guarantees as their model-based counterparts. On the other hand, we illustrate with a numerical example that the direct application of data-driven MPC without any terminal ingredients, as, e.g., suggested in the papers discussed in Section 1.2, need not be stable in closed loop and may even destabilize an open-loop stable system. Since we assume exact input-output measurements, the results in Chapter 3 can also be interpreted on a technical level as a framework for model-based output-feedback MPC with positive semidefinite cost functions.

The results of Chapter 3 have been previously presented in [JB7, JB9, JB12].

Chapter 4: Data-driven MPC for linear systems with noisy data

In Chapter 4, we consider the design and analysis of data-driven MPC schemes for linear systems in the presence of noisy data. To be precise, we assume that the output measurements, that are used to predict future trajectories via the Fundamental Lemma, are perturbed by bounded measurement noise.

In Section 4.1, we present a data-driven MPC scheme using terminal equality constraints, which is essentially a robust modification of the scheme from Section 3.1. To be precise, the optimization problem contains a slack variable as well as suitable regularization terms, similar to existing open-loop approaches discussed in Section 1.2. We prove that the proposed scheme is recursively feasible and practically exponentially stabilizes the setpoint, i.e., the closed loop converges to a neighborhood of the setpoint whose size depends on the noise level. Our results show various connections between the data quality and the resulting closed-loop performance, which improves, e.g., if the data length increases, the amplitude of the input data increases, or the noise level decreases.

In Section 4.2, we develop a constraint tightening based on noisy input-output data, which supplements the approach from Section 4.1. We prove that the suggested MPC scheme is recursively feasible and ensures that the actual output robustly satisfies the output constraints. Further, we derive procedures to estimate system constants related to controllability and observability which are required to construct the constraint tightening.

In Section 4.3, we propose a robust data-driven tracking MPC scheme for affine systems, which is a combination of the nominal tracking MPC from Section 3.3 with the robust stabilizing MPC from Section 4.1. The underlying optimization problem contains both an artificial setpoint optimized online as well as a slack variable and regularization terms in the cost. We prove that the optimal reachable equilibrium is practically exponentially stable in closed loop under the developed MPC scheme. The theoretical analysis relies on a novel separation argument, which translates noisy data of data-driven MPC into an input disturbance for nominal MPC with noise-free data and analyzes the robustness of the latter.

In Section 4.4, we apply the robust data-driven MPC scheme with terminal equality constraints from Section 4.1 to a linearized four-tank system. We explore the influence of different parameters on the closed-loop performance, where we observe satisfactory behavior over a wide range of parameters. Moreover, we showcase the applicability of the constraint tightening from Section 4.2 with a simple numerical example from the literature.

The results in Chapter 4 provide the robust counterpart to the results in Chapter 3. Loosely speaking, we show that, after slight modifications, the nominal data-driven MPC schemes in Chapter 3 still admit (practical) stability guarantees if the data are affected by noise. Moreover, the separation-type results in Section 4.3 provide a generic framework for handling noise in data-driven MPC, which can also be applied to different data-driven MPC formulations and yields open-loop performance bounds which are similar to existing results in the literature.

The results of Chapter 4 have been previously presented in [JB7, JB11, JB13].

Chapter 5: Linear tracking MPC for nonlinear systems

In contrast to the previous chapters, Chapter 5 focuses on the analysis and design of MPC approaches for *nonlinear* systems. We present two MPC schemes, both employing local linearization arguments to control nonlinear systems, one relying on explicit model knowledge and the other one only requiring input-output data.

In Section 5.1, we present a tracking MPC scheme to control nonlinear systems using the linearized dynamics at the current state for prediction, i.e., a *model-based* MPC scheme. The key idea relies on the fact that, if the system does not move too rapidly, then the linearization provides a good approximation of the nonlinear dynamics such that the prediction model is sufficiently accurate. In closed loop, this can be implicitly enforced by selecting a certain cost matrix in the MPC problem small enough. We prove that the proposed MPC scheme exponentially stabilizes the optimal reachable equilibrium for the nonlinear system if this cost matrix is sufficiently small and the initial state is close to the steady-state manifold. Thus, we show that a linear tracking MPC scheme, which only requires solving a convex quadratic program (QP) online, can be used to control a nonlinear system with stability guarantees. This result is interesting in its own right since it provides an alternative to existing linearization-based approaches for (model-based) nonlinear MPC such as the real-time iteration scheme [35, 157]. Yet, the main motivation for the results in Section 5.1 is their role in the analysis of data-driven MPC in the following section.

In Section 5.2, we propose a tracking MPC scheme to control unknown nonlinear systems using only measured input-output data. Similar to the linear case, the scheme relies on (the affine version of) the Fundamental Lemma. However, in contrast to Chapters 3 and 4, the data used for prediction are updated at every time step by a window of past measurements. This allows us to adapt to the current operating conditions of the nonlinear system by approximating the local linear dynamics, similar to re-computing the linearized dynamics at every time step in the model-based case (Section 5.1). The scheme is efficient, requiring only the solution of strictly convex QPs, and simple to apply, since no nonlinear system identification step is required. At the same time, it yields strong theoretical guarantees for the closed loop. To be precise, we show that the proposed data-driven MPC practically exponentially stabilizes the optimal reachable equilibrium for the unknown nonlinear system. Similar to Section 4.3, the proof relies on a separation argument, combining the continuity of data-driven MPC w.r.t. perturbations with the model-based results from Section 5.1. For simplicity, the technical exposition in Section 5.2 assumes that the data are not affected by noise, but an extension to robust stability guarantees in case of noisy data is straightforward.

In Section 5.3, we demonstrate the practicality of the above MPC schemes with two well-known nonlinear examples. First, in Sections 5.3.1 and 5.3.2, we apply both the model-based and the data-driven MPC scheme to a continuous stirred tank reactor (CSTR). We perform a comparison to existing model-based MPC schemes as well as MPC based on recursive linear system identification, and we study the effect of different parameters on the closed-loop performance. Next, in Section 5.3.3, we apply the data-driven MPC approach from Section 5.2 to a nonlinear four-tank system in simulation and, again, explore the influence of parameter variations on the closed loop. Finally, in Section 5.3.4, we apply the data-driven MPC scheme to control a nonlinear four-tank system in a real-world experiment. In summary, the results in Section 5.3 show that our MPC framework not only admits strong theoretical guarantees but is also useful in practical applications, yielding satisfactory performance for a wide range of parameters and relying on a simple and efficient implementation.

The results of Chapter 5 have been previously presented in [JB8, JB10, JB11].

Chapter 6: Conclusions

In Chapter 6, we summarize and discuss the findings of this thesis. Further, we provide an outlook on interesting questions for future research.

Appendix A: Technical proofs

Appendix A contains technical proofs of some of the theoretical results.

Chapter 2

Background

In this chapter, we provide a brief overview of preliminary background required for the results in this thesis. After recalling basic concepts related to discrete-time LTI systems (Section 2.1), we state the Fundamental Lemma which yields a data-driven parametrization of all trajectories of an unknown LTI system (Section 2.2). Finally, in Section 2.3, we provide an introduction to model-based MPC for linear systems.

2.1 Discrete-time LTI systems

In the following, we introduce concepts from discrete-time LTI systems theory which are relevant throughout this thesis. Let us consider systems of the form

$$\begin{aligned} x_{k+1} &= Ax_k + Bu_k, \\ y_k &= Cx_k + Du_k, \end{aligned} \tag{2.1}$$

where $x_k \in \mathbb{R}^n$ is the state, $u_k \in \mathbb{R}^m$ is the input, and $y_k \in \mathbb{R}^p$ is the output, all at time $k \in \mathbb{I}_{\geq 0}$. If (A, C) is observable, we can define the *lag* of the system (2.1).

Definition 2.1. *The lag* $\underline{l}$ *of* (2.1) *is the smallest* $l \in \mathbb{I}_{[1,n]}$ *such that the following observability matrix* Φ_l *has rank* n*:*

$$\Phi_l := \begin{bmatrix} C \\ CA \\ \vdots \\ CA^{l-1} \end{bmatrix}. \tag{2.2}$$

Note that $\underline{l} \leq n$ in general and $\underline{l} = n$ for single-input single-output systems. In practical applications, measurements of a minimal state such as x are often not accessible, in particular in a data-driven control context when a model is not available. This motivates the use of the following (non-minimal) *extended* state ξ, which can be constructed whenever input-output measurements of (2.1) are available: For any upper bound $l \in \mathbb{I}_{\geq \underline{l}}$ on the lag, we define

$$\xi_k := \begin{bmatrix} u_{[k-l,k-1]} \\ y_{[k-l,k-1]} \end{bmatrix} \in \mathbb{R}^{(m+p)l} \text{ for } k \in \mathbb{I}_{\geq l}. \tag{2.3}$$

We denote the dimension of ξ by $n_\xi := (m+p)l$. Observability of (A, C) implies the existence of a matrix $T_{x,\xi} \in \mathbb{R}^{n \times n_\xi}$ such that

$$x_k = T_{x,\xi}\xi_k \tag{2.4}$$

for any $k \in \mathbb{I}_{\geq l}$. This condition means that ξ is final state observable, compare [118, Definition 4.29]. It is straightforward (see, e.g., [46], [JB19, Lemma 2]) to derive a corresponding state-space realization whose input-output behavior is equivalent to (2.1), i.e., matrices $(\tilde{A}, \tilde{B}, \tilde{C}, \tilde{D})$ of appropriate dimensions such that

$$\begin{aligned} \xi_{k+1} &= \tilde{A}\xi_k + \tilde{B}u_k, \\ y_k &= \tilde{C}\xi_k + \tilde{D}u_k, \ k \in \mathbb{I}_{\geq 0}. \end{aligned} \tag{2.5}$$

These matrices take the following form with suitable matrices F_i, G_i, $i \in \mathbb{I}_{[1,l]}$ depending on (A, B, C, D)

$$\begin{bmatrix} u_{k-l+1} \\ \vdots \\ u_{k-1} \\ u_k \\ \hline y_{k-l+1} \\ \vdots \\ y_{k-1} \\ y_k \end{bmatrix} = \left[\begin{array}{cccc|cccc} 0 & I & \dots & 0 & 0 & \dots & \dots & 0 \\ \vdots & \ddots & \ddots & \vdots & \vdots & \ddots & \ddots & \vdots \\ 0 & \ddots & \ddots & I & \vdots & \ddots & \ddots & \vdots \\ 0 & \dots & \dots & 0 & 0 & \dots & \dots & 0 \\ \hline 0 & \dots & \dots & 0 & 0 & I & \dots & 0 \\ \vdots & \ddots & \ddots & \vdots & \vdots & \ddots & \ddots & \vdots \\ 0 & \dots & \dots & 0 & 0 & \dots & \dots & I \\ G_l & \dots & \dots & G_1 & F_l & \dots & \dots & F_1 \end{array}\right] \begin{bmatrix} u_{k-l} \\ \vdots \\ u_{k-2} \\ u_{k-1} \\ \hline y_{k-l} \\ \vdots \\ y_{k-2} \\ y_{k-1} \end{bmatrix} + \begin{bmatrix} 0 \\ \vdots \\ 0 \\ I \\ \hline 0 \\ \vdots \\ 0 \\ D \end{bmatrix} u_k, \tag{2.6}$$

compare [46], [JB19, Lemma 2]. The state-space realization (2.5) has the following interesting properties.

Lemma 2.1. *Suppose (A, B) is controllable and (A, C) is observable. Then, the matrices $(\tilde{A}, \tilde{B}, \tilde{C}, \tilde{D})$ in (2.5) satisfy the following properties:*

1. *$(\tilde{A}, \tilde{B})$ is stabilizable,*
2. *$(\tilde{A}, \tilde{B})$ is controllable if and only if Φ_l is a square matrix, i.e., $pl = n$,*
3. *$(\tilde{A}, \tilde{C})$ is detectable, and*
4. *$(\tilde{A}, \tilde{C})$ is not observable.*

Proof. **1. Stabilizability**
Since (A, B) is controllable, there exists a stabilizing state-feedback law

$$u_k = Kx_k \overset{(2.4)}{=} KT_{x,\xi}\xi_k$$

with $K \in \mathbb{R}^{m\times n}$ such that u_k, x_k, y_k all converge to zero for $k \to \infty$ from an arbitrary initial condition. The definition of ξ in (2.3) then implies that $\xi_k \to 0$ as $k \to \infty$ and hence, $(\tilde{A}, \tilde{B})$ is stabilizable.

2. Controllability
The following proof is adopted from [JB12, Lemma 13] with slight modifications.
If: According to the system dynamics (2.1), we have for any $k \geq n_\xi$

$$\xi_k = \begin{bmatrix} u_{[k-l,k-1]} \\ y_{[k-l,k-1]} \end{bmatrix} = \begin{bmatrix} u_{[k-l,k-1]} \\ \Phi_l x_{k-l} + \Gamma_l u_{[k-l,k-1]} \end{bmatrix} \tag{2.7}$$

for a suitably defined matrix Γ_l and with Φ_l as in (2.2). By observability of (A, C), i.e., full column rank of Φ_l in (2.2), it holds that $pl \geq p\underline{l} \geq n$ and hence, $k - l \geq n_\xi - l \geq n$. Thus, using controllability of (A, B), the system (2.1) can be steered to an arbitrary state x_{k-l} from any initial condition x_0 by selecting an appropriate input sequence $u_{[0,k-l-1]}$. Hence, an arbitrary state $\xi_k \in \mathbb{R}^{n_\xi}$ is reachable if there exist an input $u_{[k-l,l-1]}$ and a state x_{k-l} such that (2.7) holds. The first block row of (2.7) is trivially satisfied. On the other hand, using $u_{[k-l,k-1]} = \begin{bmatrix} I & 0 \end{bmatrix} \xi_k$, the second block row of (2.7) is equivalent to

$$\begin{bmatrix} -\Gamma_l & I \end{bmatrix} \xi_k = \Phi_l x_{k-l}. \tag{2.8}$$

Since Φ_l is square and has full column rank, it is invertible. Hence, for an arbitrary $\xi_k \in \mathbb{R}^{n_\xi}$, a state x_{k-l} satisfying (2.7) can be computed as $x_{k-l} = \Phi_l^{-1} \begin{bmatrix} -\Gamma_l & I \end{bmatrix} \xi_k$.

This proves that $(\tilde{A}, \tilde{B})$ is controllable.

Only if: Suppose $(\tilde{A}, \tilde{B})$ is controllable, i.e., any state $\xi_k \in \mathbb{R}^{n_\xi}$ is reachable. This means that, for any $\xi_k \in \mathbb{R}^{n_\xi}$, there exist $u_{[k-l,k-1]}$, x_{k-l} satisfying (2.7). Following the same steps as in the "if"-direction above, we can use (2.8) to infer that $\begin{bmatrix} -\Gamma_l & I \end{bmatrix} \xi_k$ must lie in the image of Φ_l for arbitrary $\xi_k \in \mathbb{R}^{n_\xi}$. Since $\begin{bmatrix} -\Gamma_l & I \end{bmatrix}$ has full row rank, this implies that Φ_l has full row rank, i.e., since Φ_l has full column rank by observability, Φ_l is square.

3. Detectability

If $u_k = 0$ and $y_k = 0$ for $k \in \mathbb{I}_{\geq 0}$, then $\xi_k = 0$ for $k \in \mathbb{I}_{\geq l}$, i.e., $\xi_k \to 0$ for $k \to \infty$ such that $(\tilde{A}, \tilde{C})$ is detectable.

4. Observability

We construct two different state vectors $\xi_n \neq \xi_n'$ at time n both leading to the same minimal state, i.e., $x_n = T_{\mathrm{x},\xi}\xi_n$ and $x_n = T_{\mathrm{x},\xi}\xi_n'$, thus contradicting observability of $(\tilde{A}, \tilde{C})$.

Suppose $x_0 \neq 0$. Then, by controllability of (A, B), there exists an input sequence $\{u_k\}_{k=0}^{n-1}$ steering the state to zero in n steps, i.e., $x_n = 0$. Letting now $l \leq n$ (the case $l > n$ follows with trivial modifications), we define ξ_n as in (2.3) based on the corresponding input-output trajectory $\{u_k, y_k\}_{k=0}^{n-1}$ and hence, $x_n = T_{\mathrm{x},\xi}\xi_n$. On the other hand, since $x_n = 0$, the choice $\xi_n' = 0$ also satisfies $x_n = T_{\mathrm{x},\xi}\xi_n'$. Although $\xi_n \neq \xi_n'$, the corresponding future input-output trajectories at times $k \geq n$ are indistinguishable, as they only depend on the new initial state $x_n = 0$. Therefore, it is not possible to determine a unique extended state at time n based on future input-output measurements, i.e., $(\tilde{A}, \tilde{C})$ is not observable. ∎

Equation (2.4) implies that any input-output trajectory of length at least $\underline{l}$ uniquely specifies the internal (minimal) state of the system. Thus, whenever, at two consecutive time steps, the last $\underline{l}$ input-output values coincide, then the corresponding internal state must be a steady-state for the dynamics (2.1). This motivates the following definition of an input-output equilibrium.

Definition 2.2. *We say that an input-output pair $(u^{\mathrm{s}}, y^{\mathrm{s}}) \in \mathbb{R}^{m+p}$ is an equilibrium of (2.1), if the sequence $\{u_k, y_k\}_{k=0}^{\underline{l}}$ with $(u_k, y_k) = (u^{\mathrm{s}}, y^{\mathrm{s}})$ for all $k \in \mathbb{I}_{[0,\underline{l}]}$ is a trajectory of (2.1).*

Finally, we recall two crucial implications of observability and controllability: First, if (A, B) is controllable and (A, C) is observable then Lemma 2.1 implies that $(\tilde{A}, \tilde{C})$ is detectable. In this case, there exists an input-output-to-state-stability (IOSS) Lyapunov function[1] $W(\xi) := \|\xi\|_{P_W}^2$ for some matrix $P_W \succ 0$ such that

$$W(\tilde{A}\xi + \tilde{B}u) - W(\xi) \leq -\|\xi\|_2^2 + c_{\text{IOSS},1}\|u\|_2^2 + c_{\text{IOSS},2}\|y\|_2^2 \tag{2.9}$$

for all $\xi \in \mathbb{R}^{n_\xi}$, $u \in \mathbb{R}^m$, $y = \tilde{C}\xi + \tilde{D}u$, and for suitable $c_{\text{IOSS},1}, c_{\text{IOSS},2} > 0$ [20].

Second, it is straightforward to show (compare, e.g., [129, Theorem 5]) that, if (A, B) is controllable, then there exists a constant $\Gamma > 0$ such that, for any initial condition $x_0 \in \mathbb{R}^n$ and any steady-state $(x^s, u^s) \in \mathbb{R}^{n+m}$, there exists an input trajectory $\{\bar{u}_k\}_{k=0}^{n-1}$ steering the system from x_0 to x^s, i.e., $\bar{x}_0 = x_0$, $\bar{x}_{k+1} = A\bar{x}_k + B\bar{u}_k$ for $k \in \mathbb{I}_{[0,n-1]}$, $\bar{x}_n = x^s$, while satisfying

$$\sum_{k=0}^{n-1} \|\bar{x}_k - x^s\|_2^2 + \|\bar{u}_k - u^s\|_2^2 \leq \Gamma\|x^s - x_0\|_2^2. \tag{2.10}$$

2.2 Willems' Fundamental Lemma

In this section, we discuss the Fundamental Lemma by Willems et al. [144], which provides the foundation for the main results of this thesis. To this end, we first introduce the following standard definition of persistence of excitation [144].

Definition 2.3. *We say that a sequence $\{u_k\}_{k=0}^{N-1}$ with $u_k \in \mathbb{R}^m$ is persistently exciting of order L if* $\text{rank}(H_L(u)) = mL$, *i.e., $H_L(u)$ has full row rank.*

Definition 2.3 requires that the entries of the sequence $\{u_k\}_{k=0}^{N-1}$ are linearly independent in a suitable sense and thus, it describes the "richness" of a sequence. Note that persistence of excitation of order L implies a lower bound on the length of the sequence $\{u_k\}_{k=0}^{N-1}$, i.e., $N \geq (m+1)L - 1$. We now formulate the *Fundamental Lemma* [144].

Theorem 2.1. *[144] Suppose (A, B) is controllable and $\{u_k^d, y_k^d\}_{k=0}^{N-1}$ is a trajectory of* (2.1), *where u^d is persistently exciting of order $L + n$. Then, $\{u_k, y_k\}_{k=0}^{L-1}$ is a trajectory of* (2.1) *if and only if there exists $\alpha \in \mathbb{R}^{N-L+1}$ such that*

$$\begin{bmatrix} H_L(u^d) \\ H_L(y^d) \end{bmatrix} \alpha = \begin{bmatrix} u \\ y \end{bmatrix}. \tag{2.11}$$

[1] While [20, Section 3.2] only addresses strictly proper systems with $y = Cx$, the result can be easily extended to systems with $y = Cx + Du$ by considering a modified input matrix $B' = B + LD$ in [20, Inequality (12)].

Theorem 2.1 states that an arbitrary sequence $\{u_k, y_k\}_{k=0}^{L-1}$ of length L is an input-output trajectory of (2.1) if and only if it lies in the image of the stacked Hankel matrix $\begin{bmatrix} H_L(u^{\mathrm{d}}) \\ H_L(y^{\mathrm{d}}) \end{bmatrix}$, assuming controllability and persistence of excitation. While the "if"-direction follows directly from linearity and time-invariance, proving "only if" is the non-trivial part of Theorem 2.1, see [144] for the original result in the behavioral framework, [JB2] for an explanation and interpretation in the state-space framework, and [139] for a proof in the state-space framework.

The Fundamental Lemma allows us to parametrize all trajectories of an LTI system, using only measured input-output data and no knowledge of the system matrices A, B, C, D in (2.1). Loosely speaking, the result shows that the Hankel matrices containing the measured data provide a non-parametric model of the underlying system. Based on Theorem 2.1, one can implement a wide range of system analysis and controller design techniques simply by leveraging the data-driven system representation (2.11) instead of a state-space model. As explained in Chapter 1, this general idea has sparked many recent contributions in the field of data-driven control. In the present thesis, we focus on the design and theoretical analysis of MPC schemes based on Theorem 2.1 and variations thereof with a focus on closed-loop stability and robustness.

Remark 2.1. *(Necessity of assumptions) We note that neither controllability nor persistence of excitation are strictly required for the "if and only if"-statement of Theorem 2.1 to be correct. As shown in [93], the equivalence stated in Theorem 2.1 holds if and only if* $\mathrm{rank}\left(\begin{bmatrix} H_L(u^{\mathrm{d}}) \\ H_L(y^{\mathrm{d}}) \end{bmatrix}\right) = mL + n$*, which need not imply controllability or a persistently exciting input as stated in Theorem 2.1. Similarly, as is common in the behavioral framework [113, 143], Theorem 2.1 does not require an input-output partitioning of the system variables, see [93] for details. Moreover, the result remains true when using matrix structures more general than Hankel matrices (e.g., Page matrix, mosaic-Hankel matrix, trajectory matrix), as long as every column of the data matrix is one trajectory of the system [93], compare also [139]. Throughout this thesis, we focus on the original version of the Fundamental Lemma in Theorem 2.1 above for simplicity, but we note that most of our results can be generalized based on [93]. Finally, weaker versions of the Fundamental Lemma have been developed under weaker assumptions (e.g., loss of controllability or persistence of excitation), see [108, 155]. Using such adaptations for the purpose of data-driven*

MPC is an interesting issue for future research.

2.3 Model predictive control

In the following, we provide an introduction to standard (model-based) MPC for linear systems. Details and results under more general assumptions can be found, e.g., in [104, 118]. Suppose we want to stabilize a steady-state x^s of (2.1) with corresponding input u^s while satisfying pointwise-in-time constraints on the input and the state, i.e., $u_t \in \mathbb{U}$, $x_t \in \mathbb{X}$ for all $t \in \mathbb{I}_{\geq 0}$ with closed sets $\mathbb{U} \subseteq \mathbb{R}^m$, $\mathbb{X} \subseteq \mathbb{R}^n$ satisfying $(x^s, u^s) \in \mathbb{X} \times \mathbb{U}$. An MPC scheme solving this problem can be constructed by solving, at time $t \in \mathbb{I}_{\geq 0}$ and for a given state x_t, the following open-loop optimal control problem:

Problem 2.1.

$$\underset{\bar{u}(t),\bar{x}(t)}{\text{minimize}} \quad \sum_{k=0}^{L-1} \left(\|\bar{x}_k(t) - x^s\|_Q^2 + \|\bar{u}_k(t) - u^s\|_R^2 \right) + \|\bar{x}_L(t) - x^s\|_P^2 \tag{2.12a}$$

subject to

$$\bar{x}_{k+1}(t) = A\bar{x}_k(t) + B\bar{u}_k(t), \; k \in \mathbb{I}_{[0,L-1]}, \tag{2.12b}$$

$$\bar{x}_0(t) = x_t, \tag{2.12c}$$

$$\bar{x}_k(t) \in \mathbb{X}, \; \bar{u}_k(t) \in \mathbb{U}, \; k \in \mathbb{I}_{[0,L-1]}, \tag{2.12d}$$

$$\bar{x}_L(t) \in \mathbb{X}_f. \tag{2.12e}$$

Here, $\bar{x}_k(t)$ and $\bar{u}_k(t)$ denote the k-th time step of the state and input trajectory predicted at time t. Problem 2.1 minimizes the deviation of the trajectory predicted over the horizon[2] $L \in \mathbb{I}_{>0}$ from the setpoint (x^s, u^s) using a quadratic stage cost with weights $Q, R \succ 0$. The linear state-space equation (2.12b) together with the initial condition (2.12c) ensures that $\bar{x}(t)$ is a state trajectory satisfying the linear system dynamics (2.1) with initial condition x_t. We write $\bar{u}^*(t) \in \mathbb{U}^L$, $\bar{x}^*(t) \in \mathbb{X}^{L+1}$ for the minimizer[3] of Problem 2.1. Moreover, the optimal cost of Problem 2.1 is denoted by $J_L^*(x_t)$.

[2] While the prediction horizon is often denoted by N in the MPC literature, we use L throughout this thesis since N is already occupied by the data length.

[3] In case Problem 2.1 does not admit a unique minimizer, any minimizer can be selected. The same holds true for all further MPC optimization problems considered in this thesis.

Further, Problem 2.1 contains state and input constraints (2.12d) as well as a terminal cost with weight $P \succeq 0$ and a terminal set constraint with $\mathbb{X}_f$, see (2.12e). These *terminal ingredients* are commonly used in MPC to ensure recursive feasibility of Problem 2.1, i.e., that Problem 2.1 is feasible at time $t+1$ if it is feasible at time t, and stability of the closed loop. Problem 2.1 can now be used to control System (2.1) via a receding-horizon strategy as described in Algorithm 2.1.

Algorithm 2.1. Model-based MPC scheme
Offline: Choose prediction horizon L, cost matrices $Q, R \succ 0$, constraint sets $\mathbb{U}$, $\mathbb{X}$, setpoint (x^s, u^s), and terminal ingredients $P \succeq 0$, $\mathbb{X}_f$.
Online: At time $t \in \mathbb{I}_{\geq 0}$, measure x_t and solve Problem 2.1. Apply the input $u_t = \bar{u}_0^*(t)$.

In the following, we introduce standard conditions under which Algorithm 2.1 is recursively feasible and stabilizes the closed loop.

Assumption 2.1. *(Terminal ingredients) There exist matrices $P = P^\top \succeq 0$, $K \in \mathbb{R}^{m \times n}$, and a set $\mathbb{X}_f \subseteq \mathbb{X}$ such that for all $x \in \mathbb{X}_f$, $u = u^s + K(x - x^s)$, we have*

i) $Ax + Bu \in \mathbb{X}_f$,

ii) $u \in \mathbb{U}$,

iii) the following inequality holds:

$$\|(A+BK)(x-x^s)\|_P^2 \leq \|x-x^s\|_P^2 - \|x-x^s\|_{K^\top RK} - \|x-x^s\|_Q^2. \tag{2.13}$$

Assumption 2.1 is a standard condition in MPC, compare [118, Assumption 2.14]. It implies that i) the terminal constraint set $\mathbb{X}_f$ is positively invariant, ii) the corresponding terminal control law satisfies the input constraints, and iii) $x^\top Px$ is a local control Lyapunov function.

Assumption 2.2. *(Value function upper bound) The optimal value function $J_L^*(x)$ of Problem 2.1 is quadratically upper bounded, i.e., there exists $c_u > 0$ such that $J_L^*(x) \leq c_u\|x\|_2^2$ for all x such that Problem 2.1 is feasible.*

Assumption 2.2 follows from Assumption 2.1 if $x^s \in \mathrm{int}(\mathbb{X}_f)$ or (2.1) satisfies a weak form of controllability, compare [118, Assumption 2.17]. Using these assumptions, the following stability result is immediate.

Theorem 2.2. *Suppose Assumptions 2.1 and 2.2 hold. If Problem 2.1 is feasible at initial time $t = 0$, then*

i) it is feasible at any $t \in \mathbb{I}_{\geq 0}$,

ii) the closed loop satisfies the constraints, i.e., $x_t \in \mathbb{X}$ and $u_t \in \mathbb{U}$ for all $t \in \mathbb{I}_{\geq 0}$,

iii) the steady-state x^s is exponentially stable for the resulting closed loop.

Based on the given assumptions, Theorem 2.2 shows that i) Algorithm 2.1 is recursively feasible, ii) the closed loop satisfies the constraints, and iii) the steady-state x^s is exponentially stable in closed loop. A proof of this result under more general assumptions (e.g., nonlinear system dynamics) can be found in [118, Theorem 2.19].

To guarantee constraint satisfaction and stability via Theorem 2.2, suitable terminal ingredients satisfying Assumption 2.1 need to be selected, compare [26, 118] for details. In the considered linear-quadratic setting, this is possible, e.g., by choosing P as the solution of the discrete-time algebraic Riccati equation and $\mathbb{X}_f$ as the corresponding maximal positively invariant set satisfying the constraints [118, Section 2.5.4]. Later in this thesis (Section 3.2), we will see that an analogous design is also possible if the system matrices in (2.1) are unknown and only measured data are available. Alternatively, a simple choice leading to closed-loop stability without any offline computations is a terminal equality constraint, i.e., choosing $\mathbb{X}_f = \{x^s\}$, $P = 0$, compare [118].

2.4 Summary

In this chapter, we provided background on discrete-time LTI systems, Willems' Fundamental Lemma, and model-based MPC, which will be required throughout this thesis. Classical MPC approaches rely on a (state-space) model to predict and optimize over future trajectories. In the remainder of the thesis, we present a framework for designing and analyzing MPC schemes based only on measured input-output data, using Willems' Fundamental Lemma for the prediction. Moreover, since no state measurements are available, the concepts introduced in Section 2.1 such as observability, the lag l, and the extended state ξ will be crucial for the closed-loop analysis.

Chapter 3

Data-driven MPC for linear systems with noise-free data

In this chapter, we present a framework for data-driven MPC with closed-loop stability guarantees. The proposed approach is a combination of ideas from model-based MPC with the Fundamental Lemma (Theorem 2.1). It only requires one (noise-free) input-output trajectory and no explicit model knowledge, except for an upper bound on the system's lag.

First, in Section 3.1, we present a simple data-driven MPC scheme with a terminal equality constraint to ensure closed-loop stability. Second, in Section 3.2, we include more general terminal ingredients (terminal cost and terminal set constraint) in order to improve the closed-loop performance. Since no model of the system is available, we also present a data-driven procedure to design terminal ingredients guaranteeing the desired closed-loop properties. Finally, the third MPC scheme proposed in Section 3.3 addresses a tracking objective for unknown systems with affine dynamics. For each of the three data-driven MPC schemes, we provide a theoretical analysis by proving recursive feasibility as well as constraint satisfaction and exponential stability of the closed loop. In Section 3.4, the schemes are then applied to a linearized four-tank system as a numerical example and compared with respect to their closed-loop performance

The results presented in this chapter are based on Berberich et al. [JB7, JB9, JB12].

3.1 Stability via terminal equality constraints

In this section, we propose a simple data-driven MPC scheme with terminal equality constraints. The scheme relies on noise-free measurements to predict future trajectories using Theorem 2.1 and is described in Section 3.1.1. Under mild assumptions, we prove recursive feasibility, constraint satisfaction, and exponential stability of the closed loop (Section 3.1.2).

This section is based on and taken in parts literally from [JB7][1].

3.1.1 Proposed MPC scheme

In this section, we consider data-driven MPC for the LTI system (2.1), where the matrices A, B, C, D are *unknown* and only input-output data are available. We assume that System (2.1) is controllable and observable.

Assumption 3.1. *(Controllability and observability) The pair* (A, B) *is controllable and the pair* (A, C) *is observable.*

Controllability is required for the Fundamental Lemma (Theorem 2.1), whereas observability is needed since the proposed MPC scheme only relies on input-output measurements.

In the following, our goal is to stabilize a given input-output equilibrium (u^s, y^s) in the sense of Definition 2.2 by penalizing the distance w.r.t. this equilibrium in the cost function $\|\bar{u} - u^s\|_R^2 + \|\bar{y} - y^s\|_Q^2$ for user-chosen matrices $Q, R \succ 0$. At the same time, we want to satisfy pointwise-in-time constraints $u_t \in \mathbb{U}$, $y_t \in \mathbb{Y}$, for $t \in \mathbb{I}_{\geq 0}$ with given closed sets $\mathbb{U} \subseteq \mathbb{R}^m$, $\mathbb{Y} \subseteq \mathbb{R}^p$. In contrast to standard (model-based) MPC approaches (cf. Section 2.3), we have no explicit model knowledge but only one input-output data trajectory $\{u_k^d, y_k^d\}_{k=0}^{N-1}$ of (2.1).

Recall that the model in the form of state-space matrices is a crucial ingredient of MPC since it ensures that the predicted sequences are in fact trajectories of the underlying LTI system. Theorem 2.1 provides an appealing alternative to such a state-space model since (2.11) captures all system trajectories based on the image of data-dependent Hankel matrices. Thus, to implement a data-driven MPC scheme,

[1] J. Berberich, J. Köhler, M. A. Müller, and F. Allgöwer. "Data-driven model predictive control with stability and robustness guarantees." In: *IEEE Trans. Automat. Control* 66.4 (2021), pp. 1702–1717.

one can simply replace the system dynamics constraint in Problem 2.1 by the constraint that the predicted input-output trajectories satisfy (2.11). Hence, at time t and for given input-output initial conditions $\xi_t = \begin{bmatrix} u_{[t-l,t-1]} \\ y_{[t-l,t-1]} \end{bmatrix}$ with $l \in \mathbb{I}_{\geq \underline{l}}$ for the lag $\underline{l}$ (Definition 2.1), we consider the following open-loop optimal control problem:

Problem 3.1.

$$\underset{\alpha(t),\bar{u}(t),\bar{y}(t)}{\text{minimize}} \quad \sum_{k=0}^{L-1} \|\bar{u}_k(t) - u^{\text{s}}\|_R^2 + \|\bar{y}_k(t) - y^{\text{s}}\|_Q^2 \tag{3.1a}$$

subject to

$$\begin{bmatrix} \bar{u}(t) \\ \bar{y}(t) \end{bmatrix} = \begin{bmatrix} H_{L+l}(u^{\text{d}}) \\ H_{L+l}(y^{\text{d}}) \end{bmatrix} \alpha(t), \tag{3.1b}$$

$$\begin{bmatrix} \bar{u}_{[-l,-1]}(t) \\ \bar{y}_{[-l,-1]}(t) \end{bmatrix} = \begin{bmatrix} u_{[t-l,t-1]} \\ y_{[t-l,t-1]} \end{bmatrix}, \tag{3.1c}$$

$$\bar{u}_k(t) \in \mathbb{U}, \ \bar{y}_k(t) \in \mathbb{Y}, \ k \in \mathbb{I}_{[0,L-1]}. \tag{3.1d}$$

As described above, the constraint (3.1b) replaces the system dynamics compared to classical model-based MPC schemes. Further, since only input-output measurements are available, we specify initial conditions in (3.1c) over l consecutive time steps to ensure that the internal state of the closed-loop trajectory aligns with the internal state of the predicted trajectory, compare [94]. Therefore, the overall length of the trajectory $(\bar{u}(t), \bar{y}(t)) = (\bar{u}_{[-l,L-1]}(t), \bar{y}_{[-l,L-1]}(t))$ is $L + l$. The initial conditions are specified until time step $t-1$, since the input at time t might already influence the output at time t, in case of a feedthrough-element of the plant.

A simple data-driven control strategy can be implemented by repeatedly solving Problem 3.1 in a receding-horizon fashion, similar to Algorithm 2.1. However, it is well-known that MPC without terminal constraints requires a sufficiently long prediction horizon to ensure stability and constraint satisfaction [48, 50]. Without such an assumption, the application of MPC can even destabilize an open-loop stable system [116]. There are two main approaches in the literature to guarantee stability: a) providing bounds on the minimal required prediction horizon [48] and b) including terminal ingredients such as terminal cost functions or terminal region constraints. Throughout this thesis, we will focus on b), i.e., using terminal ingredients to ensure closed-loop stability. Results regarding a), i.e., deriving lower bounds on the prediction horizon to guarantee stability, can be found in [JB17] for both noise-free as well as noisy data.

In this section, we consider a simple terminal equality constraint to guarantee exponential stability of the closed loop. To this end, at time t and for given input-output initial conditions $\xi_t = \begin{bmatrix} u_{[t-l,t-1]} \\ y_{[t-l,t-1]} \end{bmatrix}$ with $l \in \mathbb{I}_{\geq \underline{l}}$, we consider the following data-driven MPC scheme with a terminal equality constraint:

Problem 3.2.

$$\underset{\alpha(t),\bar{u}(t),\bar{y}(t)}{\text{minimize}} \quad \sum_{k=0}^{L-1} \|\bar{u}_k(t) - u^{\mathrm{s}}\|_R^2 + \|\bar{y}_k(t) - y^{\mathrm{s}}\|_Q^2 \tag{3.2a}$$

subject to

$$\begin{bmatrix} \bar{u}(t) \\ \bar{y}(t) \end{bmatrix} = \begin{bmatrix} H_{L+l}(u^{\mathrm{d}}) \\ H_{L+l}(y^{\mathrm{d}}) \end{bmatrix} \alpha(t), \tag{3.2b}$$

$$\begin{bmatrix} \bar{u}_{[-l,-1]}(t) \\ \bar{y}_{[-l,-1]}(t) \end{bmatrix} = \begin{bmatrix} u_{[t-l,t-1]} \\ y_{[t-l,t-1]} \end{bmatrix}, \tag{3.2c}$$

$$\begin{bmatrix} \bar{u}_{[L-l,L-1]}(t) \\ \bar{y}_{[L-l,L-1]}(t) \end{bmatrix} = \begin{bmatrix} \mathbb{1}_l \otimes u^{\mathrm{s}} \\ \mathbb{1}_l \otimes y^{\mathrm{s}} \end{bmatrix}, \tag{3.2d}$$

$$\bar{u}_k(t) \in \mathbb{U}, \; \bar{y}_k(t) \in \mathbb{Y}, \; k \in \mathbb{I}_{[0,L-1]}. \tag{3.2e}$$

We denote the optimal open-loop cost of Problem 3.2 by $J_L^*(\xi_t)$. Further, the minimizer of Problem 3.2 at time t is denoted by $\alpha^*(t)$, and the corresponding input-output trajectory by $\bar{u}^*(t)$, $\bar{y}^*(t)$. On the other hand, closed-loop measurements will be denoted by u_t, y_t, $t \in \mathbb{I}_{\geq 0}$.

The main difference of Problem 3.2 if compared to Problem 3.1 is that the former includes the *terminal equality constraint* (3.2d) which enforces that the last l predicted input-output values are equal to the setpoint. This implies that $\bar{\xi}_L(t)$, which is the extended state (cf. (2.3)) predicted L steps ahead corresponding to the predicted input-output trajectory, aligns with the steady-state ξ^{s} corresponding to $(u^{\mathrm{s}}, y^{\mathrm{s}})$, i.e., $\bar{\xi}_L(t) = \xi^{\mathrm{s}}$, similar to terminal equality constraints in model-based MPC, compare Section 2.3. Problem 3.2 is solved in a receding-horizon fashion, which is summarized in Algorithm 3.1. In the next section, we analyze the closed loop under Algorithm 3.1.

Algorithm 3.1. Data-driven MPC scheme with terminal equality constraints
Offline: Choose upper bound on lag $l \geq \underline{l}$, prediction horizon L, cost matrices $Q, R \succ 0$, constraint sets $\mathbb{U}$, $\mathbb{Y}$, setpoint $(u^{\text{s}}, y^{\text{s}})$, and generate data $\{u_k^{\text{d}}, y_k^{\text{d}}\}_{k=0}^{N-1}$.
Online: At time $t \in \mathbb{I}_{\geq 0}$, take the past l measurements $\{u_k, y_k\}_{k=t-l}^{t-1}$ and solve Problem 3.2. Apply the input $u_t = \bar{u}_0^*(t)$.

3.1.2 Closed-loop guarantees

Without loss of generality, we assume for the analysis that $u^{\text{s}} = 0$, $y^{\text{s}} = 0$, and thus $\xi^{\text{s}} = 0$. To prove exponential stability of the proposed scheme, we make the common assumption that the optimal value function of Problem 3.2 is quadratically upper bounded, compare Assumption 2.2. This is, e.g., satisfied in the present linear-quadratic setting if the constraints are polytopic[2] [13], in which case Problem 3.2 is a convex QP.

Assumption 3.2. *(Value function upper bound) The optimal value function $J_L^*(\xi)$ of Problem 3.2 is quadratically upper bounded, i.e., there exists $c_{\text{u}} > 0$ such that $J_L^*(\xi) \leq c_{\text{u}}\|\xi\|_2^2$ for all ξ such that Problem 3.2 is feasible.*

Moreover, we assume that the input u^{d} generating the data used for prediction is sufficiently rich in the following sense.

Assumption 3.3. *(Persistence of excitation) The sequence $\{u_k^{\text{d}}\}_{k=0}^{N-1}$ is persistently exciting of order $L + l + n$.*

Note that we assume persistence of excitation of order $L + l + n$, although Theorem 2.1 requires only an order of $L + n$. This is due to the fact that the reconstructed trajectories in Problem 3.2 are of length $L + l$ (compared to length L in Theorem 2.1), since l components are used to fix the initial conditions. Furthermore, due to the terminal constraints (3.2d), we assume that the prediction horizon is at least as long as the upper bound on the lag l.

Assumption 3.4. *(Length of prediction horizon) The prediction horizon satisfies $L \geq l$.*

[2]While [13] considered model-based linear-quadratic MPC, the result applies similarly to the present data-driven MPC setting since (3.2b) (together with the initial conditions (3.2c)) describes the input-output behavior of the system exactly and thus, both settings are equivalent.

Note that $L \geq l$ is not strictly required for our theoretical results, but typically needed to guarantee a non-trivial region of attraction. The following result shows that the MPC scheme based on Problem 3.2 (Algorithm 3.1) is recursively feasible, ensures constraint satisfaction, and leads to an exponentially stable closed loop.

Theorem 3.1. *Suppose Assumptions 3.1–3.4 hold. If Problem 3.2 is feasible at initial time $t = 0$, then*

i) it is feasible at any $t \in \mathbb{I}_{\geq 0}$,

ii) the closed loop satisfies the constraints, i.e., $u_t \in \mathbb{U}$ and $y_t \in \mathbb{Y}$ for all $t \in \mathbb{I}_{\geq 0}$,

iii) the equilibrium $\xi^s = 0$ is exponentially stable for the resulting closed loop.

Proof. Recursive feasibility and constraint satisfaction follow from standard MPC arguments, i.e., by defining a candidate solution as the shifted, previously optimal solution and appending zero (compare [118]). For the proof of exponential stability, we denote this standard candidate solution by $\bar{u}'(t+1)$, $\bar{y}'(t+1)$, $\alpha'(t+1)$. The cost of this solution is

$$\begin{aligned}\sum_{k=0}^{L-1} \|\bar{u}'_k(t+1)\|_R^2 + \|\bar{y}'_k(t+1)\|_Q^2 =& \sum_{k=1}^{L-1} \|\bar{u}^*_k(t)\|_R^2 + \|\bar{y}^*_k(t)\|_Q^2 \\ =& J_L^*(\xi_t) - \|\bar{u}^*_0(t)\|_R^2 - \|\bar{y}^*_0(t)\|_Q^2.\end{aligned}$$

Hence, it holds that

$$J_L^*(\xi_{t+1}) \leq J_L^*(\xi_t) - \|\bar{u}^*_0(t)\|_R^2 - \|\bar{y}^*_0(t)\|_Q^2. \tag{3.3}$$

Since (A, C) is observable (Assumption 3.1), the pair $(\tilde{A}, \tilde{C})$ is detectable by Lemma 2.1. Hence, there exists a matrix $P_W \succ 0$ such that $W(\xi) = \|\xi\|_{P_W}^2$ is an IOSS Lyapunov function, compare (2.9). Define the Lyapunov function candidate $V(\xi) = J_L^*(\xi) + \gamma W(\xi)$ for some $\gamma > 0$. Note that V is quadratically lower bounded, i.e., $V(\xi) \geq \gamma W(\xi) \geq \gamma \lambda_{\min}(P_W)\|\xi\|_2^2$ for all ξ such that Problem 3.2 is feasible. Further, J_L^* is quadratically upper bounded by Assumption 3.2. Hence, we have

$$V(\xi) = J_L^*(\xi) + \gamma W(\xi) \leq (c_u + \gamma \lambda_{\max}(P_W)) \|\xi\|_2^2,$$

for all ξ such that Problem 3.2 is feasible, i.e., V is quadratically upper bounded. We consider now

$$\gamma = \frac{\lambda_{\min}(Q,R)}{\max\{c_{\text{IOSS},1}, c_{\text{IOSS},2}\}} > 0.$$

Along the closed-loop trajectories, using both (2.9) as well as (3.3), it holds that

$$\begin{aligned} V(\xi_{t+1}) - V(\xi_t) &\leq \gamma\left(-\|\xi_t\|_2^2 + c_{\text{IOSS},1}\|u_t\|_2^2 + c_{\text{IOSS},2}\|y_t\|_2^2\right) - \|u_t\|_R^2 - \|y_t\|_Q^2 \\ &\leq -\gamma\|\xi_t\|_2^2. \end{aligned}$$

It follows from standard Lyapunov arguments with the Lyapunov function V (cf. [118]) that the equilibrium $\xi^s = 0$ is exponentially stable in closed loop. ■

The proof of Theorem 3.1 applies standard arguments from model-based MPC with terminal constraints (compare [118]) to analyze the closed loop under the data-driven MPC scheme in Algorithm 3.1. To handle the fact that the stage cost is merely positive *semi*-definite in the state, detectability of the stage cost is exploited via an IOSS Lyapunov function [20], similar to [47]. This leads to a combined Lyapunov function which is equal to the weighted sum of the optimal cost of Problem 3.2 and an IOSS Lyapunov function.

Summary

In this section, we presented a data-driven MPC scheme with terminal equality constraints (Algorithm 3.1). The proposed approach combined concepts from model-based MPC [118] with the Fundamental Lemma (Theorem 2.1) which is used to parametrize trajectories of the unknown LTI system directly from noise-free data. We proved that our MPC scheme exponentially stabilizes a given equilibrium while satisfying input-output constraints in closed loop. The main advantage of the presented approach is its simplicity, requiring only one input-output trajectory for its implementation but admitting closed-loop stability guarantees. In Section 3.4, we demonstrate the practical applicability of the data-driven MPC scheme developed in this section with a numerical example.

3.2 Stability via general terminal ingredients

While terminal equality constraints as in Section 3.1 are the simplest approach to ensure closed-loop stability, they have multiple potential drawbacks such as a small region of attraction and poor robustness properties. In this section, to alleviate these issues, we present a data-driven MPC scheme which includes general terminal ingredients, i.e., a terminal cost and a (non-trivial) terminal region constraint (Section 3.2.1). In Section 3.2.2, we then prove that this scheme exponentially stabilizes the closed loop. Since no explicit model is available, designing terminal ingredients ensuring closed-loop stability is non-trivial. Therefore, in Section 3.2.3, we show how the corresponding terminal ingredients can be computed using only measured data based on recent results on data-driven output-feedback design.

This section is based on and taken in parts literally from [JB12][3].

3.2.1 Proposed MPC scheme

The problem setting in this section is analogous to that in Section 3.1: Given an input-output data trajectory $\{u_k^{\mathrm{d}}, y_k^{\mathrm{d}}\}_{k=0}^{N-1}$, we want to stabilize an equilibrium $(u^{\mathrm{s}}, y^{\mathrm{s}})$ for the minimal LTI system (2.1) (cf. Assumption 3.1) while satisfying constraints $u_t \in \mathbb{U}$, $y_t \in \mathbb{Y}$, $t \in \mathbb{I}_{\geq 0}$, with closed sets $\mathbb{U} \in \mathbb{R}^m$ and $\mathbb{Y} \in \mathbb{R}^p$. To this end, at time $t \in \mathbb{I}_{\geq 0}$ and for initial conditions $\{u_k, y_k\}_{k=t-l}^{t-1}$ with $l \in \mathbb{I}_{\geq \underline{l}}$, we consider the following open-loop optimal control problem:

Problem 3.3.

$$\underset{\alpha(t),\bar{u}(t),\bar{y}(t)}{\text{minimize}} \quad \sum_{k=0}^{L-1} \left(\|\bar{u}_k(t) - u^{\mathrm{s}}\|_R^2 + \|\bar{y}_k(t) - y^{\mathrm{s}}\|_Q^2 \right) + \|\bar{\xi}_L(t) - \xi^{\mathrm{s}}\|_P^2 \tag{3.4a}$$

---

[3] J. Berberich, J. Köhler, M. A. Müller, and F. Allgöwer. "On the design of terminal ingredients for data-driven MPC." In: *IFAC-PapersOnLine* 54.6 (2021), pp. 257–263.

subject to

$$\begin{bmatrix} \bar{u}(t) \\ \bar{y}(t) \end{bmatrix} = \begin{bmatrix} H_{L+l}(u^{\mathrm{d}}) \\ H_{L+l}(y^{\mathrm{d}}) \end{bmatrix} \alpha(t), \tag{3.4b}$$

$$\begin{bmatrix} \bar{u}_{[-l,-1]}(t) \\ \bar{y}_{[-l,-1]}(t) \end{bmatrix} = \begin{bmatrix} u_{[t-l,t-1]} \\ y_{[t-l,t-1]} \end{bmatrix}, \tag{3.4c}$$

$$\bar{\xi}_L(t) = \begin{bmatrix} \bar{u}_{[L-l,L-1]}(t) \\ \bar{y}_{[L-l,L-1]}(t) \end{bmatrix} \in \Xi_{\mathrm{f}}, \tag{3.4d}$$

$$\bar{u}_k(t) \in \mathbb{U}, \ \bar{y}_k(t) \in \mathbb{Y}, \ k \in \mathbb{I}_{[0,L-1]}. \tag{3.4e}$$

Analogous to Section 3.1, we write $\bar{u}^*(t)$, $\bar{y}^*(t)$, $\alpha^*(t)$ for the optimal solution of Problem 3.3 at time $t \in \mathbb{I}_{\geq 0}$, whereas u_t, y_t denote the closed-loop input and output at time $t \in \mathbb{I}_{\geq 0}$. Further, $J_L^*(\xi_t)$ denotes the optimal cost of Problem 3.3 with initial condition $\xi_t = \begin{bmatrix} u_{[t-l,t-1]}^\top, y_{[t-l,t-1]}^\top \end{bmatrix}^\top$.

The tracking cost in (3.4a) as well as the constraints (3.4b), (3.4c), and (3.4e) are analogous to Problem 3.2. In contrast to Problem 3.2 and inspired from model-based MPC, Problem 3.3 contains general terminal ingredients in the form of a terminal cost $\|\bar{\xi}_L(t) - \xi^{\mathrm{s}}\|_P^2$ for some $P \succ 0$ and a terminal set constraint $\bar{\xi}_L(t) \in \Xi_{\mathrm{f}}$ for some $\Xi_{\mathrm{f}} \subseteq \mathbb{R}^{n_\xi}$, which involve the predicted extended state $\bar{\xi}_L(t)$ at time L. We assume that these terminal ingredients satisfy the following assumption.

Assumption 3.5. *(Terminal ingredients) There exist matrices $P = P^\top \succ 0$, $K \in \mathbb{R}^{m \times n_\xi}$, and a set $\Xi_{\mathrm{f}} \subseteq \mathbb{U}^l \times \mathbb{Y}^l$ such that for all $\xi \in \Xi_{\mathrm{f}}$, $u = u^{\mathrm{s}} + K(\xi - \xi^{\mathrm{s}})$, and $y = (\tilde{C} + \tilde{D}K)\xi$, we have*

i) $\tilde{A}\xi + \tilde{B}u \in \Xi_{\mathrm{f}}$,

ii) $u \in \mathbb{U}$ *and* $y \in \mathbb{Y}$,

iii) the following inequality holds:

$$\|(\tilde{A} + \tilde{B}K)(\xi - \xi^{\mathrm{s}})\|_P^2 \leq \|\xi - \xi^{\mathrm{s}}\|_P^2 - \|\xi - \xi^{\mathrm{s}}\|_{K^\top RK}^2 - \|y - y^{\mathrm{s}}\|_Q^2. \tag{3.5}$$

Assumption 3.5 is a standard condition in model-based MPC to ensure closed-loop stability, compare Assumption 2.1. In Section 3.2.3, we show how a matrix P and a set Ξ_{f} as in Assumption 3.5 can be constructed using only measured data and no model knowledge.

Algorithm 3.2 summarizes the MPC scheme based on repeatedly solving Problem 3.3.

Algorithm 3.2. Data-driven MPC scheme with terminal ingredients

Offline: Choose upper bound on lag $l \geq \underline{l}$, prediction horizon L, cost matrices $Q, R \succ 0$, constraint sets $\mathbb{U}$, $\mathbb{Y}$, setpoint $(u^{\mathrm{s}}, y^{\mathrm{s}})$, and generate data $\{u_k^{\mathrm{d}}, y_k^{\mathrm{d}}\}_{k=0}^{N-1}$. Compute terminal ingredients P and Ξ_{f} satisfying Assumption 3.5, e.g., via Proposition 3.1.

Online: At time $t \in \mathbb{I}_{\geq 0}$, take the past l measurements $\{u_k, y_k\}_{k=t-l}^{t-1}$ and solve Problem 3.3. Apply the input $u_t = \bar{u}_0^*(t)$.

3.2.2 Closed-loop guarantees

Without loss of generality, we assume $u^{\mathrm{s}} = 0$, $y^{\mathrm{s}} = 0$, $\xi^{\mathrm{s}} = 0$ for the following theoretical analysis.

Assumption 3.6. *(Value function upper bound) The optimal value function $J_L^*(\xi)$ of Problem 3.3 is quadratically upper bounded, i.e., there exists $c_{\mathrm{u}} > 0$ such that $J_L^*(\xi) \leq c_{\mathrm{u}}\|\xi\|_2^2$ for all ξ such that Problem 3.3 is feasible.*

Similar to Assumption 3.2, Assumption 3.6 is, e.g., satisfied if the sets $\mathbb{U}$, $\mathbb{Y}$, Ξ_{f} are polytopes [13]. Moreover, if $0 \in \mathrm{int}(\Xi_{\mathrm{f}})$, it also holds if $\mathbb{U}$ and $\mathbb{Y}$ are compact or, in case only $\mathbb{U}$ is compact and $\mathbb{Y}$ is closed, it holds with a non-quadratic upper bound, thus leading only to *asymptotic* stability guarantees for the closed loop [118, Proposition 2.16]. The following result proves desirable closed-loop properties of Algorithm 3.2.

Theorem 3.2. *Suppose Assumptions 3.1 and 3.3–3.6 hold. If Problem 3.3 is feasible at initial time $t = 0$, then*

i) it is feasible at any $t \in \mathbb{I}_{\geq 0}$,

ii) the closed loop satisfies the constraints, i.e., $u_t \in \mathbb{U}$ and $y_t \in \mathbb{Y}$ for all $t \in \mathbb{I}_{\geq 0}$,

iii) the equilibrium $\xi^{\mathrm{s}} = 0$ is exponentially stable for the resulting closed loop.

Proof. At time $t+1$, we construct a feasible candidate solution $(\bar{u}'(t+1), \bar{y}'(t+1))$ of Problem 3.3 by shifting the previously optimal solution $\bar{u}^*(t)$, $\bar{y}^*(t)$ of Problem 3.3 and appending the input $\bar{u}'_{L-1}(t+1) = K \begin{bmatrix} \bar{u}^*_{[L-l,L-1]}(t) \\ \bar{y}^*_{[L-l,L-1]}(t) \end{bmatrix}$, with K as in Assumption 3.5. This implies that the initial conditions (3.4c) and the input-output constraints (3.4e) are fulfilled and, by Assumption 3.5 iii), the terminal region constraint (3.4d) is satisfied. Since $(\bar{u}'(t+1), \bar{y}'(t+1))$ is a trajectory of (2.1), there exists $\alpha'(t+1)$ satisfying (3.4b) by Theorem 2.1, i.e., all constraints of Problem 3.3 are satisfied. Denoting the cost of Problem 3.3 for this candidate solution by $J_L'(\xi_{t+1})$, we have

$$\begin{aligned}
&J_L^*(\xi_{t+1}) - J_L^*(\xi_t) \\
\leq &J_L'(\xi_{t+1}) - J_L^*(\xi_t) \\
= &\|\bar{u}_{L-1}'(t+1)\|_R^2 + \|\bar{y}_{L-1}'(t+1)\|_Q^2 - \|\bar{u}_0^*(t)\|_R^2 - \|\bar{y}_0^*(t)\|_Q^2 \\
&+ \|\bar{\xi}_L'(t+1)\|_P^2 - \|\bar{\xi}_L^*(t)\|_P^2 \\
\overset{(3.5)}{\leq} & - \|u_t\|_R^2 - \|y_t\|_Q^2.
\end{aligned} \tag{3.6}$$

By detectability of $(\tilde{A}, \tilde{C})$ (cf. Lemma 2.1), there exists an IOSS Lyapunov function $W(\xi) = \|\xi\|_{P_W}^2$ with $P_W \succ 0$, compare (2.9). We consider as a Lyapunov function candidate the weighted sum of J_L^* and W, i.e., $V(\xi_t) = J_L^*(\xi_t) + \gamma W(\xi_t)$ with $\gamma = \frac{\lambda_{\min}(Q,R)}{\max\{c_{\text{IOSS},1}, c_{\text{IOSS},2}\}} > 0$. Combining (2.9) and (3.6), we infer $V(\xi_{t+1}) - V(\xi_t) \leq -\gamma\|\xi_t\|_2^2$. As in the proof of Theorem 3.1, the Lyapunov function candidate $V(\xi_t)$ is quadratically lower bounded and quadratically upper bounded (Assumption 3.6) such that ξ^{s} is exponentially stable in closed loop due to standard Lyapunov arguments [118]. ∎

Theorem 3.2 shows that the proposed MPC scheme based on Problem 3.3 exponentially stabilizes the desired setpoint ξ^{s} and hence, the closed-loop input-output trajectory exponentially converges to $(u^{\text{s}}, y^{\text{s}})$.

Remark 3.1. *(Stability with $\Xi_{\text{f}} = \mathbb{R}^{n_\xi}$) We note that, if the terminal set constraint in* (3.4d) *is dropped, i.e., $\Xi_{\text{f}} = \mathbb{R}^{n_\xi}$, closed-loop stability as shown in Theorem 3.2 holds locally with some region of attraction as long as the terminal cost matrix P satisfies the conditions in Assumption 3.5, compare [81]. For the special case of open-loop stable systems with no output constraints, i.e., $\mathbb{Y} = \mathbb{R}^p$, Assumption 3.5 can always be satisfied with $K = 0$ and without the terminal set constraint in* (3.4d)*, i.e., $\Xi_{\text{f}} = \mathbb{R}^{n_\xi}$, compare [119]. In this case, Problem 3.3 is globally feasible and Theorem 3.2 ensures global exponential stability.*

3.2.3 Data-driven design of terminal ingredients

In this section, we show how terminal ingredients satisfying Assumption 3.5 can be computed using only measured data and no model knowledge, based on recent results on data-driven controller design from [JB14]. To this end, we write (2.5) as a linear fractional transformation

$$\begin{bmatrix} \xi_{k+1} \\ \hline z_k \end{bmatrix} = \left[\begin{array}{c|cc} A' & B_{\mathrm{w}} & B' \\ \hline \begin{bmatrix} I \\ 0 \end{bmatrix} & 0 & \begin{bmatrix} 0 \\ I \end{bmatrix} \end{array}\right] \begin{bmatrix} \xi_k \\ \hline w_k \\ u_k \end{bmatrix}, \tag{3.7}$$
$$w_k = \Delta z_k,$$

where $w \mapsto z$ represents an additional uncertainty channel capturing all unknown elements of (2.5) (compare [160]). Here, A', B', $B_{\mathrm{w}} = \begin{bmatrix} 0 & I \end{bmatrix}^\top$ are known matrices according to the structure in (2.6), and

$$\Delta = \begin{bmatrix} G_l & \dots & G_1 & F_l & \dots & F_1 & D \end{bmatrix}$$

contains the unknown system parameters. To be precise, it holds that $\tilde{A} = A' + B_{\mathrm{w}}\Delta \begin{bmatrix} I \\ 0 \end{bmatrix}$ and $\tilde{B} = B' + B_{\mathrm{w}}\Delta \begin{bmatrix} 0 \\ I \end{bmatrix}$. We factorize $T_{\mathrm{y}}^\top Q T_{\mathrm{y}} = Q_{\mathrm{r}}^\top Q_{\mathrm{r}}$, $R = R_{\mathrm{r}}^\top R_{\mathrm{r}}$ with T_{y} such that $y_{t-1} = T_{\mathrm{y}}\xi_t$, i.e., $T_{\mathrm{y}} = \begin{bmatrix} 0 & \dots & 0 & I \end{bmatrix}$. Further, we define the data-dependent matrices

$$\Xi := \begin{bmatrix} \xi_l^{\mathrm{d}} & \xi_{l+1}^{\mathrm{d}} & \dots & \xi_{N-1}^{\mathrm{d}} \end{bmatrix}, \; \Xi_+ := \begin{bmatrix} \xi_{l+1}^{\mathrm{d}} & \xi_{l+2}^{\mathrm{d}} & \dots & \xi_N^{\mathrm{d}} \end{bmatrix},$$
$$U := \begin{bmatrix} u_l^{\mathrm{d}} & u_{l+1}^{\mathrm{d}} & \dots & u_{N-1}^{\mathrm{d}} \end{bmatrix}, \; Z := \begin{bmatrix} \Xi \\ U \end{bmatrix}, \; F := \Xi_+ - A'\Xi - B'U,$$

where $\{\xi_k^{\mathrm{d}}\}_{k=l}^{N-1}$ is the extended state trajectory corresponding to the available input-output measurements $\{u_k^{\mathrm{d}}, y_k^{\mathrm{d}}\}_{k=0}^{N-1}$. Using that $B_{\mathrm{w}}^\top$ is the Moore-Penrose inverse of B_{w}, [JB14, Lemma 2] implies the following data-dependent bound on the "uncertainty" Δ:

$$\begin{bmatrix} \Delta^\top \\ I \end{bmatrix}^\top \underbrace{\begin{bmatrix} -ZZ^\top & ZF^\top B_{\mathrm{w}} \\ B_{\mathrm{w}}^\top F Z^\top & -B_{\mathrm{w}}^\top F F^\top B_{\mathrm{w}} \end{bmatrix}}_{P_\Delta^{\mathrm{w}} :=} \begin{bmatrix} \Delta^\top \\ I \end{bmatrix} \succeq 0. \tag{3.8}$$

Moreover, we abbreviate

$$\bar{P}_{\Delta}^{\mathrm{w}} := \begin{bmatrix} 0 & I \\ B_{\mathrm{w}}^{\top} & 0 \end{bmatrix}^{\top} P_{\Delta}^{\mathrm{w}} \begin{bmatrix} 0 & I \\ B_{\mathrm{w}}^{\top} & 0 \end{bmatrix}.$$

Proposition 3.1. *Suppose there exist* $\mathcal{X} \succ 0$, $\Gamma \succ 0$, M, $\tau \geq 0$, $\gamma > 0$ *such that* $\mathrm{trace}(\Gamma) < \gamma^2$, $\begin{bmatrix} \Gamma & I \\ I & \mathcal{X} \end{bmatrix} \succ 0$, *and*

$$\begin{bmatrix} \left(\tau \bar{P}_{\Delta}^{\mathrm{w}} - \begin{bmatrix} \mathcal{X} & 0 \\ 0 & 0 \end{bmatrix}\right) & \begin{bmatrix} A'\mathcal{X} + B'M \\ \mathcal{X} \\ M \end{bmatrix} & 0 \\ \star & -\mathcal{X} & \begin{bmatrix} Q_{\mathrm{r}}\mathcal{X} \\ R_{\mathrm{r}}M \end{bmatrix}^{\top} \\ \star & \star & -I \end{bmatrix} \prec 0, \tag{3.9}$$

define $P := \mathcal{X}^{-1} - T_{\mathrm{y}}^{\top} Q T_{\mathrm{y}}$, $K := M\mathcal{X}^{-1}$, *and choose* $\beta > 0$ *such that*

$$\Xi_{\mathrm{f}} := \{\xi \in \mathbb{R}^{n_{\xi}} \mid \|\xi - \xi^{\mathrm{s}}\|_{P}^{2} \leq \beta\} \subseteq \mathbb{U}^{l} \times \mathbb{Y}^{l}. \tag{3.10}$$

Then, Assumption 3.5 holds with P, K, *and* Ξ_{f}.

Proof. By applying the Schur complement to the right-lower block of (3.9) and subsequently left- and right-multiplying $\mathrm{diag}(I, \mathcal{X}^{-1})$, we obtain

$$\begin{bmatrix} \tau \bar{P}_{\Delta}^{\mathrm{w}} - \begin{bmatrix} \mathcal{X} & 0 \\ 0 & 0 \end{bmatrix} & \begin{bmatrix} A' + B'K \\ I \\ K \end{bmatrix} \\ \star & T_{\mathrm{y}}^{\top} Q T_{\mathrm{y}} + K^{\top} R K - \mathcal{X}^{-1} \end{bmatrix} \prec 0. \tag{3.11}$$

Applying the Schur complement to the right-lower block of (3.11) and re-arranging terms leads to $\mathcal{Q} := \mathcal{X}^{-1} - T_{\mathrm{y}}^{\top} Q T_{\mathrm{y}} - K^{\top} R K \succ 0$ as well as

$$\left[\begin{array}{cc} I & 0 \\ (A' + B'K)^{\top} & \begin{bmatrix} I \\ K \end{bmatrix}^{\top} \\ \hline 0 & I \\ B_{\mathrm{w}}^{\top} & 0 \end{array}\right]^{\top} \left[\begin{array}{cc|c} -\mathcal{X} & 0 & 0 \\ 0 & \mathcal{Q}^{-1} & \\ \hline & 0 & \tau P_{\Delta}^{\mathrm{w}} \end{array}\right] \left[\begin{array}{c} \star \\ \star \\ \hline \star \\ \star \end{array}\right] \prec 0.$$

Using the full-block S-procedure [126] and $\tilde{A} = A' + B_w\Delta \begin{bmatrix} I \\ 0 \end{bmatrix}$, $\tilde{B} = B' + B_w\Delta \begin{bmatrix} 0 \\ I \end{bmatrix}$, where Δ satisfies (3.8), the latter inequality implies

$$-\mathcal{X} + (\tilde{A} + \tilde{B}K)\mathcal{Q}^{-1}(\tilde{A} + \tilde{B}K)^\top \prec 0. \tag{3.12}$$

Applying the Schur complement twice, this in turn implies

$$(\tilde{A} + \tilde{B}K)^\top \mathcal{X}^{-1}(\tilde{A} + \tilde{B}K) - \mathcal{X}^{-1} + T_y^\top Q T_y + K^\top R K \prec 0. \tag{3.13}$$

Let now $\xi^+ = (\tilde{A} + \tilde{B}K)\xi$, $y = (\tilde{C} + \tilde{D}K)\xi$ for some $\xi \in \mathbb{R}^{n_\xi}$. We then have

$$\begin{aligned} \|\xi^+ - \xi^s\|_P^2 &\overset{(3.13)}{\leq} \|\xi - \xi^s\|_{\mathcal{X}^{-1}}^2 - \|\xi - \xi^s\|_{T_y^\top Q T_y + K^\top R K}^2 - \|\xi^+ - \xi^s\|_{T_y^\top Q T_y}^2 \\ &= \|\xi - \xi^s\|_P^2 - \|y - y^s\|_Q^2 - \|\xi - \xi^s\|_{K^\top R K}^2, \end{aligned}$$

which proves Part (iii) of Assumption 3.5. Part (i) is a simple consequence of Part (iii). Finally, Part (ii) follows from invariance of $\Xi_f \subseteq \mathbb{U}^l \times \mathbb{Y}^l$. ∎

Proposition 3.1 provides a procedure to compute terminal ingredients satisfying Assumption 3.5, using no model knowledge but only input-output data. The proof relies on the full-block S-procedure [126], which is used to robustify the model-based inequality (3.5) against the unknown parameters Δ satisfying the known quadratic bound (3.8) derived by [JB14], compare also [137] for a similar approach to data-driven $\mathcal{H}_2$-state-feedback design. The design procedure in Proposition 3.1 is an LMI feasibility problem which can be solved efficiently in practice. Although any feasible solution $\mathcal{X} \succ 0$, M, $\tau \geq 0$ satisfying (3.9) leads to terminal ingredients satisfying Assumption 3.5, the additional conditions $\operatorname{trace}(\Gamma) < \gamma^2$, $\begin{bmatrix} \Gamma & I \\ I & \mathcal{X} \end{bmatrix} \succ 0$ for some $\Gamma \succ 0$, $\gamma > 0$ lead to a smaller terminal cost. To be precise, the optimal solution minimizing γ subject to the conditions in Proposition 3.1 (provided it exists) is equivalent to a linear quadratic regulator for the system (2.5), compare [JB14] for details. Hence, this optimal solution provides a standard choice for terminal ingredients, compare [26], leading to a good closed-loop performance. For a reliable numerical implementation, we perform a bisection algorithm over $\gamma > 0$ subject to the conditions in Proposition 3.1.

In the following, we discuss a crucial necessary condition for feasibility of the conditions in Proposition 3.1. Since (3.9) is a strict LMI, its feasibility requires

$\tau \bar{P}_\Delta^{\mathrm{w}} - \begin{bmatrix} \mathcal{X} & 0 \\ 0 & 0 \end{bmatrix} \prec 0$, which implies $-ZZ^\top \prec 0$. This in turn requires that the data matrix $Z = \begin{bmatrix} \Xi \\ U \end{bmatrix}$ has full row rank, which is related to persistence of excitation (Definition 2.3). In particular, [144, Corollary 2 (ii)] implies that Z has full row rank if $(\tilde{A}, \tilde{B})$ is controllable and u^{d} is persistently exciting of a sufficiently high order. Thus, one possibility to ensure that (3.9) is feasible via a sufficiently rich input signal is to require that $(\tilde{A}, \tilde{B})$ is controllable. As shown in Lemma 2.1, this is equivalent to the observability matrix Φ_l being square, i.e., $pl = n$. In fact, the latter condition is even necessary for Ξ and hence Z to have full row rank: By the Hankel matrix structure of

$$\Xi = \begin{bmatrix} \xi_l^{\mathrm{d}} & \xi_{l+1}^{\mathrm{d}} & \dots & \xi_{N-1}^{\mathrm{d}} \end{bmatrix} = \begin{bmatrix} u_0^{\mathrm{d}} & u_1^{\mathrm{d}} & \dots & u_{N-l-1}^{\mathrm{d}} \\ u_1^{\mathrm{d}} & u_2^{\mathrm{d}} & \dots & u_{N-l}^{\mathrm{d}} \\ \vdots & \vdots & \ddots & \vdots \\ u_{l-1}^{\mathrm{d}} & u_l^{\mathrm{d}} & \dots & u_{N-2}^{\mathrm{d}} \\ y_0^{\mathrm{d}} & y_1^{\mathrm{d}} & \dots & y_{N-l-1}^{\mathrm{d}} \\ y_1^{\mathrm{d}} & y_2^{\mathrm{d}} & \dots & y_{N-l}^{\mathrm{d}} \\ \vdots & \vdots & \ddots & \vdots \\ y_{l-1}^{\mathrm{d}} & y_l^{\mathrm{d}} & \dots & y_{N-2}^{\mathrm{d}} \end{bmatrix},$$

the matrix Ξ always has $\mathrm{rank}(\Xi) \leq ml + n$, i.e., it can only have full row rank if $pl = n$. This means that $pl = n$ is a necessary condition for feasibility of the presented design of terminal ingredients. For single-output systems with $p = 1$, this condition requires that $l = n$, i.e., the system order is known exactly. For multiple-output systems on the other hand, the condition $pl = n$ does in general not necessarily hold, even if both the system order n and the lag $\underline{l}$ are known exactly and we choose $l = \underline{l}$. Addressing this issue and designing data-driven controllers for the extended system (2.5) in case Z does not have full row rank is an interesting issue for future research, independently of its application to data-driven MPC.

Remark 3.2. *(Related work) The work by [38] is conceptually related to the above results since it provides a terminal set constraint for (model-based) MPC with input-output models using an implicit characterization of the maximal invariant set. Moreover, [1] propose an LMI-based procedure to construct stabilizing terminal ingredients for linear parameter-varying input-output models. Compared to these model-based results, the proposed MPC* *scheme and the computation of the terminal ingredients require only one input-output trajectory of an LTI system and no explicit model knowledge.*

Summary

In this section, we presented a data-driven MPC scheme and proved closed-loop constraint satisfaction and stability. In contrast to Section 3.1, this scheme utilizes general terminal ingredients, including a terminal cost and a terminal region constraint on the predicted extended state ξ. We showed how terminal ingredients ensuring closed-loop stability can be constructed from measured data based on recent results on data-driven output-feedback design. As we will see with a numerical example in Section 3.4, the degrees of freedom gained via the more general terminal ingredients lead to significant improvements of the closed-loop behavior if compared to the simple terminal equality constraints from Section 3.1.

3.3 Tracking MPC for affine systems

In this section, we propose a data-driven MPC scheme for setpoint tracking. The scheme relies on a data-driven system parametrization as in the Fundamental Lemma (Theorem 2.1) to predict future trajectories and contains a standard tracking cost as well as terminal equality constraints w.r.t. an input-output setpoint. In contrast to Section 3.1, this setpoint is optimized online and its deviation from the setpoint reference is penalized in the cost, analogous to model-based tracking MPC [69, 82]. Moreover, we consider plants with *affine* system dynamics instead of linear dynamics. This is motivated by the tracking objective, which does not allow to consider zero setpoints without loss of generality. Moreover, the offset terms can also represent (constant) disturbances, similar to offset-free MPC [109] which addresses tracking with (asymptotically) constant disturbances. Finally, in Chapter 5, we will address tracking MPC for nonlinear systems, for which the local linearization is in general an affine system.

Thus, we first present an extension of the Fundamental Lemma to affine systems in Section 3.3.1. Thereafter, we discuss the problem setup in Section 3.3.2. After stating the data-driven MPC scheme in Section 3.3.3, we then prove recursive feasibility, closed-loop constraint satisfaction, and exponential stability of the optimal reachable equilibrium in Section 3.3.4.

This section is based on and taken in parts literally from [JB9][4] and [JB11][5].

3.3.1 Willems' Fundamental Lemma for affine systems

In the following, we provide a data-driven parametrization of unknown systems with affine dynamics, i.e.,

$$\begin{aligned} x_{k+1} &= Ax_k + Bu_k + e, \\ y_k &= Cx_k + Du_k + r \end{aligned} \tag{3.14}$$

with state $x_k \in \mathbb{R}^n$, input $u_k \in \mathbb{R}^m$, and output $y_k \in \mathbb{R}^p$, all at time $k \in \mathbb{I}_{\geq 0}$. We assume that the matrices A, B, C, D and the offsets e, r are unknown, but one input-output trajectory $\{u_k^{\mathrm{d}}, y_k^{\mathrm{d}}\}_{k=0}^{N-1}$ of (3.14) is available. Recall that, by the Fundamental Lemma (Theorem 2.1), if $e = 0$, $r = 0$ (i.e., the system is linear) and certain persistence of excitation and controllability conditions hold, then a sequence $\{u_k, y_k\}_{k=0}^{L-1}$ is a trajectory of (3.14) if and only if there exists a vector $\alpha \in \mathbb{R}^{N-L+1}$ such that

$$\begin{bmatrix} H_L(u^{\mathrm{d}}) \\ H_L(y^{\mathrm{d}}) \end{bmatrix} \alpha = \begin{bmatrix} u \\ y \end{bmatrix}. \tag{3.15}$$

In the following, we provide an extension of this result to the class of affine systems (3.14). We note that such an extension is not trivial since, without knowledge of the vectors e and r, they cannot be set to zero without loss of generality.

Let us write $\{x_k^{\mathrm{d}}\}_{k=0}^{N-1}$ for the state sequence corresponding to the available input-output data $\{u_k^{\mathrm{d}}, y_k^{\mathrm{d}}\}_{k=0}^{N-1}$. The persistence of excitation condition used for the original Fundamental Lemma requires that $H_{L+n}(u^{\mathrm{d}})$ has full row rank which in turn, assuming controllability, implies that

$$\operatorname{rank}\left(\begin{bmatrix} H_L(u^{\mathrm{d}}) \\ H_1(x_{[0,N-L]}^{\mathrm{d}}) \end{bmatrix}\right) = mL + n. \tag{3.16}$$

[4] J. Berberich, J. Köhler, M. A. Müller, and F. Allgöwer. "Data-driven tracking MPC for changing setpoints." In: *IFAC-PapersOnLine*, 53.2 (2020), pp. 6923–6930.

[5] J. Berberich, J. Köhler, M. A. Müller, and F. Allgöwer. "Linear tracking MPC for nonlinear systems part II: the data-driven case." In: *IEEE Trans. Automat. Control* (2022). doi: 10.1109/TAC.2022.3166851. ©2021 IEEE

As we will see in the following, Condition (3.16) does not suffice to formulate the Fundamental Lemma for affine systems with $e \neq 0$, $r \neq 0$. Instead, we will require the rank condition

$$\operatorname{rank}\left(\begin{bmatrix} H_L(u^{\mathrm{d}}) \\ H_1(x^{\mathrm{d}}_{[0,N-L]}) \\ \mathbb{1}^\top_{N-L+1} \end{bmatrix}\right) = mL + n + 1. \tag{3.17}$$

In the recent paper [99], it is shown that Condition (3.17) can be enforced if (A, B) is controllable by choosing the input data sufficiently rich in the sense that $H_{L+n+1}(u^{\mathrm{d}})$ has full row rank, i.e., $\operatorname{rank}(H_{L+n+1}(u^{\mathrm{d}})) = m(L+n+1)$.

The following result extends the Fundamental Lemma (Theorem 2.1) to systems with affine dynamics.

Theorem 3.3. *Suppose* (3.17) *holds. Then,* $\{u_k, y_k\}_{k=0}^{L-1}$ *is a trajectory of* (3.14) *if and only if there exists* $\alpha = \begin{bmatrix} \alpha_0 & \alpha_1 & \dots & \alpha_{N-L} \end{bmatrix}^\top \in \mathbb{R}^{N-L+1}$ *such that*

$$\sum_{i=0}^{N-L} \alpha_i = 1, \quad \begin{bmatrix} H_L(u^{\mathrm{d}}) \\ H_L(y^{\mathrm{d}}) \end{bmatrix} \alpha = \begin{bmatrix} u \\ y \end{bmatrix}. \tag{3.18}$$

Proof. **Proof of "if":**
It clearly holds that

$$y^{\mathrm{d}}_{[i,i+L-1]} = \Phi_L x^{\mathrm{d}}_i + \Gamma_{\mathrm{u},L} u^{\mathrm{d}}_{[i,i+L-1]} + \Gamma_{\mathrm{e},L} e_L + r_L \tag{3.19}$$

for suitably defined matrices Φ_L, $\Gamma_{\mathrm{e},L}$, and $\Gamma_{\mathrm{u},L}$ depending on A, B, C, D, and for $e_L := \mathbb{1}_L \otimes e$, $r_L := \mathbb{1}_L \otimes r$. Hence,

$$\begin{aligned} y_{[0,L-1]} &\overset{(3.18)}{=} \sum_{i=0}^{N-L} y^{\mathrm{d}}_{[i,i+L-1]} \alpha_i \overset{(3.19)}{=} \sum_{i=0}^{N-L} \alpha_i \left(\Phi_L x^{\mathrm{d}}_i + \Gamma_{\mathrm{u},L} u^{\mathrm{d}}_{[i,i+L-1]} + \Gamma_{\mathrm{e},L} e_L + r_L \right) \\ &\overset{(3.18)}{=} \Phi_L \sum_{i=0}^{N-L} \alpha_i x^{\mathrm{d}}_i + \Gamma_{\mathrm{u},L} u_{[0,L-1]} + \Gamma_{\mathrm{e},L} e_L + r_L. \end{aligned} \tag{3.20}$$

This implies that $\{u_k, y_k\}_{k=0}^{L-1}$ is a trajectory of (3.14) with initial condition $x_0 = \sum_{i=0}^{N-L} \alpha_i x^d_i$.
Proof of "only if":

Let $\{u_k, y_k\}_{k=0}^{L-1}$ be a trajectory of (3.14) with initial condition x_0. Using (3.17), there exists $\alpha \in \mathbb{R}^{N-L+1}$ such that

$$\begin{bmatrix} H_L(u^{\mathrm{d}}) \\ H_1(x^{\mathrm{d}}_{[0,N-L]}) \\ \mathbb{1}_{N-L+1}^\top \end{bmatrix} \alpha = \begin{bmatrix} u_{[0,L-1]} \\ x_0 \\ 1 \end{bmatrix}. \tag{3.21}$$

Note that the last row implies $\sum_{i=0}^{N-L} \alpha_i = 1$. Moreover,

$$\begin{aligned} y_{[0,L-1]} &= \Phi_L x_0 + \Gamma_{\mathrm{u},L} u_{[0,L-1]} + \Gamma_{\mathrm{e},L} e_L + r_L \\ &\overset{(3.21)}{=} \sum_{i=0}^{N-L} \alpha_i \left(\Phi_L x_i^{\mathrm{d}} + \Gamma_{\mathrm{u},L} u^{\mathrm{d}}_{[i,i+L-1]} + \Gamma_{\mathrm{e},L} e_L + r_L \right) \\ &\overset{(3.19)}{=} \sum_{i=0}^{N-L} \alpha_i y^{\mathrm{d}}_{[i,i+L-1]} = H_L(y^{\mathrm{d}}) \alpha. \end{aligned}$$

■

Theorem 3.3 extends the Fundamental Lemma (Theorem 2.1) to systems with affine dynamics. The key difference to the linear case is the condition $\sum_{i=0}^{N-L} \alpha_i = 1$ in (3.18), i.e., the vector α sums up to one. Loosely speaking, this implies that the offsets e and r in (3.14) are carried through from the data $(u^{\mathrm{d}}, y^{\mathrm{d}})$ to the new trajectory (u, y) in (3.18). We note that a condition similar to $\sum_{i=0}^{N-L} \alpha_i = 1$ also appears in [123], albeit in a different problem setting with the objective of offset-free data-driven control. Further, instead of considering this additional condition, it is also possible to utilize the fact that the trajectory $\{u_{k+1} - u_k, y_{k+1} - y_k\}_{k=0}^{N-2}$ corresponds to an LTI system, compare, e.g., [15].

3.3.2 Problem setting

The problem setting in this section is similar to that in Section 3.1: Using only input-output data, we want to steer the affine system (3.14) towards a setpoint reference $(u^{\mathrm{r}}, y^{\mathrm{r}}) \in \mathbb{R}^{m+p}$, while the input and output satisfy pointwise-in-time constraints, i.e., $u_t \in \mathbb{U}, y_t \in \mathbb{Y}$, for all $t \in \mathbb{I}_{\geq 0}$. We assume that System (3.14) is controllable and observable, i.e., it satisfies Assumption 3.1. Moreover, the sets $\mathbb{U}$ and $\mathbb{Y}$ are assumed to be convex and compact.

Assumption 3.7. *(Constraint sets) The sets $\mathbb{U}$ and $\mathbb{Y}$ are convex and compact.*

In contrast to the previous sections, we do not require that the target setpoint $(u^{\mathrm{r}}, y^{\mathrm{r}})$ is an equilibrium of the system or satisfies the constraints, but our scheme will achieve convergence to the *optimal reachable equilibrium*. For matrices $S \succ 0, T \succeq 0$, the optimal reachable equilibrium $(u^{\text{sr}}, y^{\text{sr}})$ is defined as the feasible equilibrium $(u^{\text{s}}, y^{\text{s}})$, which minimizes $\|u^{\text{s}} - u^{\text{r}}\|_T^2 + \|y^{\text{s}} - y^{\text{r}}\|_S^2$. Due to a local controllability argument in the proof of our main result, only equilibria which are strictly inside the constraints can be considered, i.e., equilibria which lie in some convex and compact set $\mathbb{U}^{\text{s}} \times \mathbb{Y}^{\text{s}} \subseteq \text{int}(\mathbb{U} \times \mathbb{Y})$. Let $\{u_k^{\text{d}}, y_k^{\text{d}}\}_{k=0}^{N-1}$ be a measured trajectory of (3.14) with corresponding state $\{x_k^{\text{d}}\}_{k=0}^{N-1}$ and suppose that $\{u_k^{\text{d}}\}_{k=0}^{N-1}$, $\{x_k^{\text{d}}\}_{k=0}^{N-l-1}$ satisfy the persistence of excitation condition (3.17). Then, $(u^{\text{sr}}, y^{\text{sr}})$ is the minimizer of

$$
\begin{aligned}
J_{\text{eq}}^* = \min_{\alpha^{\text{s}}, u^{\text{s}}, y^{\text{s}}} \quad & \|u^{\text{s}} - u^{\text{r}}\|_T^2 + \|y^{\text{s}} - y^{\text{r}}\|_S^2 \\
\text{s.t.} \quad & \begin{bmatrix} H_{l+1}(u^{\text{d}}) \\ H_{l+1}(y^{\text{d}}) \\ \mathbb{1}_{N-l}^\top \end{bmatrix} \alpha^{\text{s}} = \begin{bmatrix} \mathbb{1}_{l+1} \otimes u^{\text{s}} \\ \mathbb{1}_{l+1} \otimes y^{\text{s}} \\ 1 \end{bmatrix}, \; u^{\text{s}} \in \mathbb{U}^{\text{s}}, \; y^{\text{s}} \in \mathbb{Y}^{\text{s}}.
\end{aligned}
\tag{3.22}
$$

Although this is not required for the implementation of the proposed scheme, Problem (3.22) can be used to compute $(u^{\text{sr}}, y^{\text{sr}})$ from measured data. We write x^{sr} and ξ^{sr} for the minimal and extended state corresponding to $(u^{\text{sr}}, y^{\text{sr}})$, respectively. Clearly, Problem (3.22) is convex. Similar to [69, 82], we require that it is even strongly convex, as captured in the following assumption.

Assumption 3.8. *(Strong convexity) The optimization problem (3.22) is strongly convex w.r.t. $(u^{\text{s}}, y^{\text{s}})$. Moreover, B has full column rank.*

The convexity condition in Assumption 3.8 implies that the optimal reachable equilibrium $(u^{\text{sr}}, y^{\text{sr}})$ is unique, and it is, e.g., satisfied if $T \succ 0$. More generally, in case that $T \succeq 0$, it is also satisfied if $D = 0$ since B has full column rank and these two conditions imply that there exists a unique equilibrium input for any equilibrium output.

Considering only systems with B having full column rank excludes over-actuated systems and guarantees the existence of a unique equilibrium input-output pair for any steady state. To be more precise, we now exploit that B has full column rank in order to derive an inequality used in the proof of our main result. For any equilibrium $(u^{\text{s}}, y^{\text{s}})$ with corresponding steady state x^{s}, it holds that

$$
(I - A)x^{\text{s}} = Bu^{\text{s}} + e, \quad y^{\text{s}} = Cx^{\text{s}} + Du^{\text{s}} + r. \tag{3.23}
$$

Since B has full column rank, (3.23) implies, for arbitrary equilibria $(u^{s,1}, y^{s,1})$, $(u^{s,2}, y^{s,2})$ with corresponding states $x^{s,1}$, $x^{s,2}$,

$$\|u^{s,1} - u^{s,2}\|_2^2 + \|y^{s,1} - y^{s,2}\|_2^2 \leq c_{x,1}\|x^{s,1} - x^{s,2}\|_2^2, \tag{3.24}$$

for $c_{x,1} = 2\|C\|_2^2 + (1 + 2\|D\|_2^2)\,\|B^\dagger(I - A)\|_2^2$, where $B^\dagger$ is the Moore-Penrose inverse of B.

We note that also a converse version of (3.24) holds: By final state observability (2.4), there exists a constant $c_{x,2} > 0$ such that

$$\|x^{s,1} - x^{s,2}\|_2^2 \leq c_{x,2}(\|u^{s,1} - u^{s,2}\|_2^2 + \|y^{s,1} - y^{s,2}\|_2^2). \tag{3.25}$$

3.3.3 Proposed MPC scheme

At time t, given past $l \in \mathbb{I}_{\geq \underline{l}}$ input-output measurements $\{u_k\}_{k=t-l}^{t-1}$, $\{y_k\}_{k=t-l}^{t-1}$ specifying initial conditions, we consider the following open-loop optimal control problem:

Problem 3.4.

$$\underset{\substack{\alpha(t),\bar{u}(t),\bar{y}(t)\\ u^s(t),y^s(t)}}{\text{minimize}} \quad \sum_{k=0}^{L}\left(\|\bar{u}_k(t) - u^s(t)\|_R^2 + \|\bar{y}_k(t) - y^s(t)\|_Q^2\right) + \|u^s(t) - u^r\|_T^2 \tag{3.26a}$$
$$+ \|y^s(t) - y^r\|_S^2$$

subject to

$$\begin{bmatrix}\bar{u}(t)\\ \bar{y}(t)\\ 1\end{bmatrix} = \begin{bmatrix}H_{L+l+1}(u^d)\\ H_{L+l+1}(y^d)\\ \mathbb{1}_{N-L-l}^\top\end{bmatrix}\alpha(t), \tag{3.26b}$$

$$\begin{bmatrix}\bar{u}_{[-l,-1]}(t)\\ \bar{y}_{[-l,-1]}(t)\end{bmatrix} = \begin{bmatrix}u_{[t-l,t-1]}\\ y_{[t-l,t-1]}\end{bmatrix}, \tag{3.26c}$$

$$\begin{bmatrix}\bar{u}_{[L-l,L]}(t)\\ \bar{y}_{[L-l,L]}(t)\end{bmatrix} = \begin{bmatrix}\mathbb{1}_{l+1} \otimes u^s(t)\\ \mathbb{1}_{l+1} \otimes y^s(t)\end{bmatrix}, \tag{3.26d}$$

$$\bar{u}_k(t) \in \mathbb{U},\ \bar{y}_k(t) \in \mathbb{Y},\ k \in \mathbb{I}_{[0,L]}, \tag{3.26e}$$

$$(u^s(t), y^s(t)) \in \mathbb{U}^s \times \mathbb{Y}^s. \tag{3.26f}$$

As before, the optimal solution of Problem 3.4 is denoted by $\alpha^*(t)$, $\bar{u}^*(t)$, $\bar{y}^*(t)$, $u^{s*}(t)$, $y^{s*}(t)$. Further, we write $J_L^*(\xi_t)$ for the optimal open-loop cost of Problem 3.4.

The constraint (3.26b) replaces the model and parametrizes all possible trajectories of the unknown affine system (3.14), compare Theorem 3.3. The initial constraint (3.26c) is analogous to the MPC schemes in Sections 3.1 and 3.2. The terminal constraint over $l+1$ steps in (3.26d) implies that the state $\bar{x}_L(t)$ corresponding to the predicted input-output trajectory $\{\bar{u}_k(t), \bar{y}_k(t)\}_{k=0}^{L}$ is equal to the steady-state $x^s(t)$ corresponding to $(u^s(t), y^s(t))$, respectively. Problem 3.4 is similar to Problem 3.2 in Section 3.1, with the main difference that the input-output setpoint is replaced by an artificial equilibrium $(u^s(t), y^s(t))$, which is optimized online, compare [69, 82]. Moreover, in addition to the tracking cost with matrices $Q, R \succ 0$, the distance of $(u^s(t), y^s(t))$ w.r.t. (u^r, y^r) is also penalized in (3.26a). Note that the terminal constraint (3.26d) and the dynamics (3.26b) imply that $(u^s(t), y^s(t))$ is indeed an equilibrium in the sense of Definition 2.2.

Problem 3.4 is solved in a receding-horizon fashion, according to Algorithm 3.3.

Algorithm 3.3. Data-driven tracking MPC scheme

Offline: Choose upper bound on lag $l \geq \underline{l}$, prediction horizon L, cost matrices $Q, R, S \succ 0$, $T \succeq 0$, constraint sets $\mathbb{U}$, $\mathbb{Y}$, $\mathbb{U}^s$, $\mathbb{Y}^s$, setpoint (u^r, y^r), and generate data $\{u_k^d, y_k^d\}_{k=0}^{N-1}$.

Online: At time $t \in \mathbb{I}_{\geq 0}$, take the past l measurements $\{u_k, y_k\}_{k=t-l}^{t-1}$ and solve Problem 3.4. Apply the input $u_t = \bar{u}_0^*(t)$.

3.3.4 Closed-loop guarantees

In this section, we prove that the MPC scheme defined via Problem 3.4 exponentially stabilizes the optimal reachable equilibrium ξ^{sr} and thus, also the input and output converge exponentially to u^{sr} and y^{sr}, respectively. For this, (u^r, y^r) is not required to satisfy the constraints or to be an equilibrium in the sense of Definition 2.2, in which case $(u^{sr}, y^{sr}) \neq (u^r, y^r)$. The latter point is particularly appealing in the context of data-driven control since verifying whether a given input-output pair is an equilibrium (which is required for the approaches in Sections 3.1 and 3.2) may be challenging in the absence of model knowledge. As in the model-based case, recursive feasibility of the scheme will be guaranteed, even if the target setpoint

$(u^{\mathrm{r}}, y^{\mathrm{r}})$ changes online. As an additional technical contribution, the result of this section extends the (model-based) setpoint tracking MPC analysis of [82] to the case that the stage cost is only positive semidefinite in the state. This is relevant in a model-based setting, e.g., if input-output models are used for prediction.

For the following theoretical analysis, we assume that the available data are persistently exciting in the sense of Section 3.3.1 and that the prediction horizon satisfies $L \geq l + n$.

Assumption 3.9. *(Persistence of excitation) The data satisfy*

$$\operatorname{rank}\left(\begin{bmatrix} H_{L+l+1}(u^{\mathrm{d}}) \\ H_1(x^{\mathrm{d}}_{[0,N-L-l-1]}) \\ \mathbb{1}^\top_{N-L-l} \end{bmatrix}\right) = m(L+l+1) + n + 1. \tag{3.27}$$

Assumption 3.10. *(Length of prediction horizon) The prediction horizon satisfies $L \geq l + n$.*

As will become clear in the proof of our main result in this section, the fact that the stage cost may not be positive definite in the state significantly complicates the analysis of the considered tracking MPC scheme. To overcome this issue, we analyze the closed loop of the proposed (1-step) MPC scheme over n consecutive time steps and show a desired Lyapunov function decay over n steps. Moreover, as in the previous sections, we exploit detectability of the stage cost via an IOSS Lyapunov function: Due to detectability of $(\tilde{A}, \tilde{C})$ (by Assumption 3.1 and Lemma 2.1), there exists an (incremental) IOSS Lyapunov function $W(\xi) = \|\xi\|^2_{P_{\mathrm{W}}}$ with $P_{\mathrm{W}} \succ 0$ which satisfies

$$W(\tilde{A}\Delta\xi + \tilde{B}\Delta u) - W(\Delta\xi) \leq -\|\Delta\xi\|_2^2 + c_{\mathrm{IOSS},1}\|\Delta u\|_2^2 + c_{\mathrm{IOSS},2}\|\Delta y\|_2^2, \tag{3.28}$$

for all $\Delta\xi = \xi_1 - \xi_2$, $\Delta u = u_1 - u_2$, $\Delta y = y_1 - y_2$, $\xi_i \in \mathbb{R}^{n_\xi}$, $u_i \in \mathbb{R}^m$, $y_i = \tilde{C}x_i + \tilde{D}u_i + r$, $i \in \mathbb{I}_{[1,2]}$, with some $c_{\mathrm{IOSS},1}, c_{\mathrm{IOSS},2} > 0$, cf. (2.9). For some $\gamma > 0$, we define a Lyapunov function candidate based on the IOSS Lyapunov function W, the optimal cost J_L^*, and the cost of the optimal reachable equilibrium J^*_{eq}, cf. (3.22), as

$$V(\xi_t) = J_L^*(\xi_t) + \gamma W(\xi_t - \xi^{\mathrm{sr}}) - J^*_{\mathrm{eq}}.$$

Using this Lyapunov function candidate, the following result proves recursive feasibility, constraint satisfaction, and exponential stability of the closed loop.

Theorem 3.4. *Suppose Assumptions 3.1 and 3.7–3.10 hold. If Problem 3.4 is feasible at initial time $t = 0$, then*

i) it is feasible at any $t \in \mathbb{I}_{\geq 0}$,

ii) the closed loop satisfies the constraints, i.e., $u_t \in \mathbb{U}$ and $y_t \in \mathbb{Y}$ for all $t \in \mathbb{I}_{\geq 0}$,

iii) the optimal reachable equilibrium ξ^{sr} is exponentially stable for the resulting closed loop.

The proof of Theorem 3.4 is provided in the appendix (Appendix A.1) and follows the lines of [69]. The main difference lies in the fact that an n-step analysis of the closed loop is performed. This can be explained by noting that the input-output behavior over $n \geq \underline{l}$ steps allows to draw conclusions on the behavior of the internal state, which is relevant for stability (compare the extended state-space system (2.5)). Further, by making an additional case distinction in Case 2b of the proof, we show a decay in the Lyapunov function for the scenario that the actual input and output values (u_{t+i}, y_{t+i}) are close to their respective artificial equilibria $(u^{s*}(t+i), y^{s*}(t+i))$ for $i \in \mathbb{I}_{[0,n-1]}$, but at least one internal state x_{t+k} is not close to $x^{s*}(t+k)$ with $k \in \mathbb{I}_{[0,n-1]}$. As is shown in Lemma A.1, this implies that, along n time steps, at least one input or output must be distant from the artificial equilibrium $(u^{s*}(t+k), y^{s*}(t+k))$. By using this insight and defining a new candidate solution as a convex combination of a simpler candidate solution and the optimal solution, a suitable decay of the optimal cost can be shown. Finally, an IOSS Lyapunov function is employed to translate this decay, which is in terms of input-output values, to a decay in the state.

Remark 3.3. *(Compact constraints) Theorem 3.4 requires compact constraints for Case 2a of the proof, which applies a local controllability argument to treat the case that the state is close to the current artificial steady-state. The proof is readily extendable to non-compact constraints, when considering initial states within some compact sublevel set of the Lyapunov function $V(\xi_t) \leq V_{\max}$, for a given level $V_{\max}$, compare Section 4.3. The only difference in this scenario is that the size of γ_2 in the proof, and hence also the exponential decay rate of V, depends on $V_{\max}$. That is, for larger initial values, the convergence rate derived in the proof decreases.*

Summary

In this section, we presented a data-driven tracking MPC scheme for affine systems. This scheme relies on an extension of the Fundamental Lemma (Theorem 3.3) to systems with affine dynamics and it admits desirable closed-loop guarantees such as recursive feasibility, constraint satisfaction, and stability. Key advantages of including an artificial equilibrium in the MPC scheme are: i) closed-loop stability guarantees in the presence of online setpoint changes, ii) the setpoint $(u^{\mathrm{r}}, y^{\mathrm{r}})$ can be arbitrary and is not required to be reachable or even an equilibrium for the unknown system dynamics, and iii) the region of attraction is significantly larger if compared to the approach without an artificial equilibrium in Section 3.1. These advantages will be illustrated with a numerical example in the following section.

3.4 Application: linearized four-tank system

In the previous sections, we developed three MPC schemes with closed-loop stability guarantees using only input-output data and no explicit model knowledge. In this section, we demonstrate the practicality of these approaches and the validity of our theoretical results by means of a numerical example. To this end, we consider a linearized version of the four-tank system from [116]. This system is well-known as a real-world example, which is open-loop stable, but can be destabilized by an MPC without terminal constraints if the prediction horizon is too short.

This section is based on and taken in parts literally from [JB7][6], [JB9][7], and [JB12][8].

[6]J. Berberich, J. Köhler, M. A. Müller, and F. Allgöwer. "Data-driven model predictive control with stability and robustness guarantees." In: *IEEE Trans. Automat. Control* 66.4 (2021), pp. 1702–1717. ©2020 IEEE

[7]J. Berberich, J. Köhler, M. A. Müller, and F. Allgöwer. "Data-driven tracking MPC for changing setpoints." In: *IFAC-PapersOnLine*, 53.2 (2020), pp. 6923–6930.

[8]J. Berberich, J. Köhler, M. A. Müller, and F. Allgöwer. "On the design of terminal ingredients for data-driven MPC." In: *IFAC-PapersOnLine* 54.6 (2021), pp. 257–263.

System dynamics and parameters

We consider a linearized version of the system from [116], which takes the form

$$x_{k+1} = \begin{bmatrix} 0.921 & 0 & 0.041 & 0 \\ 0 & 0.918 & 0 & 0.033 \\ 0 & 0 & 0.924 & 0 \\ 0 & 0 & 0 & 0.937 \end{bmatrix} x_k + \begin{bmatrix} 0.017 & 0.001 \\ 0.001 & 0.023 \\ 0 & 0.061 \\ 0.072 & 0 \end{bmatrix} u_k, \tag{3.29}$$

$$y_k = \begin{bmatrix} 1 & 0 & 0 & 0 \\ 0 & 1 & 0 & 0 \end{bmatrix} x_k.$$

For the following application of the proposed data-driven MPC schemes, the system matrices are *unknown* and only measured input-output data are available. We assume knowledge of the lag $\underline{l} = 2$ of the system, i.e., we choose, $l = \underline{l} = 2$, but we note that an upper bound (e.g., $l = 4$) leads to comparable performance. The control goal is tracking of the following input-output equilibrium of the linearized system

$$(u^{\mathrm{s}}, y^{\mathrm{s}}) = \left(\begin{bmatrix} 1 \\ 1 \end{bmatrix}, \begin{bmatrix} 0.65 \\ 0.77 \end{bmatrix} \right),$$

which is readily shown to satisfy the dynamics. To set up the data-driven MPC schemes, an input-output trajectory $\{u_k^{\mathrm{d}}, y_k^{\mathrm{d}}\}_{k=0}^{N-1}$ of length $N = 150$ is generated by applying a (persistently exciting) input trajectory sampled uniformly from $u_k^{\mathrm{d}} \in [-1, 1]^2$.

For a given trajectory, there exist infinitely many (arbitrarily large) vectors α satisfying (2.11) and, therefore, a direct implementation of the MPC schemes proposed in this chapter can be numerically ill-conditioned. Hence, we include a quadratic penalty of the form $10^{-4} \cdot \|\alpha(t)\|_2^2$ in the stage cost. In Chapter 4, we provide a theoretical analysis of the effect of such a regularization on the closed-loop performance and robustness in the presence of noisy data.

Stability via terminal equality constraints

We first apply the data-driven MPC scheme with terminal equality constraints from Section 3.1 (Algorithm 3.1). We consider an unconstrained setting, i.e., $\mathbb{U} = \mathbb{R}^m$, $\mathbb{Y} = \mathbb{R}^p$. Further, the prediction horizon is chosen as $L = 30$ and the cost matrices

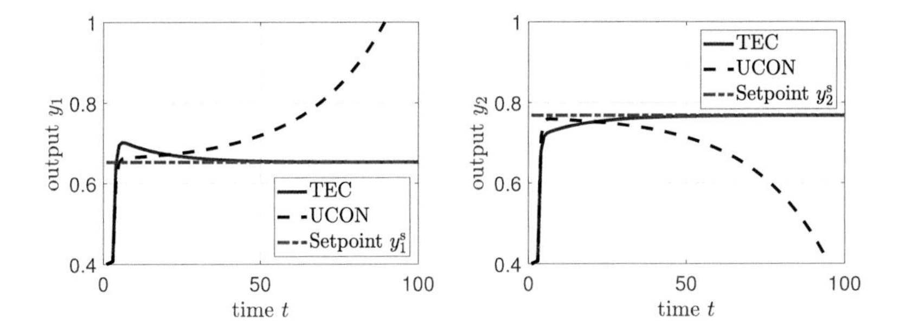

Figure 3.1. Closed-loop output trajectory resulting from the application of the data-driven MPC scheme (Algorithm 3.1) to the linearized four-tank system with terminal equality constraints (TEC, solid) and without terminal constraints (UCON, dashed).

are equal to $Q = I$, $R = 10^{-4}I$. The closed-loop trajectory under Algorithm 3.1 with these parameters can be seen in Figure 3.1. Note that the control goal is fulfilled and the closed-loop input-output trajectory converges to the setpoint (u^s, y^s). On the other hand, if the same scheme without the terminal constraint (3.2d) is applied to the system, then the closed loop is unstable and diverges since the prediction horizon L is too short. This confirms our motivation that rigorous guarantees are indeed desirable for data-driven MPC methods, in particular when they are applied to practical systems. We note that, if the horizon L or the input cost R is increased, then also an MPC scheme without terminal constraints stabilizes the desired equilibrium.

Stability via general terminal ingredients

Next, we demonstrate the advantages of general terminal ingredients as in Section 3.2 over the simple terminal equality constraints from Section 3.1. To this end, we consider the prediction horizon $L = 15$, cost matrices $Q = I$, $R = 10^{-4}I$, the input constraint set $\mathbb{U} = [-2, 2]^2$, and no output constraints, i.e., $\mathbb{Y} = \mathbb{R}^p$. First, we note that the observability matrix Φ_l of this system is square, i.e., the extended system is controllable (compare Lemma 2.1), for $l = \underline{l} = 2$, i.e., if l is chosen as the

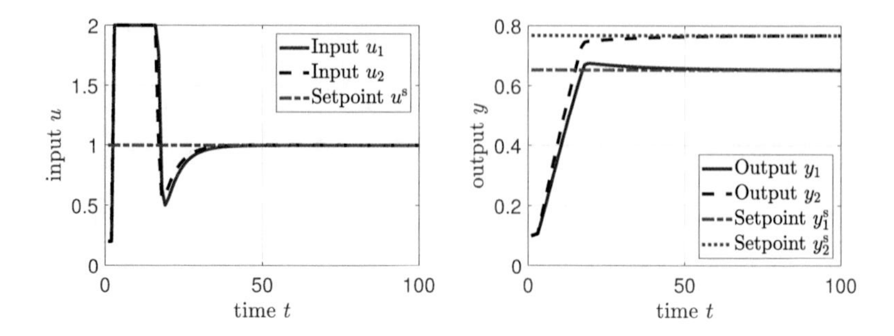

Figure 3.2. Closed-loop trajectory resulting from the application of the data-driven MPC scheme with terminal ingredients (Algorithm 3.2) to the linearized four-tank system.

lag of the system. Based on these ingredients, we apply Proposition 3.1 to compute terminal ingredients satisfying Assumption 3.5. To be precise, using Yalmip [86] with the solver MOSEK [10], we obtain a feasible solution of the conditions in Proposition 3.1 with $K = 0$ and $\gamma = 50$. We then apply the MPC scheme in Algorithm 3.2 with $\Xi_f = \mathbb{R}^{n_\xi}$. Since the above system is open-loop stable, the closed loop under this MPC scheme is globally exponentially stable (compare Remark 3.1). The closed-loop input and output trajectories are displayed in Figure 3.2. It can be seen that the MPC scheme tracks the desired setpoint. On the other hand, an MPC scheme based on terminal equality constraints as in Algorithm 3.1 is not initially feasible for the considered initial condition $x_0 = \begin{bmatrix} 0.1 & 0.1 & 0.2 & 0.2 \end{bmatrix}^\top$ with any prediction horizon $L \leq 24$ due to the input constraints. This means that x_0 does not lie in the (bounded) region of attraction guaranteed by Theorem 3.1, in contrast to the global stability guarantees of Algorithm 3.2 under the above conditions. Further, also for the above parameter configuration, an MPC scheme without any terminal ingredients (i.e., choosing $P = 0$ and $\Xi_f = \mathbb{R}^{n_\xi}$) leads to an unstable closed loop. Finally, we note that it is also possible to design a non-zero K based on Proposition 3.1 resulting in a less conservative terminal penalty. However, in this case, non-trivial terminal set constraints need to be included which reduces the region of attraction.

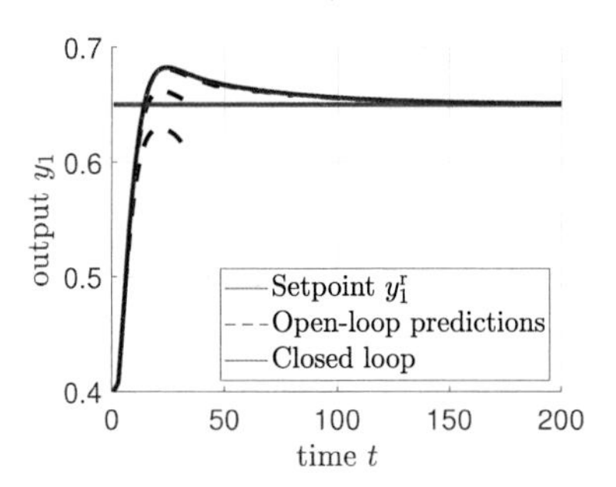

Figure 3.3. First component of the closed-loop output trajectory resulting from the application of the data-driven tracking MPC scheme (Algorithm 3.3) to the linearized four-tank system, along with open-loop predictions $\bar{y}^*(t)$ at multiple time instants $t \in \{0, 5, 15, 32, 47, 77\}$.

Tracking MPC for affine systems

In the following, we apply the data-driven tracking MPC scheme from Section 3.3 to System (3.29). In contrast to the previous sections, we assume that only the output reference $y^{\mathrm{r}} = \begin{bmatrix} 0.65 \\ 0.77 \end{bmatrix}$ is given, whereas the corresponding input setpoint is unknown, i.e., we set the input weight to $T = 0$. We impose input constraints of the form $\mathbb{U} = [-2, 2]^2$, $\mathbb{U}^{\mathrm{s}} = 0.99\mathbb{U}$, and no output constraints, i.e., $\mathbb{Y} = \mathbb{Y}^{\mathrm{s}} = \mathbb{R}^p$. The prediction horizon is chosen as $L = 30$ and the cost matrices are defined as $Q = I$, $R = 10^{-2}I$, $S = 20I$. Since the above system is linear, we drop the constraint $\sum_i \alpha_i = 1$ in (3.26b).

Figure 3.3 illustrates the closed-loop behavior of the first component of the output as well as multiple exemplary open-loop predictions. It can be seen that the artificial equilibrium $y^{s*}(t)$ is updated continuously and converges to the desired setpoint reference, which is in this case equal to the optimal reachable equilibrium, i.e., $y^{sr} = y^{\mathrm{r}}$. Thus, also the closed-loop output converges to the setpoint.

Next, we showcase the ability of the data-driven tracking MPC to handle systems with affine dynamics. To this end, suppose the dynamics of the four-tank system in (3.29) are perturbed by offsets $e = \mathbb{1}_n \otimes 0.02$, $r = \mathbb{1}_p \otimes 0.1$, compare (3.14). We now apply Algorithm 3.3 with and without the constraint that $\sum_i \alpha_i = 1$ in

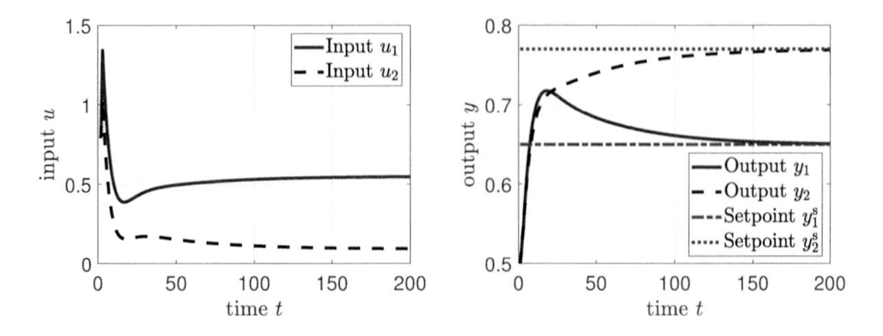

Figure 3.4. Closed-loop trajectory resulting from the application of the data-driven tracking MPC scheme (Algorithm 3.3) to the linearized four-tank system with offset.

Problem 3.4 which is required for handling affine system dynamics according to Theorem 3.3. The closed-loop trajectory under Algorithm 3.3 including the constraint $\sum_i \alpha_i = 1$ can be seen in Figure 3.4. Again, the setpoint can be tracked reliably, although the offset changes the optimal input corresponding to the output reference y^r. On the other hand, when omitting the constraint $\sum_i \alpha_i = 1$, the closed-loop output converges to $\begin{bmatrix} 0.6 \\ 0.76 \end{bmatrix} \neq \begin{bmatrix} 0.65 \\ 0.77 \end{bmatrix} = y^{sr}$, thus confirming the necessity of the constraint $\sum_i \alpha_i = 1$ for the parametrization of trajectories of unknown affine systems based on input-output data.

Comparison of the data-driven MPC approaches

Compared to the simple scheme with terminal equality constraints and without online optimization of an artificial setpoint (Algorithm 3.1), Algorithms 3.2 and 3.3 exhibit a crucial advantage: General terminal ingredients or online optimization of an artificial setpoint both significantly increase the region of attraction. This is confirmed by the above example when choosing a smaller prediction horizon or initial conditions further away from the setpoint, in which case Algorithm 3.1 quickly becomes initially infeasible.

The main advantage of the scheme with general terminal ingredients (Algorithm 3.2) over the tracking MPC scheme (Algorithm 3.3) is that, for the former, *global* stability guarantees can be provided under certain assumptions (Remark 3.1). The region of attraction under Algorithm 3.3 is a neighborhood of the steady-state manifold, which can be smaller or larger than that of Algorithm 3.2 depending on the system dynamics and design parameters. The main drawback of Algorithm 3.2 is that it requires an additional offline design of terminal ingredients, compare Proposition 3.1.

A crucial advantage of the tracking MPC scheme (Algorithm 3.3) over the other two MPC schemes is that it does not require knowledge of the equilibrium input leading to a given output setpoint. In particular, Algorithm 3.3 "finds" u^{sr} automatically, without an additional step of pre-computation. On the other hand, in case only an output setpoint is given, the optimal reachable equilibrium input $u^{\mathrm{sr}} = \begin{bmatrix} 1 \\ 1 \end{bmatrix}$ needs to be computed explicitly for Algorithms 3.1 and 3.2, which can be challenging without model knowledge.

Finally, it is also possible to combine the advantages of general terminal ingredients and an artificial setpoint, i.e., designing a data-driven MPC scheme which contains both. We conjecture that the theoretical guarantees in Theorem 3.4 remain true in this case, i.e., if the terminal equality constraints in Problem 3.4 are replaced by general terminal ingredients as in Section 3.2. This is due to the fact that the stability proof of the data-driven tracking MPC scheme (Theorem 3.4) relies on model-based tracking MPC results from [69] which also consider terminal ingredients.

3.5 Summary and discussion

3.5.1 Summary

In this section, we developed a framework for designing and analyzing data-driven MPC schemes with closed-loop guarantees based only on input-output data and without explicit model knowledge. Our approach combined concepts from standard (model-based) MPC with the data-driven prediction model provided by the Fundamental Lemma. We presented three data-driven MPC schemes, one with terminal equality constraints (Section 3.1), one with general terminal ingredients (Section 3.2), and one with online optimization of an artificial setpoint (Section 3.3).

For each of these schemes, we proved that the closed loop is exponentially stable, where the main technical challenge was posed by the output-dependent and hence (in the state) positive semidefinite cost. Finally, we compared the different MPC schemes and demonstrated their practicality with a numerical example in Section 3.4. We note that additional stability results of data-driven MPC without any terminal ingredients for a sufficiently long prediction horizon are derived in the recent paper [JB17].

3.5.2 Discussion

Let us now discuss important aspects which concern all MPC schemes in this section.

Computational complexity

We would like to emphasize the simplicity of the proposed MPC framework. Without any prior identification step, one measured data trajectory can be used directly to set up an MPC scheme for a linear system. The only prior knowledge about the system required for the implementation is a (potentially rough) upper bound $l \geq \underline{l}$ on the lag (Definition 2.1). Compared to other adaptive or learning-based MPC approaches such as [2, 11, 14, 87, 132, 158], which require initial model knowledge as well as an online estimation process, the complexity of Problem 3.2 is similar to classical MPC schemes, which rely on full model knowledge. To analyze this point in more detail, we focus on the MPC scheme in Section 3.1 but the same considerations hold for Sections 3.2 and 3.3. Note that Problem 3.2 contains in total $N-L-l+1$ decision variables since the decision variables $\bar{u}(t)$, $\bar{y}(t)$ can be replaced by $\alpha(t) \in \mathbb{R}^{N-L-l+1}$ via (3.2b) (using a condensed formulation). For u^{d} to be persistently exciting of order $L+l+n$ (Assumption 3.3), it needs to hold that $N-L-l-n+1 \geq m(L+l+n)$. Assuming equality, Problem 3.2 hence has $m(L+l+n)+n$ free parameters. On the contrary, a condensed model-based MPC optimization problem contains mL decision variables for the input trajectory (assuming that state measurements are available). Thus, the online complexity of the proposed data-driven MPC approach is slightly larger ($m(l+n)+n$ additional decision variables) than that of model-based MPC, but it does not require an a priori (offline) identification step. It is worth noting that the difference in complexity is independent of the horizon L.

Further, note that, for convex polytopic (ellipsoidal) constraint sets $\mathbb{U}$, $\mathbb{Y}$, $\mathbb{U}^s$, $\mathbb{Y}^s$, Ξ_f, Problems 3.2, 3.3, and 3.4 all are convex (quadratically constrained) QPs which can be solved efficiently. In Section 5.3, we compare the computational complexity of model-based and data-driven MPC with a numerical example.

Output-feedback MPC

Note that all MPC schemes proposed in this chapter are inherently output-feedback MPC schemes, i.e., no state measurements are required for their implementation. This is due to the fact that the Fundamental Lemma (Theorem 2.1) directly provides a parametrization of all input-output trajectories from data.

Model-based MPC with positive semidefinite stage cost

Throughout this chapter, we assumed that noise-free data are available such that the Fundamental Lemma provides an exact parametrization of system trajectories. In this case, predicting trajectories via the Fundamental Lemma is equivalent to using a model, e.g., an input-output model or a state-space model, assuming that state measurements are available for the latter. Therefore, all theoretical results in this chapter apply analogously to model-based MPC schemes with an input-output cost and, hence, our results can also be interpreted as an extension of standard MPC results to positive semidefinite stage costs. Further, the theoretical guarantees also apply to subspace predictive control [41], which employs a multi-step input-output model identified from data for predictive control. We note that similar tools (e.g., IOSS properties) have also been used in model-based MPC with positive semidefinite stage cost focusing on stability without terminal constraints [47] and output regulation [70].

Chapter 4

Data-driven MPC for linear systems with noisy data

In this chapter, we consider a data-driven MPC framework, where, in contrast to Chapter 3, the available data are not assumed to be exact measurements of the underlying system, but they can be perturbed by noise. For this scenario, we develop different MPC schemes with guarantees on robust stability and constraint satisfaction based on noisy input-output data.

First, in Section 4.1, we present a modification of the scheme with terminal equality constraints in Section 3.1 to account for noisy output measurements and we prove practical exponential stability of the closed loop. Next, in Section 4.2, this scheme is refined by adding a constraint tightening such that output constraints are satisfied robustly. Further, in Section 4.3, we present a tracking MPC scheme for affine systems with guaranteed robustness w.r.t. disturbed initial conditions and noisy output data, and we show that this scheme practically exponentially stabilizes the optimal reachable equilibrium. Finally, the proposed schemes are validated with numerical examples in Section 4.4.

The results presented in this chapter are based on Berberich et al. [JB7, JB11, JB13].

4.1 Stability via terminal equality constraints

In this section, we propose a robust data-driven MPC scheme with terminal equality constraints and we prove practical exponential stability of the closed loop in the presence of bounded additive output measurement noise. Section 4.1.1 contains the scheme, which is essentially a modification to robustify the nominal scheme in Section 3.1, as well as detailed explanations of the key ingredients. In Sections 4.1.2 and 4.1.3, we prove two technical lemmas, which will be required for our main theoretical results. Recursive feasibility of the closed loop is proven in Section 4.1.4. In Section 4.1.5, we show that, under suitable assumptions, the closed loop resulting from the application of the developed MPC scheme leads to a practically exponentially stable closed loop.

This section is based on and taken in parts literally from [JB7][1].

4.1.1 Proposed MPC scheme

Problem setting

In practical applications, the output of the unknown LTI system (2.1) is often not available exactly. Instead, one typically has access to *noisy* data, which are, e.g., perturbed by measurement noise. In this case, the stacked data-dependent Hankel matrices in Theorem 2.1 do not span the system's trajectory space exactly and thus, the output trajectories cannot be predicted accurately. Moreover, noisy output measurements enter the initial conditions in (3.2c), which deteriorates the prediction accuracy even further. Therefore, a direct application of the MPC schemes of Chapter 3 may lead to feasibility issues or it may render the closed loop unstable.

In this section, we tackle the issue of noisy measurements with a robust data-driven MPC scheme with terminal constraints. Our goal is to stabilize a given input-output equilibrium (u^s, y^s) for System (2.1) while satisfying input constraints $u_t \in \mathbb{U}$, $t \in \mathbb{I}_{\geq 0}$, for a given closed set $\mathbb{U} \subseteq \mathbb{R}^m$.

Assumption 4.1. *(Input constraint set) The input constraint set $\mathbb{U}$ satisfies $u^s \in \mathrm{int}(\mathbb{U})$.*

In this section, we only address input constraints, and robust output constraint satisfaction will be considered separately in Section 4.2. To achieve the above control goal, we only have noisy data and no knowledge of the matrices in (2.1). To be precise, we have a noisy input-output trajectory $\{u_k^d, \tilde{y}_k^d\}_{k=0}^{N-1}$ of (2.1), where

[1] J. Berberich, J. Köhler, M. A. Müller, and F. Allgöwer. "Data-driven model predictive control with stability and robustness guarantees." In: *IEEE Trans. Automat. Control* 66.4 (2021), pp. 1702–1717.

the output measurements are affected by bounded additive noise, i.e., $\tilde{y}_k^{\mathrm{d}} = y_k^{\mathrm{d}} + \varepsilon_k^{\mathrm{d}}$ for $k \in \mathbb{I}_{[0,N-1]}$. Similarly, the online output measurements used to specify initial conditions are perturbed as $\tilde{y}_k = y_k + \varepsilon_k$ for $k \in \mathbb{I}_{\geq 0}$.

Assumption 4.2. *(Noise bound) It holds that $\|\varepsilon_k^{\mathrm{d}}\|_\infty \leq \bar{\varepsilon}$, $k \in \mathbb{I}_{[0,N-1]}$, and $\|\varepsilon_k\|_\infty \leq \bar{\varepsilon}$, $k \in \mathbb{I}_{\geq 0}$, with known $\bar{\varepsilon} > 0$.*

Note that the present setting includes two types of noise. The data used for the prediction via Hankel matrices is perturbed by ε^{d}, which can thus be interpreted as a multiplicative model uncertainty. On the other hand, ε perturbs the online measurements and hence, the overall control goal is a noisy output-feedback problem. Finally, we assume that System (2.1) is controllable and observable (compare Assumption 3.1).

Assumption 4.3. *(Controllability and observability) The pair (A, B) is controllable and the pair (A, C) is observable.*

MPC scheme

Given a noisy initial input-output trajectory $\{u_k, \tilde{y}_k\}_{k=t-n}^{t-1}$ of length n, and noisy data $\{u_k^{\mathrm{d}}, \tilde{y}_k^{\mathrm{d}}\}_{k=0}^{N-1}$, we define the following optimal control problem:

Problem 4.1.

$$\underset{\substack{\alpha(t),\sigma(t)\\ \bar{u}(t),\bar{y}(t)}}{\text{minimize}} \quad \sum_{k=0}^{L-1} \left(\|\bar{u}_k(t) - u^{\mathrm{s}}\|_R^2 + \|\bar{y}_k(t) - y^{\mathrm{s}}\|_Q^2 \right) + \lambda_\alpha \bar{\varepsilon} \|\alpha(t)\|_2^2 + \frac{\lambda_\sigma}{\bar{\varepsilon}} \|\sigma(t)\|_2^2 \tag{4.1a}$$

subject to

$$\begin{bmatrix} \bar{u}(t) \\ \bar{y}(t) + \sigma(t) \end{bmatrix} = \begin{bmatrix} H_{L+n}(u^{\mathrm{d}}) \\ H_{L+n}(\tilde{y}^{\mathrm{d}}) \end{bmatrix} \alpha(t), \tag{4.1b}$$

$$\begin{bmatrix} \bar{u}_{[-n,-1]}(t) \\ \bar{y}_{[-n,-1]}(t) \end{bmatrix} = \begin{bmatrix} u_{[t-n,t-1]} \\ \tilde{y}_{[t-n,t-1]} \end{bmatrix}, \tag{4.1c}$$

$$\begin{bmatrix} \bar{u}_{[L-n,L-1]}(t) \\ \bar{y}_{[L-n,L-1]}(t) \end{bmatrix} = \begin{bmatrix} \mathbb{1}_n \otimes u^{\mathrm{s}} \\ \mathbb{1}_n \otimes y^{\mathrm{s}} \end{bmatrix}, \tag{4.1d}$$

$$\bar{u}_k(t) \in \mathbb{U}, \; k \in \mathbb{I}_{[0,L-1]}. \tag{4.1e}$$

Similar to Chapter 3, we denote the optimal solution of Problem 4.1 by $\alpha^*(t)$, $\sigma^*(t)$, $\bar{u}^*(t)$, $\bar{y}^*(t)$ and the optimal cost by $J_L^*(\tilde{\xi}_t)$, where

$$\tilde{\xi}_t := \begin{bmatrix} u_{[t-n,t-1]} \\ \tilde{y}_{[t-n,t-1]} \end{bmatrix} = \begin{bmatrix} u_{[t-n,t-1]} \\ y_{[t-n,t-1]} + \varepsilon_{[t-n,t-1]} \end{bmatrix}$$

denotes the (noisy) initial extended state.

Problem 4.1 is analogous to the nominal MPC problem (Problem 3.2) with the key difference that the output data trajectory y^{d} as well as the initial output $y_{[t-n,t-1]}$, which is obtained via online measurements, have been replaced by their noisy counterparts. Further, the following ingredients have been added:

a) A slack variable σ to account for the noisy online measurements $\tilde{y}_{[t-n,t-1]}$ and for the noisy data $\tilde{y}^{\mathrm{d}}$ used for prediction, which can be interpreted as a multiplicative model uncertainty,

b) Quadratic regularization (i.e., *ridge regularization*) of α and σ with weights $\lambda_\alpha\bar{\varepsilon} > 0$, $\frac{\lambda_\sigma}{\bar{\varepsilon}} > 0$, i.e., the regularization parameters depend on the noise level.

The above ℓ_2-norm regularization for $\alpha(t)$ implies that small values of $\|\alpha(t)\|_2^2$ are preferred. Since the noisy Hankel matrix $H_{L+n}(\tilde{y}^{\mathrm{d}})$ is multiplied by $\alpha(t)$ in (4.1b), this implicitly reduces the influence of the noise on the prediction accuracy. Loosely speaking, for increasing λ_α, the term $\lambda_\alpha\bar{\varepsilon}\|\alpha(t)\|_2^2$ reduces the "complexity" of the data-driven system description (4.1b), similar to regularization methods in linear regression, thus allowing for a trade-off between tracking performance and the avoidance of overfitting. The term $\frac{\lambda_\sigma}{\bar{\varepsilon}}\|\sigma(t)\|_2^2$ yields small values for the slack variable $\sigma(t)$, thus improving the prediction accuracy. An alternative to the present regularization terms are general quadratic regularization kernels, i.e., costs of the form $\|\alpha(t)\|_{P_\alpha}^2$, $\|\sigma(t)\|_{P_\sigma}^2$ for suitable matrices P_α, $P_\sigma \succ 0$, or norm penalties $\|\alpha(t)\|_{p_1}$, $\|\sigma(t)\|_{p_2}$ with arbitrary $p_1, p_2 = 1, 2, \ldots, \infty$. Throughout this thesis, we consider simple quadratic penalty terms since this simplifies the arguments, but we conjecture that most of our theoretical results remain to hold for different regularization terms. Finally, we note that similar forms of regularization have been used in related works on data-driven optimal control, see [92, Section 5.2.2] for a detailed exposition.

Throughout this chapter, we consider input-output initial conditions (4.1c) as well as the extended state in (2.3) over n time steps, i.e., we choose the upper bound on the lag as $l = n \geq \underline{l}$. This is mainly done for notational simplicity. In particular, all results directly apply if n is replaced by an upper bound on the system order and they can be further relaxed to upper bounds $l \geq \underline{l}$ on the lag with minor modifications.

The proposed MPC scheme relies on repeatedly solving Problem 4.1 in a multi-step fashion (compare [49, 146]). To be more precise, we consider the scenario that, after solving Problem 4.1 online, the first n computed inputs are applied to the system. Thereafter, the horizon is shifted by n steps, before the whole scheme is repeated (compare Algorithm 4.1).

Algorithm 4.1. Robust data-driven MPC scheme with terminal equality constraints

Offline: Choose upper bound on system order n, prediction horizon L, cost matrices $Q, R \succ 0$, regularization parameters $\lambda_\alpha, \lambda_\sigma > 0$, constraint set $\mathbb{U}$, noise bound $\bar{\varepsilon}$, setpoint (u^s, y^s), and generate data $\{u_k^d, \tilde{y}_k^d\}_{k=0}^{N-1}$.

Online: At time $t = n \cdot i$, $i \in \mathbb{I}_{\geq 0}$, take the past n measurements $\{u_k, \tilde{y}_k\}_{k=t-n}^{t-1}$ and solve Problem 4.1. Apply the input $u_{[t,t+n-1]} = \bar{u}^*_{[0,n-1]}(t)$ over the next n time steps.

Remark 4.1. *(Multi-step MPC) As we will see in the remainder of this section, for the considered setting with output measurement noise, the multi-step MPC scheme described in Algorithm 4.1 has superior theoretical properties if compared to its corresponding 1-step version. This is mainly due to the terminal equality constraints (4.1d), which complicate the proof of recursive feasibility, similar as in model-based robust MPC with terminal equality constraints and model mismatch. We show in this section that, for an n-step MPC scheme with a terminal equality constraint, practical exponential stability can be proven. On the other hand, we comment on the differences for the corresponding 1-step MPC scheme in Section 4.1.4 (Remark 4.4). In particular, for a 1-step MPC scheme relying on Problem 4.1, recursive feasibility holds only locally around (u^s, y^s) and thus, only local stability can be guaranteed. Finally, we note that a ν-step MPC scheme, where $\nu \leq n$ denotes the controllability index, admits the same theoretical guarantees as the n-step MPC scheme in Algorithm 4.1.*

Remark 4.2. *(Special case of noise-free data) In the nominal case of Chapter 3, i.e., for $\bar{\varepsilon} \to 0$, the regularization terms fulfill $\lambda_\alpha \bar{\varepsilon} \to 0$ and $\frac{\lambda_\sigma}{\bar{\varepsilon}} \to \infty$, which implies $\sigma(t) \to 0$.*

Hence, the system dynamics (4.1b) as well as the initial conditions (4.1c) approach their nominal counterparts. Thus, for $\bar{\varepsilon} = 0$, Problem 4.1 reduces to the nominal Problem 3.2.

Remark 4.3. *(Non-convex constraint on $\sigma(t)$) The theoretical stability results for robust data-driven MPC in the earlier work [JB7] required an additional (non-convex) constraint $\|\sigma(t)\|_\infty \leq \bar{\varepsilon}(1 + \|\alpha(t)\|_1)$ to bound the slack variable. Motivated by [JB17], the regularization of $\sigma(t)$ in Problem 4.1 depends inversely on the noise level which ensures small values of $\sigma(t)$, provided that the noise level $\bar{\varepsilon}$ is sufficiently small. With this regularization, the additional non-convex constraint from [JB7] is not required such that, for a polytopic (ellipsoidal) input constraint set $\mathbb{U}$, Problem 4.1 is a strictly convex (quadratically constrained) QP.*

As in the previous chapter (Assumption 3.3), we require that the measured input u^{d} is persistently exciting of order $L + l + n = L + 2n$.

Assumption 4.4. *(Persistence of excitation) The sequence $\{u_k^{\mathrm{d}}\}_{k=0}^{N-1}$ is persistently exciting of order $L + 2n$.*

We denote the state trajectory corresponding to $(u^{\mathrm{d}}, y^{\mathrm{d}})$ by x^{d}. According to [144, Corollary 2], Assumption 4.4 implies that the matrix

$$H_{\mathrm{ux}} = \begin{bmatrix} H_{L+n}(u^{\mathrm{d}}) \\ H_1(x^{\mathrm{d}}_{[0,N-L-n]}) \end{bmatrix} \tag{4.2}$$

has full row rank and thus admits a right-inverse $H_{\mathrm{ux}}^\dagger = H_{\mathrm{ux}}^\top \left(H_{\mathrm{ux}} H_{\mathrm{ux}}^\top\right)^{-1}$. Define the squared maximum singular value

$$c_{\mathrm{pe}} := \left\| H_{\mathrm{ux}}^\dagger \right\|_2^2. \tag{4.3}$$

For our stability results, we will require that $c_{\mathrm{pe}}\bar{\varepsilon}$ is bounded from above by a sufficiently small number. Essentially, this corresponds to a quantitative "persistence-of-excitation-to-noise"-bound. To be more precise, abbreviate in the following $U = H_{L+n}(u^{\mathrm{d}})$ and suppose that

$$\rho I_{m(L+n)} \preceq UU^\top \preceq \nu I_{m(L+n)} \tag{4.4}$$

for scalar constants $\rho, \nu > 0$. Further, define the quantity

$$c_{\mathrm{pe}}^{\mathrm{u}} = \|U^\dagger\|_2^2 = \|U^\top (UU^\top)^{-1}\|_2^2.$$

Then, it holds that

$$c_{\text{pe}}^{\text{u}} \leq \|U^\top\|_2^2 \|(UU^\top)^{-1}\|_2^2 \leq \frac{\lambda_{\max}(UU^\top)}{\lambda_{\min}(UU^\top)^2} \overset{(4.4)}{\leq} \frac{\nu}{\rho^2}. \tag{4.5}$$

Thus, if a persistently exciting input u^{d} is multiplied by a constant $c > 1$, then c_{pe}^{u} decreases proportionally to $\frac{1}{c^2}$. Further, the constant ρ can typically be chosen larger if the data length N increases. The same arguments can be carried out when assuming a bound of the form (4.4) for the matrix (4.2), but finding a suitable input which generates data achieving such a bound is less obvious. It is well-known for classical definitions of persistence of excitation that larger excitation of the input implies larger excitation of the state. Therefore, we conjecture (and we have observed for various practical simulation examples) that c_{pe} decreases with increasing data horizons N and with multiplications of a persistently exciting input data trajectory u^{d} by a scalar constant greater than one. To summarize, for a given noise level $\bar{\varepsilon}$, robust stability as guaranteed in the following sections can be obtained by choosing a large enough persistently exciting input u^{d} and/or a sufficiently large data horizon N, both of which ensure a large signal-to-noise ratio.

4.1.2 Local upper bound of Lyapunov function

In this section, we show that the optimal cost of Problem 4.1 admits a quadratic upper bound, similar to the nominal case (cf. Assumption 3.2). It is straightforward to see that such an upper bound cannot be quadratic in the minimal state x: The optimal cost J_L^* depends explicitly on $\alpha^*(t)$ via $\lambda_\alpha\bar{\varepsilon}\|\alpha^*(t)\|_2^2$, which in turn depends on the past n inputs and outputs $(u_{[t-n,t-1]}, y_{[t-n,t-1]})$ through (4.1b) and (4.1c). Even if the current state is zero, i.e., $x_t = 0$, these may in general be arbitrarily large and hence, α and therefore also J_L^* may be arbitrarily large. Thus, J_L^* does not admit an upper bound in the minimal state x_t. Therefore, similar to Section 3.2, we consider the extended (not minimal) state ξ defined in (2.3) for our analysis. Since ξ is the state of a detectable state-space realization (Lemma 2.1), there exists an IOSS Lyapunov function $W(\xi) = \|\xi\|_{P_{\text{W}}}^2$, $P_{\text{W}} \succ 0$, compare (2.9). For some $\gamma > 0$, define $V_t := J_L^*(\tilde{\xi}_t) + \gamma W(\xi_t)$.

We assume for our theoretical analysis that $(u^{\text{s}}, y^{\text{s}}) = (0, 0)$. For the presented robust data-driven MPC scheme, setpoints $(u^{\text{s}}, y^{\text{s}}) \neq (0, 0)$ change mainly one quantitative constant in Lemma 4.1 below. We comment on the main differences in the case $(u^s, y^s) \neq (0,0)$ in Section 4.1.4 (Remark 4.5).

Further, similar to the results in Chapter 3, we assume that the prediction horizon is sufficiently long (compare Assumption 3.10).

Assumption 4.5. *(Length of prediction horizon) The prediction horizon satisfies $L \geq 2n$.*

The following result shows that, under the given assumptions, we can derive a meaningful quadratic upper bound on V depending on the state ξ.

Lemma 4.1. *Suppose Assumptions 4.1–4.5 hold. Then, there exist constants $c_3, \delta > 0$ such that, for all ξ_t with $\|\xi_t\|_2 \leq \delta$, Problem 4.1 is feasible and V is bounded as*

$$\gamma\lambda_{\min}(P_W)\|\xi_t\|_2^2 \leq V_t \leq c_3\|\xi_t\|_2^2 + c_4\bar{\varepsilon}, \tag{4.6}$$

where $c_4 = 2np\lambda_\sigma$.

Proof. The lower bound is trivial. For the upper bound, we construct a feasible candidate solution to Problem 4.1 which brings the state x and thus the output y to zero in L steps. Obviously, we have $\bar{u}_{[-n,-1]}(t) = u_{[t-n,t-1]}$ as well as $\bar{y}_{[-n,-1]}(t) = \tilde{y}_{[t-n,t-1]}$ by (4.1c). By assumption, we have $L \geq 2n$ as well as $0 \in \text{int}(\mathbb{U})$. Thus, by controllability, there exists a $\delta > 0$ such that for any x_t with $\frac{1}{\|T_{x,\xi}\|_2}\|x_t\|_2 \overset{(2.4)}{\leq} \|\xi_t\|_2 \leq \delta$, there exists an input trajectory $\bar{u}_{[0,L-1]}(t) \in \mathbb{U}^L$, which steers the corresponding state $\bar{x}_{[0,L-1]}(t)$ and output $\bar{y}_{[0,L-1]}(t)$ to the origin in $L - n \geq n$ steps while satisfying

$$\sum_{k=0}^{L-1} \|\bar{u}_k(t)\|_2^2 + \|\bar{y}_k(t)\|_2^2 \leq \Gamma_{\text{uy}}\|x_t\|_2^2 \tag{4.7}$$

for a suitable constant $\Gamma_{\text{uy}} > 0$, compare (2.10). Moreover, $\alpha(t)$ is chosen as

$$\alpha(t) = H_{\text{ux}}^\dagger \begin{bmatrix} \bar{u}(t) \\ x_{t-n} \end{bmatrix}, \tag{4.8}$$

where H_{ux} is defined in (4.2). As described in more detail in [94], the output of an LTI system is a linear combination of its initial condition and the input and, therefore, the above choice of $\alpha(t)$ implies

$$\begin{bmatrix} H_{L+n}(u^d) \\ H_{L+n}(y^d) \end{bmatrix} \alpha(t) = \begin{bmatrix} \bar{u}(t) \\ \bar{y}_{[-n,-1]}(t) - \varepsilon_{[t-n,t-1]} \\ \bar{y}_{[0,L-1]}(t) \end{bmatrix},$$

where $\varepsilon_{[t-n,t-1]}$ is the true noise instance. For the slack variable $\sigma(t)$, we choose

$$\begin{aligned}\sigma_{[-n,-1]}(t) &= H_n(\varepsilon^{\mathrm{d}}_{[0,N-L-1]})\alpha(t) - \varepsilon_{[t-n,t-1]},\\ \sigma_{[0,L-1]}(t) &= H_L(\varepsilon^{\mathrm{d}}_{[n,N-1]})\alpha(t),\end{aligned} \tag{4.9}$$

which implies that (4.1b)–(4.1d) are satisfied. In the following, we employ the above candidate solution to bound the optimal cost and, thereby, the function V. Due to observability of the pair (A, C), it holds that

$$\begin{bmatrix} \bar{u}_{[-n,-1]}(t) \\ x_{t-n} \end{bmatrix} = \underbrace{\begin{bmatrix} I_{mn} & 0 \\ M_1 & \Phi_n^{\dagger} \end{bmatrix}}_{M:=} \xi_t, \tag{4.10}$$

where $\Phi_n^{\dagger}$ is the Moore-Penrose inverse of the observability matrix Φ_n, cf. (2.2). The lower block of (4.10) follows from observability and the linear system dynamics $x_{k+1} = Ax_k + Bu_k$, $y_k = Cx_k + Du_k$ for $k \in \mathbb{I}_{[t-n,t-1]}$, which can be used to compute the matrix M_1 depending on A, B, C, D. Hence, $\alpha(t)$ can be bounded as

$$\begin{aligned}\|\alpha(t)\|_2^2 &\overset{(4.8)}{\leq} \|H_{\mathrm{ux}}^{\dagger}\|_2^2 \left(\|\bar{u}(t)\|_2^2 + \|x_{t-n}\|_2^2\right) \\ &= \|H_{\mathrm{ux}}^{\dagger}\|_2^2 \left(\|\bar{u}_{[0,L-1]}(t)\|_2^2 + \left\|\begin{bmatrix} \bar{u}_{[-n,-1]}(t) \\ x_{t-n} \end{bmatrix}\right\|_2^2\right) \\ &\overset{(4.7),(4.10)}{\leq} \underbrace{\|H_{\mathrm{ux}}^{\dagger}\|_2^2}_{c_{\mathrm{pe}}=} \left(\Gamma_{\mathrm{uy}}\|x_t\|_2^2 + \|M\|_2^2\|\xi_t\|_2^2\right).\end{aligned} \tag{4.11}$$

Using standard norm equivalence properties, it holds for arbitrary $k \in \mathbb{N}$ that

$$\|H_k(\varepsilon^{\mathrm{d}}_{[0,N-L-n+k-1]})\|_2^2 \leq c_5 k \bar{\varepsilon}^2, \tag{4.12}$$

where $c_5 := p(N-L-n+1)$. Based on the definition of $\sigma(t)$ in (4.9), and using (4.12) as well as the inequality $(a+b)^2 \leq 2(a^2+b^2)$, we can bound $\sigma(t)$ in terms of $\alpha(t)$ as

$$\|\sigma(t)\|_2^2 \leq 2np\bar{\varepsilon}^2 + c_5(L+2n)\bar{\varepsilon}^2\|\alpha(t)\|_2^2. \tag{4.13}$$

Combining the above inequalities and using $W(\xi_t) \leq \lambda_{\max}(P_{\mathrm{W}})\|\xi_t\|_2^2$, V is upper bounded as

$$\begin{aligned}V_t \leq& \lambda_{\max}(Q,R)\Gamma_{\mathrm{uy}}\|x_t\|_2^2 + \gamma\lambda_{\max}(P_{\mathrm{W}})\|\xi_t\|_2^2 + 2np\bar{\varepsilon}\lambda_{\sigma} \\ &+ (\lambda_{\alpha} + c_5(L+2n)\lambda_{\sigma})\, c_{\mathrm{pe}}\bar{\varepsilon}\left(\Gamma_{\mathrm{uy}}\|x_t\|_2^2 + \|M\|_2^2\|\xi_t\|_2^2\right).\end{aligned}$$

Finally, x_t is bounded by ξ_t as $\|x_t\|_2^2 \overset{(2.4)}{\leq} \|T_{\mathrm{x},\xi}\|_2^2\|\xi_t\|_2^2$, which leads to $V_t \leq c_3\|\xi_t\|_2^2 + c_4\bar{\varepsilon}$, where

$$\begin{aligned} c_3 =& \lambda_{\max}(Q,R)\Gamma_{\mathrm{uy}}\|T_{\mathrm{x},\xi}\|_2^2 + \gamma\lambda_{\max}(P_{\mathrm{W}}) \\ &+ (\lambda_\alpha + c_5(L+2n)\lambda_\sigma)\, c_{\mathrm{pe}}\bar{\varepsilon}\left(\Gamma_{\mathrm{uy}}\|T_{\mathrm{x},\xi}\|_2^2 + \|M\|_2^2\right), \\ c_4 =& 2np\lambda_\sigma. \end{aligned}$$

■

In Section 3.1, we assumed that the optimal cost is quadratically upper bounded (cf. Assumption 3.2), which is not restrictive in the nominal linear-quadratic setting. Lemma 4.1 proves that, under mild assumptions, the optimal cost of Problem 4.1 admits (locally) a similar upper bound and can thus be seen as the robust counterpart of Assumption 3.2.

The term c_4 is solely due to the cost of the slack variable σ. This can be explained by noting that, for $\xi_t = 0$, $\alpha(t)$, $\bar{u}_{[0,L-1]}(t)$, $\bar{y}_{[0,L-1]}(t)$ can all be chosen to be zero, as long as σ compensates the noise, i.e., $\sigma_{[-n,-1]}(t) = -\varepsilon_{[t-n,t-1]}$.

4.1.3 Prediction error bound

We denote the output trajectory resulting from an open-loop application of $\bar{u}^*(t)$ to System (2.1) with initial condition x_t by $\hat{y}$. To be precise, $\hat{y}$ is defined as

$$\hat{y}_{t+k} = CA^k x_t + \sum_{j=0}^{k-1} CA^{k-1-j}B\bar{u}_j^*(t) + D\bar{u}_k^*(t). \tag{4.14}$$

One of the reasons why it is difficult to analyze the presented MPC scheme is the non-trivial relation between the predicted output $\bar{y}^*(t)$ and the "actual" output $\hat{y}$. In the following, we derive a bound on the difference between the two quantities, which will play an important role in proving recursive feasibility and practical stabiliy of the proposed scheme. For some $k \in \mathbb{I}_{\geq 0}$, choose constants $\rho_{2,k}$, $\rho_{\infty,k}$ such that

$$\rho_{2,k} \geq \|CA^k\Phi_n^\dagger\|_2^2, \quad \rho_{\infty,k} \geq \|CA^k\Phi_n^\dagger\|_\infty, \tag{4.15}$$

where $\Phi_n^\dagger$ is the Moore-Penrose inverse of the observability matrix Φ_n, cf. (2.2).

Lemma 4.2. *Suppose Assumptions 4.2 and 4.3 hold. If Problem 4.1 is feasible at time t, then the following inequalities hold for all $k \in \mathbb{I}_{[0,L-1]}$*

$$\begin{aligned} \|\hat{y}_{t+k} - \bar{y}_k^*(t)\|_2^2 \leq & 8c_5\bar{\varepsilon}^2\|\alpha^*(t)\|_2^2 + 2\|\sigma_k^*(t)\|_2^2 \\ & + \rho_{2,n+k}\left(16n\bar{\varepsilon}^2\left(c_5\|\alpha^*(t)\|_2^2 + p\right) + 4\|\sigma_{[-n,-1]}^*(t)\|_2^2\right), \end{aligned} \tag{4.16}$$

$$\begin{aligned} \|\hat{y}_{t+k} - \bar{y}_k^*(t)\|_\infty \leq & \bar{\varepsilon}\|\alpha^*(t)\|_1 + \|\sigma_k^*(t)\|_\infty \\ & + \rho_{\infty,n+k}\left(\bar{\varepsilon}\left(\|\alpha^*(t)\|_1 + 1\right) + \|\sigma_{[-n,-1]}^*(t)\|_\infty\right), \end{aligned} \tag{4.17}$$

with c_5 from (4.12).

Proof. We show only (4.17) and note that (4.16) can be derived following the same steps, using (4.12) as well as the inequality $(a+b)^2 \leq 2a^2 + 2b^2$. As written above, $\hat{y}$ is the trajectory which results from an open-loop application of $\bar{u}^*(t)$ and with initial conditions specified by x_t. On the other hand, according to (4.1b), $\bar{y}^*(t)$ is comprised as

$$\bar{y}^*(t) = H_{L+n}(\varepsilon^{\mathrm{d}})\alpha^*(t) + H_{L+n}(y^{\mathrm{d}})\alpha^*(t) - \sigma^*(t).$$

It follows directly from (4.1b) and (4.1c) that the second term on the right-hand side $H_{L+n}(y^{\mathrm{d}})\alpha^*(t)$ is the trajectory which results from an open-loop application of $\bar{u}^*(t)$ and with initial output conditions

$$\tilde{y}_{[t-n,t-1]} + \sigma_{[-n,-1]}^*(t) - H_n(\varepsilon_{[0,N-L-1]}^{\mathrm{d}})\alpha^*(t).$$

Define

$$y_{[t-n,t+L-1]}^- = \hat{y}_{[t-n,t+L-1]} - H_{L+n}(y^{\mathrm{d}})\alpha^*(t).$$

Due to the LTI dynamics and since y^- contains the difference between two trajectories with the same input, we can assume $\bar{u}^*(t) = 0$ for the following arguments without loss of generality. Hence, y^- is equal to the output component of a trajectory (u^-, y^-) with zero input and with initial trajectory

$$\begin{bmatrix} u_{[t-n,t-1]}^- \\ y_{[t-n,t-1]}^- \end{bmatrix} = \begin{bmatrix} 0 \\ H_n(\varepsilon_{[0,N-L-1]}^{\mathrm{d}})\alpha^*(t) - \varepsilon_{[t-n,t-1]} - \sigma_{[-n,-1]}^*(t) \end{bmatrix}. \tag{4.18}$$

The relation to the internal state x^- can be derived as

$$y_{[t-n,t-1]}^- = \Phi_n x_{t-n}^-,$$

with the observability matrix Φ_n. This leads to the corresponding output at time $t+k$

$$y_{t+k}^{-} = CA^{n+k}\Phi_n^{\dagger} y_{[t-n,t-1]}^{-}. \tag{4.19}$$

Further, writing e_i for a row vector whose i-th component is equal to 1 and which is zero otherwise, we obtain

$$\|H_{L+n}(\varepsilon^{\mathrm{d}})\alpha(t)\|_\infty = \max_{i\in\mathbb{I}_{[1,p(L+n)]}} |e_i H_{L+n}(\varepsilon^{\mathrm{d}})\alpha(t)| \leq \bar{\varepsilon}\|\alpha(t)\|_1. \tag{4.20}$$

Combining (4.19) with the expression for $y_{[t-n,t-1]}^{-}$ in (4.18) as well as Inequality (4.20), $\|y_{t+k}^{-}\|_\infty$ can be bounded as

$$\|y_{t+k}^{-}\|_\infty \leq \rho_{\infty,n+k}\left(\bar{\varepsilon}\left(\|\alpha^*(t)\|_1+1\right)+\|\sigma_{[-n,-1]}^*(t)\|_\infty\right).$$

Note that

$$\|\hat{y}_{t+k} - \bar{y}_k^*(t)\|_\infty \leq \|y_{t+k}^{-}\|_\infty + \bar{\varepsilon}\|\alpha^*(t)\|_1 + \|\sigma_k^*(t)\|_\infty,$$

which concludes the proof. ∎

Essentially, Lemma 4.2 gives a bound on the mismatch between the predicted output and the actual output resulting from the open-loop application of $\bar{u}^*(t)$, depending on the optimal solutions α^*, σ^*, and on system parameters. In model-based robust MPC schemes, similar bounds are typically used to propagate uncertainty, where the role of the weighting vector α to account for multiplicative uncertainty is replaced by the state x and a model-based uncertainty description (compare [74] for details). The main difference in the proposed MPC scheme is that the predicted trajectory $\bar{y}^*(t)$ is in general not a trajectory of the LTI system (2.1) corresponding to the input $\bar{u}^*(t)$. On the contrary, in model-based robust MPC, the predicted trajectory usually satisfies the dynamics of a (nominal) model of the system.

4.1.4 Recursive feasibility

The following result shows that, if the proposed robust MPC scheme is feasible at time t, then it is also feasible at time $t+n$, assuming that the noise level is sufficiently small.

Proposition 4.1. *Suppose Assumptions 4.1–4.5 hold. Then, for any $V_{\max} > 0$, there exists an $\bar{\varepsilon}_{\max} > 0$ such that for all $\bar{\varepsilon} \leq \bar{\varepsilon}_{\max}$, if $V_t \leq V_{\max}$ for some $t \in \mathbb{I}_{\geq 0}$, then Problem 4.1 is feasible at time $t+n$.*

Proof. Suppose Problem 4.1 is feasible at time t with $V_t \leq V_{\max}$. As in Lemma 4.2, the trajectory resulting from an open-loop application of $\bar{u}^*(t)$ and with initial conditions specified by $(u_{[t-n,t-1]}, y_{[t-n,t-1]})$ is denoted by $\hat{y}$. For $k \in \mathbb{I}_{[-n,L-2n-1]}$, we choose for the candidate input the shifted previously optimal solution, i.e., $\bar{u}'_k(t+n) = \bar{u}^*_{k+n}(t)$. Over the first n steps, the candidate output must satisfy $\bar{y}'_{[-n,-1]}(t+n) = \tilde{y}_{[t,t+n-1]}$ due to (4.1c). Further, for $k \in \mathbb{I}_{[0,L-2n-1]}$, the output is chosen as $\bar{y}'_k(t+n) = \hat{y}_{t+n+k}$. Since $\bar{y}^*_{[L-n,L-1]}(t) = 0$ by (4.1d), the prediction error bound of Lemma 4.2 implies that, for any $k \in \mathbb{I}_{[L-n,L-1]}$, it holds that

$$\|\hat{y}_{t+k}\|_\infty \leq \bar{\varepsilon}\|\alpha^*(t)\|_1 + \|\sigma^*(t)\|_\infty + \rho_{\infty,n+k}\left(\bar{\varepsilon}\left(\|\alpha^*(t)\|_1 + 1\right) + \|\sigma^*_{[-n,-1]}(t)\|_\infty\right).$$

Using that $\lambda_\alpha\bar{\varepsilon}\|\alpha^*(t)\|_2^2 + \frac{\lambda_\sigma}{\bar{\varepsilon}}\|\sigma^*(t)\|_2^2 \leq J_L^*(\tilde{\xi}_t) \leq V_{\max}$, we can bound $\alpha^*(t)$ and $\sigma^*(t)$ as

$$\|\alpha^*(t)\|_1 \leq \sqrt{N-L-n+1}\|\alpha^*(t)\|_2 \leq \sqrt{N-L-n+1}\sqrt{\frac{V_{\max}}{\lambda_\alpha\bar{\varepsilon}}}, \tag{4.21}$$

$$\|\sigma^*(t)\|_\infty \leq \|\sigma^*(t)\|_2 \leq \sqrt{\frac{\bar{\varepsilon}V_{\max}}{\lambda_\sigma}}. \tag{4.22}$$

Hence, if $\bar{\varepsilon}_{\max}$ is sufficiently small, then $\hat{y}_{t+k}$ becomes arbitrarily small at the above time instants. This implies that the internal state in some minimal realization corresponding to the trajectory $(\bar{u}^*(t), \hat{y})$ at time $t+L-n$, i.e., $\hat{x}_{t+L-n} = \Phi_+\hat{y}_{[t+L-n,t+L-1]}$, approaches zero for $\bar{\varepsilon} \to 0$. Thus, similar to the proof of Lemma 4.1, there exists an input trajectory $\bar{u}'_{[L-2n,L-n-1]}(t+n)$, which brings the state and the corresponding output $\bar{y}'_{[L-2n,L-n-1]}(t+n)$ to zero in n steps, while satisfying $\bar{u}'_k(t+n) \in \mathbb{U}$ for $k \in \mathbb{I}_{[L-2n,L-n-1]}$ as well as

$$\sum_{k=L-2n}^{L-n-1} \|\bar{u}'_k(t+n)\|_2^2 + \|\bar{y}'_k(t+n)\|_2^2 \leq \Gamma_{\mathrm{uy}}\|\hat{x}_{t+L-n}\|_2^2. \tag{4.23}$$

Moreover, in the interval $\mathbb{I}_{[L-n,L-1]}$, we choose $\bar{u}'_{[L-n,L-1]}(t+n) = 0$, $\bar{y}'_{[L-n,L-1]}(t+n) = 0$, i.e., (4.1d) is satisfied. The above arguments imply that

$$\left(\bar{u}'(t+n), \begin{bmatrix} \hat{y}_{[t,t+n-1]} \\ \bar{y}'_{[0,L-1]}(t+n) \end{bmatrix}\right)$$

is a trajectory of System (2.1) and we denote the corresponding state by $\bar{x}'(t+n)$. We choose $\alpha'(t+n)$ as a corresponding solution to (2.11), i.e., as

$$\alpha'(t+n) = H_{\mathrm{ux}}^{\dagger} \begin{bmatrix} \bar{u}'_{[-n,L-1]}(t+n) \\ x_t \end{bmatrix} \tag{4.24}$$

with H_{ux} from (4.2). Finally, we fix

$$\sigma'(t+n) = H_{L+n}(\tilde{y}^{\mathrm{d}})\alpha'(t+n) - \bar{y}'(t+n), \tag{4.25}$$

which implies that (4.1b) holds. ■

Proposition 4.1 shows that, for any sublevel set of the Lyapunov function V, there exists a sufficiently small noise bound $\bar{\varepsilon}_{\max}$ such that, for any $\bar{\varepsilon} \leq \bar{\varepsilon}_{\max}$ and any state starting in the sublevel set at time t, the n-step MPC scheme is feasible at time $t+n$. In particular, the required noise bound decreases if the size of the sublevel set, i.e., $V_{\max}$, increases and vice versa. This can be explained by noting that the noise in (4.1b) corresponds to a multiplicative uncertainty, which affects the prediction accuracy more strongly if the current state is further away from the origin and hence the Lyapunov function V_t is larger. We note that this does not imply recursive feasibility of the n-step MPC scheme in the standard sense since it remains to be shown that the sublevel set $V_t \leq V_{\max}$ is invariant, which will be proven in Section 4.1.5. In our main result, the set of initial states for which $V_0 \leq V_{\max}$ will play the role of the guaranteed region of attraction of the closed-loop system.

The input candidate solution used to prove recursive feasibility in Proposition 4.1 is analogous to a candidate solution one would use to show robust recursive feasibility in model-based robust MPC with terminal equality constraints. The output candidate solution is sketched in Figure 4.1. Up to time $L-2n-1$, $\bar{y}'(t+n)$ is equal to $\hat{y}$ (shifted by n times steps), which is the output resulting from an open-loop application of $\bar{u}^*(t)$. This choice together with the prediction error bound of Lemma 4.2 implies that the internal state corresponding to $\bar{y}'(t+n)$ at time $L-2n$ is close to zero. Thus, by controllability, there exists an input trajectory satisfying the input constraints, which brings the state and the output to zero in n steps. In the interval $\mathbb{I}_{[L-2n,L-n-1]}$, the candidate output is chosen as this trajectory. This also implies that the choice $\bar{y}'_{[L-n,L-1]}(t+n) = 0$ makes the candidate solution between

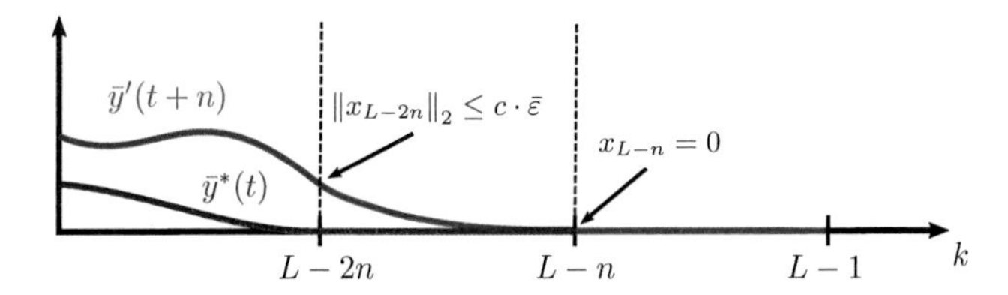

Figure 4.1. Sketch of the candidate output for recursive feasibility. Due to the terminal equality constraints (4.1d), the last n steps of the optimal predicted output $\bar{y}^*(t)$ are equal to zero. According to the prediction error bound derived in Lemma 4.2, this implies that the state resulting from an open-loop application of the optimal input $\bar{u}^*(t)$ is small at time $L-2n$, provided that $\bar{\varepsilon}$ is sufficiently small. Therefore, a candidate solution $\bar{y}'(t+n)$ can be constructed by appending the open-loop output $\hat{y}$ by a local deadbeat controller, which steers the state to the origin in n steps. ©2020 IEEE

0 and $L-1$, i.e., $(\bar{u}'_{[0,L-1]}(t+n), \bar{y}'_{[0,L-1]}(t+n))$, a trajectory[2] of System (2.1). Finally, the suggested candidate input is also similar to [154], where inherent robustness of quasi-infinite horizon (model-based) MPC is shown.

Note that, due to the flexibility gained via the slack variable $\sigma(t)$, recursive feasibility of Problem 4.1 can alternatively be proven using a trivial candidate solution: Simply choose the input $\bar{u}_k(t) = u^{\mathrm{s}}$ and the output $\bar{y}_k(t) = y^{\mathrm{s}}$ for $k \in \mathbb{I}_{[0,L-1]}$, the vector $\alpha(t)$ such that $H_{L+n}(u^{\mathrm{d}})\alpha(t) = \bar{u}(t)$ holds, and the slack variable as $\sigma(t) = H_{L+n}(\tilde{y}^{\mathrm{d}})\alpha(t) - \bar{y}(t)$. However, this candidate solution does not lead to any useful bounds on the Lyapunov function decay. This motivates the more sophisticated candidate solution constructed in Proposition 4.1, which we employ in Section 4.1.5 to prove closed-loop stability.

Remark 4.4. *(1-step MPC) For a 1-step MPC scheme, a similar argument to prove recursive feasibility can be applied, given that $\bar{u}^*_{[L-2n,L-n-1]}(t)$ and $\bar{y}^*_{[L-2n,L-n-1]}(t)$ (and hence $\hat{y}_{[t+L-2n,t+L-n-1]}$) are close to zero. This is required to construct a feasible input which steers the state and the corresponding output to zero, similar to the proof of Proposition 4.1,*

[2] In most practical cases, $(\bar{u}^*(t), \bar{y}^*(t))$ are not trajectories of the system due to the slack variable σ and the noise.

and it is, e.g., the case if the initial state x_t is close to zero. That is, the result of Proposition 4.1 holds locally for a 1-step MPC scheme, as expected based on model-based MPC with terminal equality constraints under disturbances using inherent robustness properties.

Remark 4.5. *(Non-zero setpoint) As mentioned in Section 4.1.1, all of our theoretical guarantees for the presented robust MPC scheme can be straightforwardly extended to the case $(u^s, y^s) \neq 0$, with the corresponding steady-state $\xi^s \neq 0$. The main difference lies in the bound (4.6), which becomes $V_t \leq \tilde{c}_3 \|\xi_t - \xi^s\|_2^2 + \tilde{c}_4 \bar{\varepsilon}$ for constants $\tilde{c}_3 \neq c_3, \tilde{c}_4 \neq c_4$, where $\tilde{c}_3$ can be made arbitrarily close to c_3. On the other hand, $\tilde{c}_4$ changes depending on ξ^s, since the right-hand side of (4.11) would need to be proportional to $\|\xi_t - \xi^s\|_2^2 + \|\xi^s\|_2^2$. The same phenomenon can be observed in a bound of $\alpha'(t+n)$ based on (4.24), which will be used in the stability proof. As will become clear later in this section, such changes in the bound of $\alpha'(t+n)$ as well as in the constant $\tilde{c}_4$ do not affect our qualitative theoretical results, but they may potentially (quantitatively) deterioriate the robustness w.r.t. the noise level $\bar{\varepsilon}$. Intuitively, this can be explained by noting that (4.1b) corresponds to a multiplicative uncertainty and thus, stabilization of the origin is simpler than stabilization of any other equilibrium. Since equilibria with $(u^s, y^s) \neq 0$ require a significantly more involved notation, we omit this extension in the present section. In Section 4.3, we study the case of non-zero setpoints in robust data-driven tracking MPC based on a regularization of the form $\|\alpha(t) - \alpha^s\|_2^2$ which is not w.r.t. zero.*

4.1.5 Closed-loop guarantees

The following is our main stability result in this section. It shows that, under the given assumptions, for a low noise amplitude and large persistence of excitation, and for suitable regularization parameters, the application of the proposed MPC scheme in Algorithm 4.1 leads to a practically exponentially stable closed loop.

Theorem 4.1. *Suppose Assumptions 4.1–4.5 hold. Then, for any $V_{\max} > 0$, there exist constants $\underline{\lambda}_\alpha, \overline{\lambda}_\alpha, \underline{\lambda}_\sigma, \overline{\lambda}_\sigma > 0$ such that, for all $\lambda_\alpha, \lambda_\sigma$ satisfying*

$$\underline{\lambda}_\alpha \leq \lambda_\alpha \leq \overline{\lambda}_\alpha, \quad \underline{\lambda}_\sigma \leq \lambda_\sigma \leq \overline{\lambda}_\sigma, \tag{4.26}$$

there exist constants $\bar{\varepsilon}_{\max}, \bar{c}_{\text{pe}} > 0$, as well as $\beta \in \mathcal{K}_\infty$ such that, for all $\bar{\varepsilon} > 0$ (cf. Assumption 4.2) and $c_{\text{pe}} > 0$ (as defined in (4.3)) satisfying

$$\bar{\varepsilon} \leq \bar{\varepsilon}_{\max}, \quad c_{\text{pe}} \bar{\varepsilon} \leq \bar{c}_{\text{pe}}, \tag{4.27}$$

the bound $V_t \leq V_{\max}$ holds recursively for all $t \in \mathbb{I}_{\geq 0}$ and V_t converges exponentially to $[0, \beta(\bar{\varepsilon})]$ in closed loop with the n-step MPC scheme (Algorithm 4.1) for all initial conditions for which $V_0 \leq V_{\max}$.

Proof. The proof consists of three parts: First, we bound the increase in the Lyapunov function V. Thereafter, we prove that, for suitably chosen bounds on the parameters, there exists a function β, which satisfies the above requirements. Finally, we show invariance of the sublevel set $V_t \leq V_{\max}$ and exponential convergence of V_t to $[0, \beta(\bar{\varepsilon})]$.

(i) Practical stability

Suppose Problem 4.1 is feasible at time t and let $V_{\max} > 0$ be arbitrary. Further, let $\bar{\varepsilon}_{\max}$ be sufficiently small such that Proposition 4.1 is applicable. Denoting $\ell(u, y) := \|u\|_R^2 + \|y\|_Q^2$, the cost of the candidate solution derived in Proposition 4.1 at time $t + n$ is

$$\sum_{k=0}^{L-1} \ell\left(\bar{u}_k'(t+n), \bar{y}_k'(t+n)\right) + \lambda_\alpha \bar{\varepsilon} \|\alpha'(t+n)\|_2^2 + \frac{\lambda_\sigma}{\bar{\varepsilon}} \|\sigma'(t+n)\|_2^2.$$

Thus, we obtain for the optimal cost

$$\begin{aligned} & J_L^*(\tilde{\xi}_{t+n}) \\ \leq & J_L^*(\tilde{\xi}_t) - \sum_{k=0}^{L-1} \ell\left(\bar{u}_k^*(t), \bar{y}_k^*(t)\right) - \lambda_\alpha \bar{\varepsilon} \|\alpha^*(t)\|_2^2 - \frac{\lambda_\sigma}{\bar{\varepsilon}} \|\sigma^*(t)\|_2^2 \\ & + \lambda_\alpha \bar{\varepsilon} \|\alpha'(t+n)\|_2^2 + \frac{\lambda_\sigma}{\bar{\varepsilon}} \|\sigma'(t+n)\|_2^2 + \sum_{k=0}^{L-1} \ell(\bar{u}_k'(t+n), \bar{y}_k'(t+n)). \end{aligned} \tag{4.28}$$

In the following key technical part of the proof (Parts (i.a)-(i.d)), we derive useful bounds for most terms on the right-hand side of (4.28). This will lead to a decay bound of the optimal cost which is then used to prove practical exponential stability of the closed loop.

(i.a) Stage cost bounds

We first bound those terms in (4.28), which involve the stage cost. The above difference can be decomposed as

$$\begin{aligned}
&\sum_{k=0}^{L-1} \ell(\bar{u}_k'(t+n), \bar{y}_k'(t+n)) - \sum_{k=0}^{L-1} \ell\left(\bar{u}_k^*(t), \bar{y}_k^*(t)\right) \\
=& \sum_{k=L-2n}^{L-n-1} \ell\left(\bar{u}_k'(t+n), \bar{y}_k'(t+n)\right) - \sum_{k=0}^{n-1} \ell\left(\bar{u}_k^*(t), \bar{y}_k^*(t)\right) \\
&+ \sum_{k=0}^{L-2n-1} \left(\ell\left(\bar{u}_k'(t+n), \bar{y}_k'(t+n)\right) - \ell\left(\bar{u}_{k+n}^*(t), \bar{y}_{k+n}^*(t)\right)\right),
\end{aligned} \tag{4.29}$$

where we use that $\bar{u}_k'(t+n)$, $\bar{y}_k'(t+n)$, $\bar{u}_k^*(t)$, $\bar{y}_k^*(t)$ are all zero for $k \in \mathbb{I}_{[L-n,L-1]}$ due to (4.1d). To bound the first term on the right-hand side of (4.29), note that

$$\|\hat{x}_{t+L-n}\|_2^2 \leq \|\Phi_n^\dagger\|_2^2 \|\hat{y}_{[t+L-n,t+L-1]}\|_2^2$$

with $\hat{x}_{t+L-n}$ as in the proof of Proposition 4.1 and $\hat{y}$ as in (4.14). Further, since $\bar{y}^*_{[L-n,L-1]}(t) = 0$, $\hat{y}$ can be bounded in the considered time interval using Lemma 4.2, i.e., Inequality (4.16), which leads to

$$\begin{aligned}
\|\hat{y}_{[t+L-n,t+L-1]}\|_2^2 \leq& 8c_5 n\bar{\varepsilon}^2 \|\alpha^*(t)\|_2^2 + 2\|\sigma^*(t)\|_2^2 \\
&+ \sum_{k=L-n}^{L-1} \rho_{2,n+k} \left(16n\bar{\varepsilon}^2 \left(c_5\|\alpha^*(t)\|_2^2 + p\right) + 4\|\sigma^*_{[-n,-1]}(t)\|_2^2\right).
\end{aligned}$$

Hence, it holds that

$$\begin{aligned}
&\sum_{k=L-2n}^{L-n-1} \ell\left(\bar{u}_k'(t+n), \bar{y}_k'(t+n)\right) \\
\overset{(4.23)}{\leq}& \lambda_{\max}(Q,R)\Gamma_{\mathrm{uy}} \|\hat{x}_{t+L-n}\|_2^2 \\
\leq& \lambda_{\max}(Q,R)\Gamma_{\mathrm{uy}} \|\Phi_+\|_2^2 \Big(8c_5 n\bar{\varepsilon}^2 \|\alpha^*(t)\|_2^2 + 2\|\sigma^*(t)\|_2^2 \\
&+ \sum_{k=L-n}^{L-1} \rho_{2,n+k} \left(16n\bar{\varepsilon}^2 \left(c_5\|\alpha^*(t)\|_2^2 + p\right) + 4\|\sigma^*_{[-n,-1]}(t)\|_2^2\right)\Big).
\end{aligned} \tag{4.30}$$

Next, we bound the difference between the third and the fourth term on the right-hand side of (4.29). The following relations are readily derived:

$$\begin{aligned}
&\|\bar{y}_k'(t+n)\|_Q^2 - \|\bar{y}_{k+n}^*(t)\|_Q^2 \\
=& \|\bar{y}_k'(t+n) - \bar{y}_{k+n}^*(t) + \bar{y}_{k+n}^*(t)\|_Q^2 - \|\bar{y}_{k+n}^*(t)\|_Q^2 \\
=& \|\bar{y}_k'(t+n) - \bar{y}_{k+n}^*(t)\|_Q^2 + 2\left(\bar{y}_k'(t+n) - \bar{y}_{k+n}^*(t)\right)^\top Q\bar{y}_{k+n}^*(t) \\
\leq& \|\bar{y}_k'(t+n) - \bar{y}_{k+n}^*(t)\|_Q^2 + 2\|\bar{y}_k'(t+n) - \bar{y}_{k+n}^*(t)\|_Q \|\bar{y}_{k+n}^*(t)\|_Q.
\end{aligned} \tag{4.31}$$

By using $2\|\bar{y}^*_{k+n}(t)\|_Q \leq 1 + \|\bar{y}^*_{k+n}(t)\|_Q^2$ as well as $\|\bar{y}^*_{k+n}(t)\|_Q^2 \leq V_{\max}$, we arrive at

$$2\|\bar{y}'_k(t+n) - \bar{y}^*_{k+n}(t)\|_Q \|\bar{y}^*_{k+n}(t)\|_Q \leq \|\bar{y}'_k(t+n) - \bar{y}^*_{k+n}(t)\|_Q\,(1+V_{\max})\,. \quad (4.32)$$

Since the inputs coincide over the considered time interval, and due to (4.31) as well as (4.32), it holds that

$$\begin{aligned}&\sum_{k=0}^{L-2n-1} \ell\left(\bar{u}'_k(t+n), \bar{y}'_k(t+n)\right) - \ell\left(\bar{u}^*_{k+n}(t), \bar{y}^*_{k+n}(t)\right) &(4.33)\\ \leq &\sum_{k=0}^{L-2n-1} \|\bar{y}'_k(t+n) - \bar{y}^*_{k+n}(t)\|_Q^2 + \|\bar{y}'_k(t+n) - \bar{y}^*_{k+n}(t)\|_Q\,(1+V_{\max})\,.\end{aligned}$$

The difference $\|\bar{y}'_k(t+n) - \bar{y}^*_{k+n}(t)\|_Q$ can be bounded via (4.17) (Lemma 4.2) and $\|a\|_Q \leq \sqrt{\lambda_{\max}(Q)}\|a\|_2 \leq \sqrt{p\lambda_{\max}(Q)}\|a\|_\infty$, which holds for arbitrary $a \in \mathbb{R}^p$. Using (4.22) to bound $\|\sigma^*(t)\|_\infty$, it can be shown that the bound is of the form

$$\|\bar{y}'_k(t+n) - \bar{y}^*_{k+n}(t)\|_Q \leq C_1\bar{\varepsilon}\|\alpha^*(t)\|_2 + \beta_1(\bar{\varepsilon}) \leq C_1\bar{\varepsilon}\left(1 + \|\alpha^*(t)\|_2^2\right) + \beta_1(\bar{\varepsilon})$$

for some $C_1 > 0$, $\beta_1 \in \mathcal{K}_\infty$. Hence, applying Lemma 4.2 to (4.33), the sum of (4.30) and (4.33) can be bounded by

$$\beta_2(\bar{\varepsilon})\|\alpha^*(t)\|_2^2 + \beta_3(\bar{\varepsilon})\|\sigma^*(t)\|_2^2 + \beta_4(\bar{\varepsilon})$$

for suitable $\beta_i \in \mathcal{K}_\infty$, $i \in \mathbb{I}_{[2,4]}$, where $\beta_2(\bar{\varepsilon}) = C_2\bar{\varepsilon}^2 + C_3\bar{\varepsilon}$ for some $C_2, C_3 > 0$. Therefore, if $\underline{\lambda}_\alpha$ and $\underline{\lambda}_\sigma$ are sufficiently large, then (4.28) implies

$$\begin{aligned}J_L^*(\tilde{\xi}_{t+n}) \leq &J_L^*(\tilde{\xi}_t) - \sum_{k=0}^{n-1} \ell\left(\bar{u}^*_k(t), \bar{y}^*_k(t)\right) &(4.34)\\ &+ \lambda_\alpha\bar{\varepsilon}\|\alpha'(t+n)\|_2^2 + \frac{\lambda_\sigma}{\bar{\varepsilon}}\|\sigma'(t+n)\|_2^2 + \beta_5(\bar{\varepsilon})\end{aligned}$$

for a suitable $\beta_5 \in \mathcal{K}_\infty$.

(i.b) Bound of $\|\sigma'(t+n)\|_2^2$

By applying standard norm bounds to the slack variable candidate $\sigma'(t+n)$ defined in (4.25), we obtain

$$\|\sigma'(t+n)\|_2^2 \leq 2np\bar{\varepsilon}^2 + c_5(L+2n)\bar{\varepsilon}^2\|\alpha'(t+n)\|_2^2, \quad (4.35)$$

with $c_5 = p(N-L-n+1)$ as in (4.12).

(i.c) Bound of $\|\alpha'(t+n)\|_2^2$

For the vector $\alpha'(t+n)$, it holds that

$$\begin{aligned}\|\alpha'(t+n)\|_2^2 &\overset{(4.3),(4.24)}{\leq} c_{\text{pe}} \left\| \begin{bmatrix} \bar{u}'_{[-n,L-1]}(t+n) \\ \bar{x}'_{-n}(t+n) \end{bmatrix} \right\|_2^2 \\ &= c_{\text{pe}} \left(\|x_t\|_2^2 + \|\bar{u}'_{[-n,L-1]}(t+n)\|_2^2 \right) \\ &= c_{\text{pe}} \|x_t\|_2^2 + c_{\text{pe}} \left(\|\bar{u}^*_{[0,L-n-1]}(t)\|_2^2 + \|\bar{u}'_{[L-2n,L-n-1]}(t+n)\|_2^2 \right).\end{aligned}$$

Similar to (4.30), we can use (4.23) to bound the last term as

$$\begin{aligned}&\|\bar{u}'_{[L-2n,L-n-1]}(t+n)\|_2^2 \\ \leq &\Gamma_{\text{uy}} \|\Phi_+\|_2^2 \Big(8c_5 n\bar{\varepsilon}^2 \|\alpha^*(t)\|_2^2 + 2\|\sigma^*(t)\|_2^2 \\ &+ \sum_{k=L-n}^{L-1} \rho_{2,n+k} \left(16n\bar{\varepsilon}^2 \left(c_5 \|\alpha^*(t)\|_2^2 + p \right) + 4\|\sigma^*_{[-n,-1]}(t)\|_2^2 \right) \Big).\end{aligned} \tag{4.36}$$

The bound (4.36) is of the same form as (4.30) and (4.33). Due to this fact, using the bound (4.35) for $\sigma'(t+n)$, and by potentially choosing $\underline{\lambda}_\alpha$ and $\underline{\lambda}_\sigma$ larger, (4.28) implies

$$\begin{aligned}J_L^*(\tilde{\xi}_{t+n}) \leq &J_L^*(\tilde{\xi}_t) - \sum_{k=0}^{n-1} \ell\left(\bar{u}_k^*(t), \bar{y}_k^*(t)\right) \\ &+ (\lambda_\alpha + \lambda_\sigma c_6) \left(\|x_t\|_2^2 + \|\bar{u}^*_{[0,L-n-1]}(t)\|_2^2 \right) c_{\text{pe}} \bar{\varepsilon} + \beta_6(\bar{\varepsilon})\end{aligned} \tag{4.37}$$

for suitable $c_6 > 0$, $\beta_6 \in \mathcal{K}_\infty$.

(i.d) IOSS bound

As in the proof of Theorem 3.1, we consider now $V_t = J_L^*(\tilde{\xi}_t) + \gamma W(\xi_t)$ with the IOSS Lyapunov function W for some $\gamma > 0$. It follows directly from (2.9), (4.37), and from $\|x_t\|_2^2 \leq \|T_{\mathrm{x},\xi}\|_2^2 \|\xi_t\|_2^2$ that

$$\begin{aligned}&V_{t+n} - V_t \\ \leq &- \sum_{k=0}^{n-1} \ell\left(\bar{u}_k^*(t), \bar{y}_k^*(t)\right) + \gamma (c_{\text{IOSS},1} \|u_{[t,t+n-1]}\|_2^2 + c_{\text{IOSS},2} \|y_{[t,t+n-1]}\|_2^2 - \|\xi_{[t,t+n-1]}\|_2^2) \\ &+ (\lambda_\alpha + \lambda_\sigma c_6) \, c_{\text{pe}} \bar{\varepsilon} \left(\|T_{\mathrm{x},\xi}\|_2^2 \|\xi_t\|_2^2 + \|\bar{u}^*_{[0,L-n-1]}(t)\|_2^2 \right) + \beta_6(\bar{\varepsilon}).\end{aligned} \tag{4.38}$$

The inequality $(a+b)^2 \leq 2(a^2+b^2)$ yields

$$\|y_{[t,t+n-1]}\|_2^2 \leq 2\|\bar{y}^*_{[0,n-1]}(t)\|_2^2 + 2\|y_{[t,t+n-1]} - \bar{y}^*_{[0,n-1]}(t)\|_2^2,$$

where the latter term can again be bounded using Lemma 4.2. Similar to the earlier steps of this proof, the components of the bound $\|y_{[t,t+n-1]} - \bar{y}^*_{[0,n-1]}(t)\|_2^2$ vanish in (4.38) if $\underline{\lambda}_\sigma, \underline{\lambda}_\alpha$ are chosen sufficiently large, except for an additive constant, which depends solely on the noise. Moreover, choosing $\gamma = \frac{\lambda_{\min}(Q,R)}{\max\{c_{\text{IOSS},1}, 2c_{\text{IOSS},2}\}}$, it holds that

$$\gamma(c_{\text{IOSS},1}\|u_{[t,t+n-1]}\|_2^2 + 2c_{\text{IOSS},2}\|\bar{y}^*_{[0,n-1]}(t)\|_2^2) \leq \sum_{k=0}^{n-1} \ell(\bar{u}_k^*(t), \bar{y}_k^*(t)).$$

Combining these facts, we arrive at

$$\begin{aligned}&V_{t+n} - V_t\\ &\leq \left((\lambda_\alpha + \lambda_\sigma c_6)\|T_{x,\xi}\|_2^2 c_{\text{pe}}\bar{\varepsilon} - \gamma\right)\|\xi_t\|_2^2 + (\lambda_\alpha + \lambda_\sigma c_6)\, c_{\text{pe}}\bar{\varepsilon}\|\bar{u}^*_{[0,L-n-1]}(t)\|_2^2 + \beta_7(\bar{\varepsilon})\end{aligned}$$

for a suitable $\beta_7 \in \mathcal{K}_\infty$. Finally, note that

$$\lambda_{\min}(R)\|\bar{u}^*_{[0,L-n-1]}(t)\|_2^2 \leq V_t,$$

which leads to

$$\begin{aligned}V_{t+n} - V_t &\leq \left((\lambda_\alpha + \lambda_\sigma c_6)\|T_{x,\xi}\|_2^2 c_{\text{pe}}\bar{\varepsilon} - \gamma\right)\|\xi_t\|_2^2 + \frac{(\lambda_\alpha + \lambda_\sigma c_6)\, c_{\text{pe}}\bar{\varepsilon}}{\lambda_{\min}(R)} V_t + \beta_7(\bar{\varepsilon})\\ &=: (\beta_8(\bar{\varepsilon}) - \gamma)\|\xi_t\|_2^2 + \beta_9(\bar{\varepsilon})V_t + \beta_7(\bar{\varepsilon}),\end{aligned} \tag{4.39}$$

where $\beta_8, \beta_9 \in \mathcal{K}_\infty$.

(ii) Construction of β

The local upper bound in Lemma 4.1, which holds for any ξ_t with $\|\xi_t\|_2 \leq \delta$, implies that the following holds for any $V_{\max} > 0$, and any ξ_t with $V_t \leq V_{\max}$:

$$V_t \leq \underbrace{\max\left\{c_3, \frac{V_{\max} - c_4\bar{\varepsilon}}{\delta^2}\right\}}_{c_{3,V_{\max}}:=}\|\xi_t\|_2^2 + c_4\bar{\varepsilon}. \tag{4.40}$$

We first consider $V_{\max} = \delta^2 c_3 + c_4\bar{\varepsilon}$, which implies $c_{3,V_{\max}} = c_3$. Further, we define $\eta(\bar{\varepsilon}) := \gamma - \beta_8(\bar{\varepsilon}) - c_3\beta_9(\bar{\varepsilon})$ as well as

$$\hat{\beta}(\bar{\varepsilon}) := \frac{\gamma c_4\bar{\varepsilon} + c_3\beta_7(\bar{\varepsilon})}{\eta(\bar{\varepsilon})}$$

for any $\bar{\varepsilon}$ for which $\eta(\bar{\varepsilon}) > 0$. It clearly holds that $\hat{\beta}(0) = 0$. Next, we show the existence of a constant $\bar{\varepsilon}_{\max}$ such that $\hat{\beta}$ is strictly increasing on $[0, \bar{\varepsilon}_{\max}]$. If

$\eta(\bar{\varepsilon}) > 0$, then $\hat{\beta}$ is strictly increasing since its numerator increases with $\bar{\varepsilon}$ whereas its denominator decreases with $\bar{\varepsilon}$. In the following, we show that $\eta(\bar{\varepsilon}) > 0$. By definition, we have

$$\begin{aligned}\eta(\bar{\varepsilon}) =& \gamma - (\lambda_\alpha + \lambda_\sigma c_6) \|T_{x,\xi}\|_2^2 c_{pe}\bar{\varepsilon} \\ &- \frac{(\lambda_\alpha + \lambda_\sigma c_6)\, c_{pe}\bar{\varepsilon}}{\lambda_{\min}(R)} \Big(\lambda_{\max}(Q,R)\Gamma_{uy}\|T_{x,\xi}\|_2^2 + \gamma\lambda_{\max}(P_W) \\ &+ (\lambda_\alpha + c_5(L+2n)\lambda_\sigma)\, c_{pe}\bar{\varepsilon}(\Gamma_{uy}\|T_{x,\xi}\|_2^2 + \|M\|_2^2)\Big).\end{aligned}$$

It can be seen directly from this expression that, if $\lambda_\alpha \leq \overline{\lambda}_\alpha, \lambda_\sigma \leq \overline{\lambda}_\sigma$, with arbitrary but fixed upper bounds $\overline{\lambda}_\alpha, \overline{\lambda}_\sigma$, and $c_{pe}\bar{\varepsilon}$ is sufficiently small, then $\eta(\bar{\varepsilon}) > 0$. Finally, if $\bar{\varepsilon}_{\max}$ is sufficiently small, then

$$\frac{\gamma c_4\bar{\varepsilon} + c_3\beta_7(\bar{\varepsilon})}{\gamma - \beta_8(\bar{\varepsilon}) - c_3\beta_9(\bar{\varepsilon})} \leq \delta^2 c_3 + c_4\bar{\varepsilon},$$

i.e., $\hat{\beta}(\bar{\varepsilon}_{\max}) \leq V_{\max}$. Thus, we have constructed $\hat{\beta} : [0, \bar{\varepsilon}_{\max}] \to [0, V_{\max}]$, which is continuous, strictly increasing, and satisfies $\beta(0) = 0$. We now define β as a continuation of $\hat{\beta}$ in $\mathcal{K}_\infty$, i.e., $\beta(\bar{\varepsilon}) = \hat{\beta}(\bar{\varepsilon})$ for all $\bar{\varepsilon} \in [0, \bar{\varepsilon}_{\max}]$ and $\beta \in \mathcal{K}_\infty$.

(iii) Invariance and exponential convergence

Suppose $V_t \leq V_{\max}$ such that Problem 4.1 is feasible and, therefore, (4.39) and (4.40) hold. Further, note that $\eta(\bar{\varepsilon}) > 0$ implies $\beta_8(\bar{\varepsilon}) < \gamma$. Defining $V_{\beta,t} := V_t - \beta(\bar{\varepsilon})$, we thus obtain

$$\begin{aligned}V_{t+n} &\overset{(4.39)}{\leq} (1 + \beta_9(\bar{\varepsilon}))\, V_t + (\beta_8(\bar{\varepsilon}) - \gamma)\, \|\xi_t\|_2^2 + \beta_7(\bar{\varepsilon}) \\ &\overset{(4.40)}{\leq} \left(1 + \beta_9(\bar{\varepsilon}) + \frac{\beta_8(\bar{\varepsilon}) - \gamma}{c_3}\right) V_t + \frac{c_4\bar{\varepsilon}}{c_3}(\gamma - \beta_8(\bar{\varepsilon})) + \beta_7(\bar{\varepsilon}) \\ &\leq \left(1 + \beta_9(\bar{\varepsilon}) + \frac{\beta_8(\bar{\varepsilon}) - \gamma}{c_3}\right) V_{\beta,t} + \beta(\bar{\varepsilon}),\end{aligned}$$

where the last inequality follows from elementary computations. This in turn implies the following contraction property

$$V_{\beta,t+n} \leq \underbrace{\left(1 + \beta_9(\bar{\varepsilon}) + \frac{\beta_8(\bar{\varepsilon}) - \gamma}{c_3}\right)}_{<1} V_{\beta,t}. \tag{4.41}$$

If the noise bound $\bar{\varepsilon}_{\max}$ is sufficiently small, then this implies invariance of the sublevel set $V_t \leq V_{\max}$ and hence, by Proposition 4.1, recursive feasibility of the

n-step MPC scheme. Applying the contraction property (4.41) recursively, we conclude that V_t converges exponentially to $[0, \beta(\bar{\varepsilon})]$.

So far, we have only considered the case $V_{\max} = \delta^2 c_3 + c_4 \bar{\varepsilon}$. It remains to show that, *for any* $V_{\max} > 0$, there exist suitable parameter bounds such that

$$\eta_{V_{\max}}(\bar{\varepsilon}) := \gamma - \beta_8(\bar{\varepsilon}) - c_{3,V_{\max}} \beta_9(\bar{\varepsilon}) > 0$$

with $c_{3,V_{\max}}$ from (4.40). It is easily seen from the above discussion that, for any *fixed* $V_{\max} > 0$ and for *fixed* bounds $\overline{\lambda}_\alpha, \overline{\lambda}_\sigma$, $\eta(\bar{\varepsilon}) > 0$ can always be ensured if $c_{\mathrm{pe}} \bar{\varepsilon}$ is sufficiently small, i.e., if the bound $\bar{c}_{\mathrm{pe}}$ is sufficiently small. ∎

Discussion

Theorem 4.1 shows that the closed loop of the proposed data-driven MPC scheme admits a practical Lyapunov function [52], which converges robustly and exponentially to a set, whose size shrinks with the noise level. Since $\|\xi_t\|_2^2 \leq \frac{1}{\gamma \lambda_{\min}(P_W)} V_t$ due to (4.6), this implies practical exponential stability of the equilibrium $\xi = 0$, compare [52]. The result requires that the noise level $\bar{\varepsilon}$ is small, the amount of persistence of excitation is large compared to the noise level (i.e., $c_{\mathrm{pe}} \bar{\varepsilon}$ is small), and the regularization parameters are chosen suitably. Concerning the latter requirement, λ_α cannot be chosen arbitrarily large, which can be explained by noting that the optimal α is usually not zero, even in the noise-free case. On the other hand, λ_α cannot be too close to zero since solutions $\alpha(t)$ of (4.1b) are not unique and large choices of $\alpha(t)$ amplify the influence of the noise in $\tilde{y}^{\mathrm{d}}$ on the prediction accuracy. Further, λ_σ has to be chosen sufficiently large to ensure stability, but not arbitrarily large for a fixed noise level. To be more precise, $\lambda_\alpha c_{\mathrm{pe}} \bar{\varepsilon}$ and $\lambda_\sigma c_{\mathrm{pe}} \bar{\varepsilon}$ have to be small, i.e., for a fixed c_{pe}, choosing the regularization parameters too large deteriorates the robustness of the scheme w.r.t. the noise level. Considering noise-dependent regularizations of $\alpha(t)$ and $\sigma(t)$, i.e., regularization parameters $\bar{\varepsilon} \lambda_\alpha$ and $\frac{\lambda_\sigma}{\bar{\varepsilon}}$ in Problem 4.1, is crucial for our theoretical analysis. We note that [28] also employs a regularization of $\alpha(t)$ which is linear in the noise bound in the context of data-driven optimal control via distributional robustness.

In the proof of Theorem 4.1, a close connection between the region of attraction, i.e., the set of initial conditions with $V_0 \leq V_{\max}$, and various parameters becomes apparent. First of all, the noise bound $\bar{\varepsilon}$ needs to be sufficiently small depending on $V_{\max}$ to allow for an application of Proposition 4.1. Moreover, if $V_{\max}$ increases, then also $c_{3,V_{\max}}$ increases and hence, $\beta_9(\bar{\varepsilon})$ must decrease to ensure $\eta_{V_{\max}}(\bar{\varepsilon}) > 0$ and thereby exponential stability. To render $\beta_9(\bar{\varepsilon})$ small, $c_{\text{pe}}\bar{\varepsilon}$ must decrease, i.e., the amount of persistence of excitation compared to the noise level must increase. Thus, for $c_{\text{pe}}\bar{\varepsilon} \to 0$ (and a sufficiently small noise bound $\bar{\varepsilon}$ due to Proposition 4.1), the region of attraction approaches the set of all initially feasible points. For a fixed c_{pe}, the size of the region of attraction increases if the noise level decreases and vice versa. A similar connection between the maximal disturbance and the region of attraction can be found in [154], which studies inherent robustness properties of quasi-infinite horizon MPC (but the result applies similarly to model-based n-step MPC with terminal equality constraints). Further, if c_{pe} decreases then so do $\beta_8(\bar{\varepsilon})$ as well as $\beta_9(\bar{\varepsilon})$ and hence also $\beta(\bar{\varepsilon})$. This implies that larger persistence of excitation (i.e., a lower $c_{\text{pe}}\bar{\varepsilon}$) does not only increase the region of attraction but it also reduces the tracking error.

Summary

In this section, we presented a robust data-driven MPC scheme (Algorithm 4.1) for unknown linear systems based on noisy input-output data. Algorithm 4.1 is an adapted version of the nominal scheme from Section 3.1, containing a slack variable and regularization terms to cope with the noise. The presented MPC scheme is simple to apply, only requiring to solve a strictly convex QP online, but admits strong theoretical guarantees. In particular, our main theoretical result (Theorem 4.1) shows that the closed loop exponentially converges to a neighborhood of the origin whose size increases with the noise level. Further, it provides insights into the interplay of system, data, and design parameters and the resulting closed-loop performance. In Section 4.4.1, we illustrate the practicality of the proposed approach and we investigate the mentioned interplay in more detail with a numerical example.

4.2 Robust constraint satisfaction

In this section, we present a robust data-driven MPC scheme based on noisy input-output data which ensures output constraint satisfaction in closed loop. To be precise, we expand the MPC scheme from Section 4.1 by an additional constraint tightening ensuring robust output constraint satisfaction. After introducing the extended MPC scheme in Section 4.2.1, the closed-loop guarantees are presented and proven in Section 4.2.2. Further, in Section 4.2.3, we discuss how certain controllability and observability constants required to set up the constraint tightening can be computed using only measured data without explicit model knowledge.

This section is based on and taken in parts literally from [JB13][3].

4.2.1 Proposed MPC scheme

Problem setting

We consider the same problem setup as in Section 4.1 with the *only* difference that, additionally, we want to satisfy output constraints $y_t \in \mathbb{Y}$ for all $t \in \mathbb{I}_{\geq 0}$ in closed loop with a given set $\mathbb{Y} \subseteq \mathbb{R}^p$. For simplicity, we consider hyperbox constraints on the output, which are w.l.o.g. assumed to be symmetric w.r.t. zero, and we assume that the input constraint set $\mathbb{U} \subseteq \mathbb{R}^m$ is a convex, compact polytope.

Assumption 4.6. *(Constraint sets) The input constraint set $\mathbb{U}$ is a convex, compact polytope and $\mathbb{Y}$ is defined as*

$$\mathbb{Y} = \{y \in \mathbb{R}^p \mid \|y\|_\infty \leq y_{\max}\}. \tag{4.42}$$

for some $y_{\max} > 0$. Further, $(u^s, y^s) \in \operatorname{int}(\mathbb{U} \times \mathbb{Y})$.

It is straightforward to extend the results of this section to constraints with different bounds $y^i_{\max}$, $i = 1, \ldots, p$, for each component of the output, e.g., by scaling all output measurements by $\frac{y^i_{\max}}{y_{\max}}$. Moreover, we conjecture that similar results can be obtained for general polytopic output constraints.

MPC scheme

Given a noisy data trajectory $\{u^d_k, \tilde{y}^d_k\}_{k=0}^{N-1}$ as well as past n (noisy) input-output measurements $\{u_k, \tilde{y}_k\}_{k=t-n}^{t-1}$ to specify initial conditions, we define the following data-driven MPC scheme:

[3] J. Berberich, J. Köhler, M. A. Müller, and F. Allgöwer. "Robust constraint satisfaction in data-driven MPC." In: *Proc. 59th Conf. Decision and Control (CDC)*. 2020, pp. 1260–1267. ©2020 IEEE

Problem 4.2.

$$\underset{\substack{\alpha(t),\sigma(t)\\ \bar{u}(t),\bar{y}(t)}}{\text{minimize}} \quad \sum_{k=0}^{L-1} \left(\|\bar{u}_k(t) - u^s\|_R^2 + \|\bar{y}_k(t) - y^s\|_2^2 \right) + \lambda_\alpha \bar{\varepsilon} \|\alpha(t)\|_2^2 + \frac{\lambda_\sigma}{\bar{\varepsilon}} \|\sigma(t)\|_2^2 \tag{4.43a}$$

subject to

$$\begin{bmatrix} \bar{u}(t) \\ \bar{y}(t) + \sigma(t) \end{bmatrix} = \begin{bmatrix} H_{L+n}\left(u^d\right) \\ H_{L+n}\left(\tilde{y}^d\right) \end{bmatrix} \alpha(t), \tag{4.43b}$$

$$\begin{bmatrix} \bar{u}_{[-n,-1]}(t) \\ \bar{y}_{[-n,-1]}(t) \end{bmatrix} = \begin{bmatrix} u_{[t-n,t-1]} \\ \tilde{y}_{[t-n,t-1]} \end{bmatrix}, \tag{4.43c}$$

$$\begin{bmatrix} \bar{u}_{[L-n,L-1]}(t) \\ \bar{y}_{[L-n,L-1]}(t) \end{bmatrix} = \begin{bmatrix} \mathbb{1}_n \otimes u^s \\ \mathbb{1}_n \otimes y^s \end{bmatrix}, \; \bar{u}_k(t) \in \mathbb{U}, \tag{4.43d}$$

$$\|\bar{y}_k(t)\|_\infty + a_{1,k}\|\bar{u}(t)\|_1 + a_{2,k}\|\alpha(t)\|_1 + a_{3,k}\|\sigma(t)\|_\infty + a_{4,k} \leq y_{\max}, \; k \in \mathbb{I}_{[0,L-n-1]}. \tag{4.43e}$$

Problem 4.2 is analogous to Problem 4.1 in Section 4.1 with the only difference being the constraint tightening (4.43e) which ensures closed-loop output constraint satisfaction despite noisy measurements. As will become clear later in this section, the output prediction error induced by the noise depends on the size of the quantities $\bar{u}(t)$, $\alpha(t)$, $\sigma(t)$, compare also Lemma 4.2, which explains their occurrence in the constraint tightening. Appropriate definitions of the constants $a_{i,k}$ to guarantee recursive feasibility and closed-loop constraint satisfaction are provided below in Equations (4.52) and (4.53). Note that all constraints in Problem 4.2 can be written as linear equality and inequality constraints and thus, Problem 4.2 is a strictly convex QP which can be solved efficiently.

Throughout this section, we denote the optimal solution of Problem 4.2 at time t by $\bar{u}^*(t)$, $\bar{y}^*(t)$, $\alpha^*(t)$, $\sigma^*(t)$, and the optimal cost by $J_L^*(\tilde{\xi}_t)$. Closed-loop values of the input and output at time t are denoted by u_t and y_t, respectively. Analogously to Section 4.1, Problem 4.2 is solved in an n-step fashion, as defined in Algorithm 4.2.

Constraint tightening

In the following, we define the constants $a_{i,k}$ involved in the constraint tightening (4.43e), based on data and system parameters. First, we assume knowledge of a

---

Algorithm 4.2. Robust data-driven MPC scheme with constraint tightening

Offline: Choose upper bound on system order n, prediction horizon L, cost matrices $Q, R \succ 0$, regularization parameters $\lambda_\alpha, \lambda_\sigma > 0$, constraint set $\mathbb{U}$, noise bound $\bar{\varepsilon}$, setpoint (u^s, y^s), compute constants $a_{i,k}$ (compare (4.52) and (4.53)), and generate data $\{u_k^d, \tilde{y}_k^d\}_{k=0}^{N-1}$.

Online: At time $t = n \cdot i$, $i \in \mathbb{I}_{\geq 0}$, take the past n measurements $\{u_k, \tilde{y}_k\}_{k=t-n}^{t-1}$ and solve Problem 4.2. Apply the input $u_{[t,t+n-1]} = \bar{u}^*_{[0,n-1]}(t)$ over the next n time steps.

---

controllability constant $\Gamma > 0$ such that for any initial trajectory $\{u_k, y_k\}_{k=-n}^{n-1}$ with $u_{[0,n-1]} = 0$, there exists an input-output trajectory $\{\bar{u}_k, \bar{y}_k\}_{k=-n}^{2n-1}$ such that

$$(\bar{u}_{[-n,-1]}, \bar{y}_{[-n,-1]}) = (u_{[-n,-1]}, y_{[-n,-1]}), \tag{4.44}$$

$$(\bar{u}_{[n,2n-1]}, \bar{y}_{[n,2n-1]}) = (0, 0), \tag{4.45}$$

$$\|\bar{u}_{[0,n-1]}\|_1 \leq \Gamma \left\| y_{[0,n-1]} \right\|_\infty. \tag{4.46}$$

This condition is similar to (2.10) and it means that, for any given initial input-output trajectory $\{u_k, y_k\}_{k=-n}^{-1}$, there exists an input that can be appended to this trajectory which steers the system to zero in n steps. Moreover, the norm of this input is bounded by the norm of the output that would result from the same initial trajectory when applying a zero input. By Assumption 4.3 (controllability), the above condition is always satisfied for *some* $\Gamma > 0$. Further, we assume that there exist (known) constants $\rho_{\infty,k}$, $k \in \mathbb{I}_{[n,L+n-1]}$, defined as

$$\begin{aligned} \rho_{\infty,k} \;=\; & \max_{y \in \mathbb{R}^{(k+1)p}} \|y_k\|_\infty \\ & \text{s.t. } \|y_{[0,n-1]}\|_\infty = 1 \text{ and } \{0, y_j\}_{j=0}^{k} \text{ is an input-output trajectory of (2.1).} \end{aligned} \tag{4.47}$$

The constants $\rho_{\infty,k}$ are related to observability of the underlying system and it is straightforward to verify that they satisfy the second inequality in (4.15). According to its definition, $\rho_{\infty,k}$ is equal to the norm of the output of the unknown LTI system (2.1) at time k, assuming that the input is zero and the initial output is norm-bounded by 1. In Section 4.2.3, we describe how Γ as well as $\rho_{\infty,k}$ can be estimated from measured data. Furthermore, since the input and output constraint sets $\mathbb{U}$, $\mathbb{Y}$ are compact, we can define

$$\xi_{\max} := \max_{\xi \in \mathbb{U}^n \times \mathbb{Y}^n} \|\xi\|_1. \tag{4.48}$$

Finally, we define

$$H_{\mathrm{u}\xi} = \begin{bmatrix} H_{L+n}(u^{\mathrm{d}}) \\ H_1\left(\xi^{\mathrm{d}}_{[0,N-L-n]}\right) \end{bmatrix}, \tag{4.49}$$

where ξ^{d} is the extended state[4] corresponding to the measured data $\{u_k^{\mathrm{d}}, y_k^{\mathrm{d}}\}_{k=0}^{N-1}$, compare (2.3). We define the constant

$$c_{\mathrm{pe}}^{\xi} := \|H_{\mathrm{u}\xi}^{\dagger}\|_1, \tag{4.50}$$

where $H_{\mathrm{u}\xi}^{\dagger}$ is the Moore-Penrose inverse of $H_{\mathrm{u}\xi}$. Note that $H_{\mathrm{u}\xi}$ does in general not have full row rank, even if the input is persistently exciting, since the state-space model with state ξ may not be controllable, compare Lemma 2.1, and both persistence of excitation as well as controllability are required to ensure that $H_{\mathrm{u}\xi}$ has full row rank [144, Corollary 2]. Nevertheless, if u^{d} is persistently exciting of order $L+2n$, then for any desired input $\{u_k\}_{k=0}^{L-1}$ and extended state ξ_0, there exists $\alpha \in \mathbb{R}^{N-L-n+1}$ such that

$$H_{\mathrm{u}\xi}\alpha = \begin{bmatrix} u \\ \xi_0 \end{bmatrix}, \tag{4.51}$$

compare [94] for details. The constant c_{pe}^{ξ} is analogous to c_{pe} in Section 4.1 (compare (4.3)), replacing the (minimal) state x by the (extended) state ξ. In this section, we consider the matrix $H_{\mathrm{u}\xi}$ instead of H_{ux} since the constant c_{pe}^{ξ} is explicitly required to set up the proposed constraint tightening and we only have measurements of the extended state ξ but not of the minimal state x. Note that the constant c_{pe}^{ξ} can be rendered arbitrarily small by choosing the norm of a sufficiently rich persistently exciting input u^{d} sufficiently large. In Section 4.1, in the absence of output constraints, it was shown that a small value of c_{pe}^{ξ} corresponds to a large region of attraction and a small tracking error for the closed loop under the presented

[4] In order to construct this state at times 0 through $n-1$, we require n additional data points $\{u_k^{\mathrm{d}}, y_k^{\mathrm{d}}\}_{k=-n}^{-1}$. For notational simplicity, we neglect this fact throughout this section and simply assume that $N+n$ overall data points are available, the first n of which are only used to construct $\{\xi_k^{\mathrm{d}}\}_{k=0}^{n-1}$.

data-driven MPC scheme. Similarly, we will see in the following that a small constant c_{pe}^{ξ} reduces the conservatism of the proposed output constraint tightening.

Now, we are in the position to state the coefficients $a_{i,k}$, $i = 1, \dots, 4$, in (4.43e). For the first n steps $k \in \mathbb{I}_{[0,n-1]}$, define

$$a_{1,k} = 0, \; a_{2,k} = \bar{\varepsilon} a_{3,k}, \; a_{3,k} = 1 + \rho_{\infty,n}^{\max}, \; a_{4,k} = \bar{\varepsilon} \rho_{\infty,n}^{\max}, \tag{4.52}$$

where $\rho_{\infty,n}^{\max} := \max_{k \in \mathbb{I}_{[0,n-1]}} \rho_{\infty,n+k}$ with $\rho_{\infty,k}$ as in (4.15). For $k \in \mathbb{I}_{[0,L-2n-1]}$, the coefficients $a_{i,k+n}$ are defined recursively as

$$\begin{aligned} a_{1,k+n} &= a_{1,k} + (a_{2,k} + a_{3,k}\bar{\varepsilon}) c_{\text{pe}}^{\xi}, \qquad (4.53) \\ a_{2,k+n} &= \bar{\varepsilon} a_{3,k+n}, \\ a_{3,k+n} &= 1 + \rho_{\infty,2n+k} + \Gamma(1 + \rho_{\infty,L}^{\max}) a_{1,k+n}, \\ a_{4,k+n} &= a_{4,k} + \bar{\varepsilon} \rho_{\infty,2n+k} + \bar{\varepsilon} a_{1,k+n} \Gamma \rho_{\infty,L}^{\max} + \bar{\varepsilon} a_{3,k} + (a_{2,k} + a_{3,k}\bar{\varepsilon}) c_{\text{pe}}^{\xi} \xi_{\max}, \end{aligned}$$

where $\rho_{\infty,L}^{\max} := \max_{k \in \mathbb{I}_{[0,n-1]}} \rho_{\infty,L+k}$. According to the above definition, $a_{2,k} = \bar{\varepsilon} a_{3,k}$ for all $k \in \mathbb{I}_{[0,L-n-1]}$ and hence, $a_{2,k}$ is arbitrarily small if $\bar{\varepsilon}$ is arbitrarily small. This implies that the same holds true for $a_{1,k}$ and $a_{4,k}$ and hence, $a_{3,k}$ becomes arbitrarily close to $1 + \rho_{\infty,n+k}$ for $k \geq n$. Thus, except for $a_{3,k}$, all constants involved in the constraint tightening can be rendered arbitrarily small for sufficiently small noise levels. Finally, the slack variable $\sigma(t)$ becomes arbitrarily small for small noise levels due to the regularization $\frac{\lambda_\sigma}{\bar{\varepsilon}} \|\sigma(t)\|_2^2$ in the cost and hence, the tightened constraints (4.43e) recover the nominal output constraints $\|\bar{y}_k(t)\|_\infty \leq y_{\max}$ if the noise level tends to zero.

In order to implement the MPC scheme, the coefficients $a_{i,k}$ need to be computed according to the above recursion. This requires knowledge of the constants $\rho_{\infty,k}$, Γ, $\xi_{\max}$, c_{pe}^{ξ}. Note that $\xi_{\max}$ can be computed directly using only the definition of the constraints $\mathbb{U}, \mathbb{Y}$. Similarly, the constant c_{pe}^{ξ} in (4.50) can be (approximately) computed based on the data $\{u_k^{\text{d}}, \tilde{y}_k^{\text{d}}\}_{k=0}^{N-1}$, i.e.,

$$c_{\text{pe}}^{\xi} \approx \|H_{u\tilde{\xi}}\|_1,$$

with $\tilde{\xi}_k = \begin{bmatrix} u_{[k-n,k-1]}^\top & \tilde{y}_{[k-n,k-1]}^\top \end{bmatrix}^\top$. On the other hand, the quantities $\rho_{\infty,k}$ and Γ cannot easily be computed without additional model knowledge. In Section 4.2.3, we propose methods to estimate these constants from measured data based on the Fundamental Lemma.

4.2.2 Closed-loop guarantees

In the following, we provide a theoretical analysis of closed-loop properties of the proposed MPC scheme. To this end, we consider the case $(u^s, y^s) = (0, 0)$, i.e., the desired setpoint is the origin, and we comment on setpoints $(u^s, y^s) \neq (0, 0)$ in Remark 4.6.

Recursive feasibility

We begin by proving that, if the proposed data-driven MPC scheme is feasible at time t, then it is feasible at time $t + n$, provided that the noise bound $\bar{\varepsilon}$ is sufficiently small. While this was proven in Section 4.1 for Problem 4.2 without any constraints on the output, i.e., without (4.43e), it is an essential contribution of the present section to prove the same recursive feasibility property for the scheme with tightened output constraints (4.43e). To this end, we use the same Lyapunov function candidate to analyze stability as in Section 4.1, i.e.,

$$V_t := J_L^*(\tilde{\xi}_t) + \gamma W(\xi_t)$$

for some $\gamma > 0$ and an IOSS Lyapunov function $W(\xi) = \|\xi\|_{P_W}^2$.

Proposition 4.2. *Suppose Assumptions 4.2–4.6 hold. Then, for any $V_{\max} > 0$, there exists an $\bar{\varepsilon}_{\max} > 0$ such that for all $\bar{\varepsilon} \leq \bar{\varepsilon}_{\max}$, if $V_t \leq V_{\max}$ for some $t \geq 0$, then Problem 4.2 is feasible at time $t + n$.*

Proof. Denote by $\alpha'(t+n)$, $\sigma'(t+n)$, $\bar{u}'(t+n)$, and $\bar{y}'(t+n)$ the candidate solution used to prove recursive feasibility of the data-driven MPC scheme without tightened output constraints in Section 4.1 (Proposition 4.1). It only remains to show that this candidate satisfies (4.43e). Note that, by definition, the optimal solution at time t satisfies

$$\|\bar{y}_k^*(t)\|_\infty + a_{1,k}\|\bar{u}^*(t)\|_1 + a_{2,k}\|\alpha^*(t)\|_1 + a_{3,k}\|\sigma^*(t)\|_\infty + a_{4,k} \leq y_{\max} \tag{4.54}$$

for $k \in \mathbb{I}_{[0,L-n-1]}$. We bound now each separate component appearing in the constraint (4.43e) at time $t + n$ to show that it is satisfied for $k \in \mathbb{I}_{[0,L-2n-1]}$ by the candidate solution. First, it holds for the output that

$$\begin{aligned}\|\bar{y}'_k(t+n)\|_\infty \leq& \|\bar{y}^*_{k+n}(t)\|_\infty + \|\hat{y}_{t+n+k} - \bar{y}^*_{k+n}(t)\|_\infty \\ \overset{(4.17),(4.54)}{\leq}& y_{\max} - a_{1,k+n}\|\bar{u}^*(t)\|_1 - a_{2,k+n}\|\alpha^*(t)\|_1 - a_{3,k+n}\|\sigma^*(t)\|_\infty - a_{4,k+n} \\ &+ \bar{\varepsilon}(1+\rho_{\infty,2n+k})\|\alpha^*(t)\|_1 + (1+\rho_{\infty,2n+k})\|\sigma^*(t)\|_\infty + \bar{\varepsilon}\rho_{\infty,2n+k}.\end{aligned} \tag{4.55}$$

We obtain for the input candidate that

$$\|\bar{u}'(t+n)\|_1 = \|\bar{u}^*_{[0,L-n-1]}(t)\|_1 + \|\bar{u}'_{[L-2n,L-n-1]}(t+n)\|_1.$$

Due to the terminal equality constraint $\bar{u}^*_{[L-n,L-1]}(t) = 0$ as well as the definition of $\hat{y}$ in (4.14), we can use (4.46) to bound the second term on the right-hand side as

$$\|\bar{u}'_{[L-2n,L-n-1]}(t+n)\|_1 \overset{(4.46)}{\leq} \Gamma\|\hat{y}_{[t+L-n,t+L-1]}\|_\infty.$$

Using now the bound (4.17) together with the fact that $\bar{y}^*_k(t) = 0$ for $k \in \mathbb{I}_{[L-n,L-1]}$ according to the terminal equality constraint (4.43d), it follows that

$$\|\bar{u}'_{[L-2n,L-n-1]}(t+n)\|_1 \leq \Gamma\big(\bar{\varepsilon}\rho^{\max}_{\infty,L} + \bar{\varepsilon}(1+\rho^{\max}_{\infty,L})\|\alpha^*(t)\|_1 + (1+\rho^{\max}_{\infty,L})\|\sigma^*(t)\|_\infty\big), \tag{4.56}$$

where $\rho^{\max}_{\infty,L} = \max_{k\in\mathbb{I}_{[0,n-1]}} \rho_{\infty,L+k}$. Note that the candidate $\alpha'(t+n)$ satisfies

$$\alpha'(t+n) = H^\dagger_{\mathrm{u}\xi}\begin{bmatrix}\bar{u}'(t+n)\\ \xi_t\end{bmatrix}, \tag{4.57}$$

where $H^\dagger_{\mathrm{u}\xi}$ is the Moore-Penrose inverse of $H_{\mathrm{u}\xi}$, compare also (4.51). Thus, it can be bounded as

$$\|\alpha'(t+n)\|_1 \leq c^{\xi}_{\mathrm{pe}}(\|\bar{u}'(t+n)\|_1 + \underbrace{\|\xi_t\|_1}_{\leq \xi_{\max}}) \tag{4.58}$$

with $c^{\xi}_{\mathrm{pe}} = \|H^\dagger_{\mathrm{u}\xi}\|_1$. Finally, similar to (4.35), it can be shown that the slack variable candidate satisfies

$$\|\sigma'(t+n)\|_\infty \leq \bar{\varepsilon}(1+\|\alpha'(t+n)\|_1). \tag{4.59}$$

It follows from straightforward algebraic manipulations, using the definition of the coefficients $a_{i,k}$ in (4.52) and (4.53) as well as the bounds (4.55), (4.56), (4.58), and (4.59), that the candidate solution satisfies

$$\|\bar{y}'_k(t+n)\|_\infty + a_{1,k}\|\bar{u}'(t+n)\|_1 + a_{2,k}\|\alpha'(t+n)\|_1 + a_{3,k}\|\sigma'(t+n)\|_\infty + a_{4,k} \leq y_{\max},$$

i.e., the constraint (4.43e) holds for $k \in \mathbb{I}_{[0,L-2n-1]}$. If $\bar{\varepsilon}$ is sufficiently small, then $\bar{y}'_k(t+n)$ is arbitrarily small for $k \in \mathbb{I}_{[L-2n,L-n-1]}$ (recall its definition in the proof of Proposition 4.1) and also the other components appearing on the left-hand side of the constraint (4.43e) become arbitrarily small (compare the discussion below (4.53)). Hence, since $y_{\max} > 0$, the constraint (4.43e) holds also for $k \in \mathbb{I}_{[L-2n,L-n-1]}$ which thus concludes the proof. ∎

Proposition 4.2 shows that the proposed n-step MPC scheme is n-step feasible in the sense that, for a given Lyapunov function sublevel set $V_t \leq V_{\max}$ with some $V_{\max} > 0$, there exists a noise bound $\bar{\varepsilon}$ sufficiently small such that the scheme is feasible at time $t+n$. The proof of Proposition 4.2 utilizes the same candidate solution as in Proposition 4.1. While it is shown in Proposition 4.1 that this candidate solution satisfies the constraints (4.43b)-(4.43d), Proposition 4.2 proves that it additionally satisfies the tightened constraints (4.43e), which is the main technical contribution of this section.

Remark 4.6. *(Non-zero setpoint) Analogously to Section 4.1, also the results in this section hold true qualitatively for non-zero setpoints $(u^s, y^s) \neq (0,0)$, but the guarantees deteriorate quantitatively, compare Remark 4.5. Additionally, the implementation of the constraint tightening* (4.43e) *changes slightly when considering $(u^s, y^s) \neq (0,0)$. This is due to the fact that, in this case, the controllability bound* (4.46) *leads to*

$$\|\bar{u}_{[0,n-1]} - \mathbb{1}_n \otimes u^s\|_1 \leq \Gamma \|y_{[0,n-1]} - \mathbb{1}_n \otimes y^s\|_\infty.$$

When modifying the bound (4.58) *accordingly, an additional additive term on the right-hand side of* (4.58) *needs to be introduced, depending on the setpoint (u^s, y^s). We do not address the case $(u^s, y^s) \neq (0,0)$ since, compared to the other terms involved in the constraint tightening, the effect of a non-zero setpoint is relatively small and considering it complicates some of the arguments.*

Robust constraint satisfaction

Together with Theorem 4.1, Proposition 4.2 already proves that the proposed MPC scheme (Algorithm 4.2) leads to a practically exponentially stable closed loop. It only remains to show that the closed-loop output satisfies the constraints $y_t \in \mathbb{Y}$ for all $t \in \mathbb{I}_{\geq 0}$.

Theorem 4.2. *Suppose Assumptions 4.2–4.6 hold. Then, for any $V_{\max} > 0$, there exist constants $\underline{\lambda}_\alpha, \overline{\lambda}_\alpha, \underline{\lambda}_\sigma, \overline{\lambda}_\sigma > 0$ such that, for all $\lambda_\alpha, \lambda_\sigma$ satisfying*

$$\underline{\lambda}_\alpha \leq \lambda_\alpha \leq \overline{\lambda}_\alpha, \quad \underline{\lambda}_\sigma \leq \lambda_\sigma \leq \overline{\lambda}_\sigma, \tag{4.60}$$

there exist constants $\bar{\varepsilon}_{\max}, \bar{c}_{\text{pe}} > 0$, as well as $\beta \in \mathcal{K}_\infty$ such that, for all $\bar{\varepsilon}$ (cf. Assumption 4.2) and c_{pe}^{ξ} (as defined in (4.50)) satisfying

$$\bar{\varepsilon} \leq \bar{\varepsilon}_{\max}, \quad c_{\text{pe}}^{\xi}\bar{\varepsilon} \leq \bar{c}_{\text{pe}}, \tag{4.61}$$

the bound $V_t \leq V_{\max}$ holds recursively for all $t \in \mathbb{I}_{\geq 0}$, the input and output constraints are satisfied, i.e., $u_t \in \mathbb{U}$ and $y_t \in \mathbb{Y}$ for all $t \in \mathbb{I}_{\geq 0}$, and V_t converges exponentially to $[0, \beta(\bar{\varepsilon})]$ in closed loop with the n-step MPC scheme (Algorithm 4.2) for all initial conditions for which $V_0 \leq V_{\max}$.

Proof. Theorem 4.1 shows that the set $V_t \leq V_{\max}$ is robustly positively invariant and that V_t converges exponentially to $[0, \beta(\bar{\varepsilon})]$ for an MPC scheme without the tightened output constraint (4.43e). As Proposition 4.2 shows, the candidate solution used for the proof in Theorem 4.1 also satisfies the tightened constraints and hence, we can apply the exact same arguments to conclude recursive feasibility as well as exponential convergence to $[0, \beta(\bar{\varepsilon})]$. To prove closed-loop output constraint satisfaction, note that $\hat{y}_{t+k} = y_{t+k}$ for $k \in \mathbb{I}_{[0,n-1]}$ due to the n-step MPC scheme, with $\hat{y}$ from (4.14). Hence, (4.17) implies

$$\|y_{t+k} - \bar{y}_k^*(t)\|_\infty \leq \bar{\varepsilon}(1 + \rho_{\infty,n}^{\max})\|\alpha^*(t)\|_1 + (1 + \rho_{\infty,n}^{\max})\|\sigma^*(t)\|_\infty + \bar{\varepsilon}\rho_{\infty,n}^{\max} \tag{4.62}$$

for any $k \in \mathbb{I}_{[0,n-1]}$, where $\rho_{\infty,n}^{\max} = \max_{k \in \mathbb{I}_{[0,n-1]}} \rho_{\infty,n+k}$. Further, since the optimal solution satisfies (4.43e), it follows from the definition of the coefficients $a_{i,k}$, $i \in \mathbb{I}_{[1,4]}$, $k \in \mathbb{I}_{[0,n-1]}$, in (4.52) that

$$\|\bar{y}_k^*(t)\|_\infty \leq y_{\max} - \bar{\varepsilon}(1 + \rho_{\infty,n}^{\max})\|\alpha^*(t)\|_1 - (1 + \rho_{\infty,n}^{\max})\|\sigma^*(t)\|_\infty - \bar{\varepsilon}\rho_{\infty,n}^{\max} \tag{4.63}$$

for any $k \in \mathbb{I}_{[0,n-1]}$. Combining (4.62) and (4.63), we thus obtain

$$\|y_{t+k}\|_\infty \leq \|y_{t+k} - \bar{y}_k^*(t)\|_\infty + \|\bar{y}_k^*(t)\|_\infty \leq y_{\max},$$

i.e., the closed-loop output satisfies the constraints. ∎

Except for output constraint satisfaction $y_t \in \mathbb{Y}$, Theorem 4.2 follows from Proposition 4.2 and the results in Section 4.1. It is the key contribution of the present section to suggest a suitable output constraint tightening, compare (4.43e) with parameters (4.52) and (4.53), and to prove recursive feasibility of this constraint tightening (Proposition 4.2) as well as closed-loop constraint satisfaction (Theorem 4.2).

Remark 4.7. *(Related work) Recently, various approaches have been developed to guarantee open-loop constraint satisfaction for data-driven optimal control. In [28], the authors enhance data-driven optimal control with a conditional value at risk constraint on the output, for which a tractable reformulation can be derived using tools from distributionally robust optimization. Stochastic constraints are also considered in [66, 111] in the context of data-driven MPC, however, without any theoretical guarantees. Moreover, [44] addresses data-driven optimal control based on the input-output parametrization [45] and includes a constraint tightening ensuring robust output constraint satisfaction. Further, [63, 147] employ min-max formulations to enforce robust constraint satisfaction in the presence of deterministic noise. In contrast to the above works, Theorem 4.2 guarantees robust constraint satisfaction not only of Problem 4.2 in an open-loop application but also in* closed loop *when applying the data-driven MPC scheme in Algorithm 4.2. Finally, [JB18] extends the results of this section by developing a robust data-driven MPC scheme with guaranteed closed-loop constraint satisfaction in the presence of process noise.*

4.2.3 Estimation of system constants

In order to define the tightened output constraints (4.43e) and thus, to implement the proposed MPC scheme in practice, appropriate values for the coefficients $a_{i,k}$ as in (4.52) and (4.53) need to be computed. This requires knowledge of the system constants Γ and $\rho_{\infty,k}$ defined in (4.46) and (4.47), respectively. In this section, we employ the Fundamental Lemma (Theorem 2.1) to derive a purely data-driven estimation procedure for these system constants. Similar to [JB21], which verifies integral quadratic constraints from measured data, we use Theorem 2.1 to optimize over all system trajectories based on a single measured trajectory in order to compute the desired constants, i.e., to verify a quantitative controllability and observability property.

While the stability and constraint satisfaction results of the previous section hold true if the data are affected by noise (assuming that upper bounds on $\rho_{\infty,k}$ and Γ are known), we assume throughout this section that a noise-free input-output trajectory $\{u_k^{\mathrm{d}}, y_k^{\mathrm{d}}\}_{k=0}^{N-1}$ of the unknown system is available. An extension of the estimation procedures presented in this section to the case of noisy measurements is an interesting issue for future research.

Controllability

In the following, we derive a data-driven procedure to compute a constant $\Gamma > 0$ according to (4.46). To be more precise, Γ is equal to the optimal value of the optimization problem

$$\begin{aligned} \max_{u,y,\alpha} \min_{\bar{u},\bar{y},\bar{\alpha}} \quad & \|\bar{u}_{[0,n-1]}\|_1 \\ \text{s.t.} \quad & (\bar{u}_{[-n,-1]}, \bar{y}_{[-n,-1]}) = (u_{[-n,-1]}, y_{[-n,-1]}), \\ & (\bar{u}_{[n,2n-1]}, \bar{y}_{[n,2n-1]}) = (0,0), \\ & \|y_{[0,n-1]}\|_\infty \leq 1, \ u_{[0,n-1]} = 0, \\ & \begin{bmatrix} H_{3n}(u^{\mathrm{d}}) \\ H_{3n}(y^{\mathrm{d}}) \end{bmatrix} \bar{\alpha} = \begin{bmatrix} \bar{u} \\ \bar{y} \end{bmatrix}, \begin{bmatrix} H_{2n}(u^{\mathrm{d}}) \\ H_{2n}(y^{\mathrm{d}}) \end{bmatrix} \alpha = \begin{bmatrix} u \\ y \end{bmatrix}. \end{aligned} \tag{4.64}$$

In (4.64), we impose $\|y_{[0,n-1]}\|_\infty \leq 1$ instead of $\|y_{[0,n-1]}\|_\infty = 1$ since the inequality constraint is always satisfied with equality due to the maximization in y, and the constraint $\|y_{[0,n-1]}\|_\infty \leq 1$ is convex. We proceed in solving (4.64) as follows: The set of all feasible (u, y) in (4.64) is a convex, compact polytope $\mathbb{Z}$ and hence, instead of maximizing over all (u, y), it suffices to solve the minimization problem for all vertices of this polytope. To be more precise, denote the set of vertices of $\mathbb{Z}$ by $\mathbb{Z}_{\mathrm{v}} = \{(u^i, y^i), i \in \mathbb{I}_{[1,v]}\}$ and note that $\mathbb{Z}_{\mathrm{v}}$ can be conveniently computed from data, e.g., using the MPT3-toolbox [55]. We compute Γ as the optimal value of

$$\begin{aligned} \max_{i\in\mathbb{I}_{[1,v]}} \min_{\bar{u},\bar{y},\bar{\alpha}} \quad & \|\bar{u}_{[0,n-1]}\|_1 \\ \text{s.t.} \quad & (\bar{u}_{[-n,-1]}, \bar{y}_{[-n,-1]}) = (u^i_{[-n,-1]}, y^i_{[-n,-1]}), \\ & (\bar{u}_{[n,2n-1]}, \bar{y}_{[n,2n-1]}) = (0,0), \\ & \begin{bmatrix} H_{3n}(u^{\mathrm{d}}) \\ H_{3n}(y^{\mathrm{d}}) \end{bmatrix} \bar{\alpha} = \begin{bmatrix} \bar{u} \\ \bar{y} \end{bmatrix}. \end{aligned} \tag{4.65}$$

Note that, for any fixed i, the inner minimization problem in (4.65) is a linear program (LP) which can be solved efficiently using standard solvers. Hence, to compute the constant Γ, we need to (i) compute the vertex set $\mathbb{Z}_{\mathrm{v}}$ and (ii) solve a single LP for each element of $\mathbb{Z}_{\mathrm{v}}$. Due to the computation of the vertices $\mathbb{Z}_{\mathrm{v}}$, the complexity of the above approach scales exponentially with the number of inputs and outputs and with the system dimension. Nevertheless, it remains practical for medium-sized problems, due to the availability of efficient LP solvers. Moreover, the computation is carried out once offline, i.e., it does not increase the online complexity of the proposed MPC scheme.

Observability

Next, we compute the constants $\rho_{\infty,k}$, $k \in \mathbb{I}_{[n,L+n-1]}$, as in (4.47) from data. Applying Theorem 2.1, the optimization problem (4.47) can be reformulated as

$$\begin{aligned} \rho_{\infty,k} \quad = \quad & \max_{y,\alpha} \|y_k\|_\infty \\ & \text{s.t. } \|y_{[0,n-1]}\|_\infty \leq 1, \\ & \begin{bmatrix} H_{k+1}(u^{\mathrm{d}}) \\ H_{k+1}(y^{\mathrm{d}}) \end{bmatrix} \alpha = \begin{bmatrix} 0 \\ y_{[0,k]} \end{bmatrix}. \end{aligned} \tag{4.66}$$

As in (4.64), we replace $\|y_{[0,n-1]}\|_\infty = 1$ by $\|y_{[0,n-1]}\|_\infty \leq 1$ since the constraint is always active for the optimal solution. The maximization of $\|y_k\|_\infty$ can be reformulated as a linear objective via additional integer variables and hence, Problem (4.66) can be directly solved using a mixed integer programming solver, which is, e.g., provided by MOSEK [10]. For systems with one output, i.e., $p = 1$, Problem (4.66) is an LP.

Summary

In this section, we presented a constraint tightening which extends the robust data-driven MPC scheme with terminal equality constraints from Section 4.1. We showed that the resulting MPC scheme is recursively feasible and that the closed-loop output robustly satisfies the output constraints. The applicability of the proposed tightening will be illustrated with a numerical example in Section 4.4.

4.3 Tracking MPC for affine systems

In this section, we present a robust data-driven tracking MPC scheme for affine systems based only on input-output data affected by noise. The presented results combine and extend the data-driven MPC results in Sections 3.3 and 4.1 regarding nominal setpoint tracking and robustness to noise, respectively.

After stating the problem setting in Section 4.3.1, we present the MPC scheme in Section 4.3.2. In Section 4.3.3, we provide a theoretical analysis of the scheme, proving closed-loop practical stability of the optimal reachable equilibrium. Finally, in Section 4.3.4, we derive and discuss performance bounds and suboptimality estimates which are analogous to existing results on data-driven (open-loop) optimal control.

This section is based on and taken in parts literally from [JB11][5].

4.3.1 Problem setting

In this section, we revisit the problem setting from Section 3.3 with the key difference that the available data are affected by noise. To be precise, we want to steer the affine system (3.14), which is assumed to be controllable and observable (Assumption 4.3), towards a given setpoint while satisfying input constraints. We only have access to a *noisy* input-output trajectory $\{u_k^{\mathrm{d}}, \tilde{y}_k^{\mathrm{d}}\}_{k=0}^{N-1}$, where the output $\tilde{y}_k^{\mathrm{d}} = y_k^{\mathrm{d}} + \varepsilon_k^{\mathrm{d}}$, $k \in \mathbb{I}_{[0,N-1]}$, is perturbed by additive measurement noise $\{\varepsilon_k^{\mathrm{d}}\}_{k=0}^{N-1}$. Additionally, the input-output measurements obtained online which are used to include initial conditions are also affected by noise $\{\varepsilon_k^{\mathrm{u}}, \varepsilon_k\}_{k=0}^{\infty}$ as $\tilde{u}_k = u_k + \varepsilon_k^{\mathrm{u}}$, $\tilde{y}_k = y_k + \varepsilon_k$, $k \in \mathbb{I}_{\geq 0}$.

Assumption 4.7. *(Noise bound) It holds that* $\|\varepsilon_k^{\mathrm{d}}\|_2 \leq \bar{\varepsilon}$, $k \in \mathbb{I}_{[0,N-1]}$, *and* $\|\varepsilon_k^{\mathrm{u}}\|_2 \leq \bar{\varepsilon}$, $\|\varepsilon_k\|_2 \leq \bar{\varepsilon}$, $k \in \mathbb{I}_{\geq 0}$, *with known* $\bar{\varepsilon} > 0$.

Note that, in contrast to Sections 4.1 and 4.2 (Assumption 4.2), Assumption 4.7 allows for noise entering both the online input and output measurements. The present section can handle noise in the online input measurements due to a more flexible theoretical analysis which relies on a continuity result for data-driven MPC

[5] J. Berberich, J. Köhler, M. A. Müller, and F. Allgöwer. "Linear tracking MPC for nonlinear systems part II: the data-driven case." In: *IEEE Trans. Automat. Control* (2022). doi: 10.1109/TAC.2022.3166851. ©2021 IEEE

(Proposition 4.3). We conjecture that noisy input measurements in the offline data u^{d} can also be handled with slight modifications, i.e., adding an extra slack variable for the input in the proposed MPC scheme.

Based on the above data, we want to track an output setpoint[6] $y^{\mathrm{r}} \in \mathbb{R}^p$, which need not be an equilibrium of (3.14), while satisfying pointwise-in-time input constraints $u_t \in \mathbb{U} \subseteq \mathbb{R}^m$, $t \in \mathbb{I}_{\geq 0}$. We assume that $\mathbb{U}$ is a convex, compact polytope.

Assumption 4.8. *(Input constraint set) The input constraint set* $\mathbb{U}$ *is a convex, compact polytope.*

We do not consider output constraints to avoid the additional challenges in the required robust constraint tightening, compare Section 4.2. Similar to Section 3.3, the proposed MPC scheme includes an artificial setpoint in the online optimization, thereby guaranteeing closed-loop stability of the optimal reachable equilibrium. Due to a local controllability argument required for our theoretical results, we consider only equilibria whose input component lies in the interior of the constraints, i.e., in some convex and compact polytope $\mathbb{U}^{\mathrm{s}} \subseteq \mathrm{int}(\mathbb{U})$. Given a matrix $S \succ 0$ and a data trajectory $\{u_k^{\mathrm{d}}, y_k^{\mathrm{d}}\}_{k=0}^{N-1}$ of (3.14) with persistently exciting input and state according to Assumption 3.9, we define the optimal reachable equilibrium $(u^{\mathrm{sr}}, y^{\mathrm{sr}})$ as the minimizer of

$$J_{\mathrm{eq}}^* := \min_{\alpha^{\mathrm{s}}, u^{\mathrm{s}}, y^{\mathrm{s}}} \|y^{\mathrm{s}} - y^{\mathrm{r}}\|_S^2 \tag{4.67}$$
$$\text{s.t.} \quad \begin{bmatrix} H_{L+n+1}(u^{\mathrm{d}}) \\ H_{L+n+1}(y^{\mathrm{d}}) \\ \mathbb{1}_{N-L-n}^\top \end{bmatrix} \alpha^{\mathrm{s}} = \begin{bmatrix} \mathbb{1}_{L+n+1} \otimes u^{\mathrm{s}} \\ \mathbb{1}_{L+n+1} \otimes y^{\mathrm{s}} \\ 1 \end{bmatrix}, \; u^{\mathrm{s}} \in \mathbb{U}^{\mathrm{s}},$$

compare the corresponding optimization problem (3.22) in Section 3.3. We note that y^{sr} is unique due to $S \succ 0$ and u^{sr} is unique due to Assumption 4.9 below. On the other hand, the corresponding solution α^{s} is in general not unique and we denote the solution with minimum 2-norm by α^{sr}. We denote the optimal reachable extended state corresponding to $(u^{\mathrm{sr}}, y^{\mathrm{sr}})$ by ξ^{sr}.

[6] For simplicity, we only consider output setpoints in this section. Input setpoints can be included via an augmented output $y' = \begin{bmatrix} y \\ u \end{bmatrix}$.

Assumption 4.9. *(Unique steady-state) The matrix* $\begin{bmatrix} A-I & B \\ C & D \end{bmatrix}$ *has full column rank.*

Assumption 4.9 is a standard condition in tracking MPC (cf. [118, Lemma 1.8], [83, Remark 1]) and it implies that the system has no transmission zeros at 1 [90, Ass. 1]. Moreover, if Assumption 4.9 holds, then there exists an affine (hence Lipschitz continuous) map $\hat{g}$ which uniquely maps output equilibria y^{s} to their corresponding input-state components $(x^{\mathrm{s}}, u^{\mathrm{s}})$, i.e., $\hat{g}(y^{\mathrm{s}}) = (x^{\mathrm{s}}, u^{\mathrm{s}})$ and

$$\left\| \begin{bmatrix} x_1^{\mathrm{s}} \\ u_1^{\mathrm{s}} \end{bmatrix} - \begin{bmatrix} x_2^{\mathrm{s}} \\ u_2^{\mathrm{s}} \end{bmatrix} \right\|_2^2 \leq c_g \|y_1^{\mathrm{s}} - y_2^{\mathrm{s}}\|_2^2 \tag{4.68}$$

with some $c_g > 0$ for any two input-output equilibria $(u_1^{\mathrm{s}}, y_1^{\mathrm{s}})$, $(u_2^{\mathrm{s}}, y_2^{\mathrm{s}})$ with corresponding steady-states x_1^{s}, x_2^{s}. Since $S \succ 0$, the cost of (4.67) is strongly convex in y^{s} and for any y^{s} satisfying the constraints of (4.67) for some u^{s}, α^{s}, we have

$$\|y^{\mathrm{s}} - y^{\mathrm{r}}\|_S^2 - J_{\mathrm{eq}}^* \geq \|y^{\mathrm{s}} - y^{\mathrm{sr}}\|_S^2, \tag{4.69}$$

compare [69, Inequality (11)]. Similar to Section 3.3 (Assumption 3.9), we assume that the data are persistently exciting such that the Fundamental Lemma for affine systems (Theorem 3.3) is applicable.

Assumption 4.10. *(Persistence of excitation) The data satisfy*

$$\operatorname{rank}\left(\begin{bmatrix} H_{L+n+1}(u^{\mathrm{d}}) \\ H_1(x^{\mathrm{d}}_{[0,N-L-n-1]}) \\ \mathbb{1}_{N-L-n}^\top \end{bmatrix}\right) = m(L+n+1)+n+1. \tag{4.70}$$

4.3.2 Proposed MPC scheme

Given data $\{u_k^{\mathrm{d}}, \tilde{y}_k^{\mathrm{d}}\}_{k=0}^{N-1}$ as well as initial conditions $\{\tilde{u}_k, \tilde{y}_k\}_{k=t-n}^{t-1}$, the following open-loop optimal control problem is the basis for our MPC scheme:

Problem 4.3.

$$\underset{\substack{\alpha(t),\sigma(t),\bar{u}(t)\\ \bar{y}(t),u^{\mathrm{s}}(t),y^{\mathrm{s}}(t)}}{\text{minimize}} \quad \sum_{k=-n}^{L} \left(\|\bar{u}_k(t) - u^{\mathrm{s}}(t)\|_R^2 + \|\bar{y}_k(t) - y^{\mathrm{s}}(t)\|_Q^2 \right) + \|y^{\mathrm{s}}(t) - y^{\mathrm{r}}\|_S^2 \tag{4.71a}$$

$$+ \lambda_\alpha \bar{\varepsilon}^{\beta_\alpha} \|\alpha(t) - \alpha^{\mathrm{sr}}\|_2^2 + \frac{\lambda_\sigma}{\bar{\varepsilon}^{\beta_\sigma}} \|\sigma(t)\|_2^2$$

subject to

$$\begin{bmatrix} \bar{u}(t) \\ \bar{y}(t) + \sigma(t) \\ 1 \end{bmatrix} = \begin{bmatrix} H_{L+n+1}(u^{\mathrm{d}}) \\ H_{L+n+1}(\tilde{y}^{\mathrm{d}}) \\ \mathbb{1}_{N-L-n}^{\top} \end{bmatrix} \alpha(t), \tag{4.71b}$$

$$\begin{bmatrix} \bar{u}_{[-n,-1]}(t) \\ \bar{y}_{[-n,-1]}(t) \end{bmatrix} = \begin{bmatrix} \tilde{u}_{[t-n,t-1]} \\ \tilde{y}_{[t-n,t-1]} \end{bmatrix}, \tag{4.71c}$$

$$\begin{bmatrix} \bar{u}_{[L-n,L]}(t) \\ \bar{y}_{[L-n,L]}(t) \end{bmatrix} = \begin{bmatrix} \mathbb{1}_{n+1} \otimes u^{\mathrm{s}}(t) \\ \mathbb{1}_{n+1} \otimes y^{\mathrm{s}}(t) \end{bmatrix}, \tag{4.71d}$$

$$\bar{u}_k(t) \in \mathbb{U},\ k \in \mathbb{I}_{[0,L]},\ u^{\mathrm{s}}(t) \in \mathbb{U}^{\mathrm{s}}. \tag{4.71e}$$

Problem 4.3 is a combination of the MPC problems considered for nominal setpoint tracking in Section 3.3 (Problem 3.4) and for robust stabilization in Section 4.1 (Problem 4.1). Inspired from tracking MPC [69, 82], the tracking cost with $Q, R \succ 0$ as well as the terminal equality constraints (4.71d) are defined w.r.t. an artificial setpoint $(u^{\mathrm{s}}(t), y^{\mathrm{s}}(t))$ which is optimized online and whose distance to y^{r} is penalized with weight $S \succ 0$. The fact that the cost (4.71a) is summed over $k \in \mathbb{I}_{[-n,L]}$ (instead of $k \in \mathbb{I}_{[0,L]}$ as, e.g., in Chapter 3 and Section 4.1) simplifies the theoretical analysis and represents a weighting of the initial (extended) state, analogously to model-based tracking MPC in [69, 83]. Additionally, Problem 4.3 contains a slack variable $\sigma(t)$ as well as quadratic penalties of $\alpha(t)$ and $\sigma(t)$ to cope with the influence of noisy data, analogously to the results in Section 4.1.

The regularization of $\alpha(t)$ and $\sigma(t)$ depends on the noise bound $\bar{\varepsilon}$ such that, in the limit $\bar{\varepsilon} \to 0$, a nominal MPC scheme with guaranteed exponential stability is recovered. In addition to $\lambda_\alpha, \lambda_\sigma > 0$, Problem 4.3 contains the parameters $\beta_\alpha, \beta_\sigma > 0$ which are assumed to satisfy $\beta_\alpha + 2\beta_\sigma < 2$. In the literature, different choices for β_α, β_σ have been considered for alternative data-driven MPC formulations, e.g., $\beta_\alpha = 1$, $\beta_\sigma = 0$ (see [JB7, 28]) or $\beta_\alpha = 1$, $\beta_\sigma = 1$ (see [JB17] or Section 4.1). In this section, the additional flexibility provided by the parameters β_α and β_σ is mainly required for an argument in the stability proof of data-driven MPC for nonlinear systems in Section 5.2. All theoretical results in the present section remain valid as long as $\beta_\alpha + 2\beta_\sigma < 2$.

Throughout this section, the optimal solution of Problem 4.3 at time t is denoted by $\bar{u}^*(t)$, $\bar{y}^*(t)$, $\alpha^*(t)$, $\sigma^*(t)$, $u^{\mathrm{s}*}(t)$, $y^{\mathrm{s}*}(t)$, and the closed-loop input, state, and output at time t are denoted by u_t, x_t, and y_t, respectively.

Remark 4.8. *(Computation of α^{sr}) Inspired by [39], the regularization of $\alpha(t)$ is not w.r.t. zero but depends on α^{sr} since we want to track the generally non-zero equilibrium $(u^{\mathrm{sr}}, y^{\mathrm{sr}})$. Note that α^{sr} can be (approximately) computed as the least-squares solution of (4.67) by inserting the available noisy input-output measurements $\{u_k, \tilde{y}_k\}_{k=0}^{N-1}$. Due to the influence of the noise, it is beneficial for practical purposes to solve the following robust version of (4.67) with parameters $\lambda_\alpha^{\mathrm{s}}, \lambda_\sigma^{\mathrm{s}} > 0$:*

$$\min_{\alpha^{\mathrm{s}},\sigma^{\mathrm{s}},u^{\mathrm{s}},y^{\mathrm{s}}} \|y^{\mathrm{s}} - y^{\mathrm{r}}\|_S^2 + \lambda_\alpha^{\mathrm{s}}\|\alpha^{\mathrm{s}}\|_2^2 + \lambda_\sigma^{\mathrm{s}}\|\sigma^{\mathrm{s}}\|_2^2 \tag{4.72}$$

$$\text{s.t.} \quad \begin{bmatrix} H_{L+n+1}(u_{[0,N-1]}) \\ H_{L+n+1}(\tilde{y}_{[0,N-1]}) \\ \mathbb{1}_{N-L-n}^\top \end{bmatrix} \alpha^{\mathrm{s}} = \begin{bmatrix} \mathbb{1}_{L+n+1} \otimes u^{\mathrm{s}} \\ \mathbb{1}_{L+n+1} \otimes y^{\mathrm{s}} + \sigma^{\mathrm{s}} \\ 1 \end{bmatrix}, \quad u^{\mathrm{s}} \in \mathbb{U}^{\mathrm{s}}.$$

If an approximation $\alpha^{\mathrm{s}\prime}$ of α^{sr} with $\|\alpha^{\mathrm{s}\prime} - \alpha^{\mathrm{sr}}\|_2 \leq c$ for some $c > 0$ is known, our theoretical results remain true, albeit with more conservative bounds which deteriorate for increasing values of c. In particular, the presented results hold qualitatively for $\alpha^{\mathrm{s}\prime} = 0$.

Algorithm 4.3. Robust data-driven tracking MPC scheme

Offline: Choose upper bound on system order n, prediction horizon L, cost matrices $Q, R, S \succ 0$, regularization parameters $\lambda_\alpha, \lambda_\sigma, \beta_\alpha, \beta_\sigma > 0$, constraint sets $\mathbb{U}$, $\mathbb{U}^{\mathrm{s}}$, noise bound $\bar{\varepsilon}$, setpoint y^{r}, and generate data $\{u_k^{\mathrm{d}}, \tilde{y}_k^{\mathrm{d}}\}_{k=0}^{N-1}$. Compute an approximation of α^{sr} by solving (4.67) or (4.72), compare Remark 4.8.

Online: At time $t = n \cdot i$, $i \in \mathbb{I}_{\geq 0}$, take the past n measurements $\{u_k, \tilde{y}_k\}_{k=t-n}^{t-1}$ and solve Problem 4.3. Apply the input $u_{[t,t+n-1]} = \bar{u}^*_{[0,n-1]}(t)$ over the next n time steps.

Algorithm 4.3 summarizes the proposed (multi-step) MPC scheme. Similar to Algorithm 4.1 in Section 4.1, considering a multi-step MPC scheme instead of a standard (one-step) MPC scheme simplifies the theoretical analysis with terminal equality constraints due to a local controllability argument in the proof. We conjecture that the results in this section hold locally close to the steady-state x^{sr} if Algorithm 4.3 is executed in a one-step fashion, cf. Remark 4.4.

Similar to the previous sections, our theoretical analysis relies on the extended state ξ_t and its noisy version

$$\tilde{\xi}_t := \begin{bmatrix} \tilde{u}_{[t-n,t-1]} \\ \tilde{y}_{[t-n,t-1]} \end{bmatrix}, \tag{4.73}$$

and we write $J_L^*(\tilde{\xi}_t)$ for the optimal cost of Problem 4.3. Further, we note that, by observability, there exist matrices T_x, T_e, T_r, such that

$$x_t = T_\mathrm{x}\xi_t + T_\mathrm{e}e + T_\mathrm{r}r. \tag{4.74}$$

Thus, similar to (2.4), for any two pairs of state and extended state vectors $(x_t^\mathrm{a}, \xi_t^\mathrm{a})$ and $(x_t^\mathrm{b}, \xi_t^\mathrm{b})$, it holds that

$$x_t^\mathrm{a} - x_t^\mathrm{b} = T_\mathrm{x}(\xi_t^\mathrm{a} - \xi_t^\mathrm{b}). \tag{4.75}$$

4.3.3 Closed-loop guarantees

In this section, we prove that Algorithm 4.3 practically exponentially stabilizes the optimal reachable equilibrium in closed loop. To this end, we employ a continuity property of Problem 4.3 which states that noisy data can be translated into an additive input disturbance for data-driven MPC with noise-free data (Proposition 4.3). We then prove that the latter scheme is robust w.r.t. input disturbances (Theorem 4.3) which we combine with Proposition 4.3 to conclude practical exponential stability of the closed loop under the robust data-driven MPC scheme in Algorithm 4.3 (Corollary 4.1).

Let us define the following *nominal* data-driven optimal control problem:

Problem 4.4.

$$\underset{\substack{\alpha(t),\bar{u}(t),\bar{y}(t)\\u^\mathrm{s}(t),y^\mathrm{s}(t)}}{\text{minimize}} \quad \sum_{k=-n}^{L}\left(\|\bar{u}_k(t)-u^\mathrm{s}(t)\|_R^2 + \|\bar{y}_k(t)-y^\mathrm{s}(t)\|_Q^2\right) + \|y^\mathrm{s}(t)-y^\mathrm{r}\|_S^2 \tag{4.76a}$$

$$+\lambda_\alpha\bar{\varepsilon}^{\beta_\alpha}\|\alpha(t)-\alpha^\mathrm{sr}\|_2^2$$

subject to

$$\begin{bmatrix}\bar{u}(t)\\\bar{y}(t)\\1\end{bmatrix} = \begin{bmatrix}H_{L+n+1}\left(u^\mathrm{d}\right)\\H_{L+n+1}\left(y^\mathrm{d}\right)\\\mathbb{1}_{N-L-n}^\top\end{bmatrix}\alpha(t), \tag{4.76b}$$

$$\begin{bmatrix}\bar{u}_{[-n,-1]}(t)\\\bar{y}_{[-n,-1]}(t)\end{bmatrix} = \begin{bmatrix}u_{[t-n,t-1]}\\y_{[t-n,t-1]}\end{bmatrix}, \tag{4.76c}$$

$$\begin{bmatrix}\bar{u}_{[L-n,L]}(t)\\\bar{y}_{[L-n,L]}(t)\end{bmatrix} = \begin{bmatrix}\mathbb{1}_{n+1}\otimes u^\mathrm{s}(t)\\\mathbb{1}_{n+1}\otimes y^\mathrm{s}(t)\end{bmatrix}, \tag{4.76d}$$

$$\bar{u}_k(t)\in\mathbb{U},\ k\in\mathbb{I}_{[0,L]},\ u^\mathrm{s}(t)\in\mathbb{U}^\mathrm{s}. \tag{4.76e}$$

We denote the optimal solution of Problem 4.4 by $\check{\alpha}^*(t)$, $\check{u}^{s*}(t)$, $\check{y}^{s*}(t)$, $\check{u}^*(t)$, $\check{y}^*(t)$, and the corresponding optimal cost by $\check{J}_L^*(\xi_t)$. For the stability analysis, we consider the Lyapunov function candidate $V(\xi_t) := \check{J}_L^*(\xi_t) - J_{\text{eq}}^*$.

Assumption 4.11. *(LICQ) Problem 4.4 satisfies a linear independence constraint qualification (LICQ), i.e., the row entries of the equality and active inequality constraints are linearly independent.*

Assumption 4.11 is required for a technical step in the following result. Such an LICQ assumption is common in linear MPC, compare [13], and we conjecture that it may be possible to relax it at the price of a more involved analysis. The following result bounds the difference between the *nominal* optimal input $\check{u}^*(t)$ corresponding to Problem 4.4 and the *perturbed* optimal input $\bar{u}^*(t)$ corresponding to Problem 4.3.

Proposition 4.3. *Suppose Assumptions 4.7, 4.8, 4.10, and 4.11 hold. Then, for any $\bar{J} > 0$, there exist $\bar{\varepsilon}_{\max} > 0$, $\beta_u \in \mathcal{K}_\infty$ such that, if $\check{J}_L^*(\xi_t) \leq \bar{J}$ and $\bar{\varepsilon} \leq \bar{\varepsilon}_{\max}$, then*

$$\|\bar{u}^*(t) - \check{u}^*(t)\|_2 \leq \beta_u(\bar{\varepsilon}). \tag{4.77}$$

The proof of Proposition 4.3 is postponed to Section 5.2 (Proposition 5.4), where we derive a more general and powerful result which contains the statement of Proposition 4.3 as a special case. Motivated by Proposition 4.3, the following result studies the closed-loop properties of the nominal MPC scheme based on repeatedly solving Problem 4.4 with additional input disturbances.

Theorem 4.3. *Suppose Assumptions 4.3, 4.5, 4.8, 4.9, and 4.10 hold, and consider System (3.14) under control with an n-step MPC scheme (cf. Algorithm 4.3) based on Problem 4.4, where the input applied to (3.14) is perturbed as*

$$u_{[t,t+n-1]} = \check{u}^*_{[0,n-1]}(t) + d_{[t,t+n-1]} \tag{4.78}$$

for $t = ni$, $i \in \mathbb{I}_{\geq 0}$, with some disturbance $\{d_t\}_{t=0}^\infty$ bounded as $\|d_t\|_2 \leq \bar{\varepsilon}$ for all $t \in \mathbb{I}_{\geq 0}$.

For any $V_{\max} > 0$, there exist $\bar{\varepsilon}_{\max}, c_l, c_u > 0$, $0 < \check{c}_V < 1$, and $\beta_d \in \mathcal{K}_\infty$ such that, if $V(\xi_0) \leq V_{\max}$, then, for all $\bar{\varepsilon} \leq \bar{\varepsilon}_{\max}$, $t = ni$, $i \in \mathbb{I}_{\geq 0}$, Problem 4.4 is feasible and the closed loop satisfies

$$c_l\|\xi_t - \xi^{sr}\|_2^2 \leq V(\xi_t) \leq c_u\|\xi_t - \xi^{sr}\|_2^2, \tag{4.79}$$

$$V(\xi_{t+n}) \leq \check{c}_V V(\xi_t) + \beta_d(\bar{\varepsilon}). \tag{4.80}$$

The proof of Theorem 4.3 is provided in the appendix (Appendix A.2). The result shows that the nominal data-driven MPC scheme with noise-free data based on repeatedly solving Problem 4.4 in an n-step fashion (compare Algorithm 4.3) practically exponentially stabilizes the optimal reachable equilibrium ξ^{sr} in the presence of input disturbances. To be precise, the decay bound (4.80) in combination with the lower and upper bounds in (4.79) implies that the closed loop converges to a neighborhood of ξ^{sr}, the size of which shrinks if the disturbance bound (denoted by $\bar{\varepsilon}$ with a slight abuse of notation) is small. The guaranteed region of attraction is then given by $V(\xi_0) \leq V_{\max}$ and, in particular, for a larger size $V_{\max}$ the maximal disturbance bound $\bar{\varepsilon}$ ensuring the closed-loop properties decreases. Moreover, the proof of Theorem 4.3 reveals that the closed-loop robustness improves (i.e., the maximal disturbance bound $\bar{\varepsilon}$ leading to practical stability increases) if the persistence of excitation condition in Assumption 4.10 is quantitatively stronger, i.e., the minimum singular value of the matrix in (4.70) increases. Recall that we observed a similar relation in Section 4.1 for linear data-driven MPC without online optimization of $(u^{\mathrm{s}}(t), y^{\mathrm{s}}(t))$.

In the following, we combine Proposition 4.3 and Theorem 4.3 to prove that the closed loop under the robust data-driven MPC scheme based on Problem 4.3 is practically exponentially stable in the presence of noisy output data and perturbed initial conditions. To this end, we employ the Lyapunov function $V(\xi_t)$ used in Theorem 4.3.

Corollary 4.1. *Suppose Assumptions 4.3, 4.5, and 4.7–4.11 hold. Then, for any* $V_{\max} > 0$*, there exist* $\bar{\varepsilon}_{\max} > 0$ *and* $\beta_{\mathrm{V}} \in \mathcal{K}_\infty$ *such that, for all initial conditions with* $V(\xi_0) \leq V_{\max}$ *and all* $\bar{\varepsilon} \leq \bar{\varepsilon}_{\max}$*, the closed-loop trajectory under Algorithm 4.3 satisfies*

$$V(\xi_{t+n}) \leq \check{c}_{\mathrm{V}} V(\xi_t) + \beta_{\mathrm{V}}(\bar{\varepsilon}) \tag{4.81}$$

for all $t = ni$*,* $i \in \mathbb{I}_{\geq 0}$*, with* $\check{c}_{\mathrm{V}}$ *as in* (4.80).

Proof. If $\bar{\varepsilon} \leq \bar{\varepsilon}_{\max}$ is sufficiently small, then Proposition 4.3 and Theorem 4.3 imply (4.81) for $t = 0$ with $\beta_{\mathrm{V}} := \beta_{\mathrm{d}} \circ \beta_{\mathrm{u}}$, where $\circ$ denotes concatenation. Further, with $\bar{\varepsilon}$ sufficiently small, we have $V(\xi_{t+n}) \leq V_{\max}$ such that the argument can be applied recursively and (4.81) holds for all $t = ni$, $i \in \mathbb{I}_{\geq 0}$. ∎

Corollary 4.1 shows that the closed loop under Algorithm 4.3 exponentially converges to a neighborhood of the optimal reachable equilibrium whose size increases with $\bar{\varepsilon}$. In particular, due to the quadratic upper and lower bounds on V (compare (4.79)), ξ^{sr} is practically exponentially stable. The result is a simple consequence of the facts that the noise in Problem 4.3 can be translated into an input disturbance for nominal data-driven MPC (Proposition 4.3) and the closed loop under the latter is practically exponentially stable w.r.t. the disturbance bound (Theorem 4.3). Thus, the analysis presented in this section reveals a separation principle of data-driven MPC with noise-free and noisy data. That is, any data-driven MPC scheme whose nominal version is robust w.r.t. input disturbances will also lead to a practically stable closed loop in the presence of noisy output measurements affecting the offline data in the Hankel matrices and perturbed input-output initial conditions. A similar approach will be pursued for deriving stability guarantees of data-driven MPC for nonlinear systems in Section 5.2. Finally, as we discuss in Section 4.5, Proposition 4.3 also allows us to extend and unify other robustness results on data-driven MPC.

4.3.4 Performance bounds and suboptimality estimates

We conclude the section by showing that our results in this chapter can be used to derive performance bounds and suboptimality estimates for the open-loop application of data-driven optimal control, which are also investigated in the recent literature. To be precise, we consider bounds on the *actual* cost, i.e., resulting from an application of the optimal input to the true system, in terms of 1) the cost of the underlying optimal control problem, as also studied by [28, 63], and 2) the optimal achievable cost with exact model knowledge / noise-free data, as also studied by [43, 44, 149].

As in Section 4.1, we write $\hat{y}_{[t,t+L]}$ for the actual output trajectory resulting from an open-loop application of $\bar{u}^*(t)$ to the affine system (3.14) with initial condition x_t, compare (4.14). Further, we write $\hat{J}_L$ for the cost of this trajectory, i.e.,

$$\hat{J}_L = \sum_{k=-n}^{L} \|\bar{u}_k^*(t) - u^{s*}(t)\|_R^2 + \|\hat{y}_{t+k} - y^{s*}(t)\|_Q^2 + \|y^{s*}(t) - y^{\text{r}}\|_S^2. \tag{4.82}$$

Combining (4.82) with $\|a+b\|_Q^2 \leq (1+\bar{\varepsilon}^{-\frac{\beta_\sigma}{2}})\|a\|_Q^2 + (1+\bar{\varepsilon}^{\frac{\beta_\sigma}{2}})\|b\|_Q^2$, which holds for any vectors a, b, we can derive

$$\hat{J}_L \overset{(4.82)}{\leq} \left(1+\bar{\varepsilon}^{\frac{\beta_\sigma}{2}}\right) J_L^*(\tilde{\xi}_t) + \left(1+\bar{\varepsilon}^{-\frac{\beta_\sigma}{2}}\right) \sum_{k=-n}^{L} \|\hat{y}_{t+k} - \bar{y}_k^*(t)\|_Q^2. \tag{4.83}$$

The term $\|\hat{y}_{t+k} - \bar{y}_k^*(t)\|_Q^2$ describes the prediction error at the k-th time step, i.e., the difference between the optimal output of Problem 4.3 and the actual output that would result from applying the optimal input $\bar{u}^*(t)$ in open loop. This term can be bounded analogously to Lemma 4.2 such that (4.83) implies

$$\hat{J}_L \leq \left(1+\bar{\varepsilon}^{\frac{\beta_\sigma}{2}}\right) J_L^*(\tilde{\xi}_t) + \hat{\beta}_1(\bar{\varepsilon}) \tag{4.84}$$

for an appropriately defined $\hat{\beta}_1 \in \mathcal{K}_\infty$. This shows that the open-loop performance achieved by the optimal input of Problem 4.3 for the actual system (3.14) is bounded in terms of the optimal cost $J_L^*(\tilde{\xi}_t)$ of Problem 4.3. Moreover, the bound approaches $J_L^*(\tilde{\xi}_t)$ as $\bar{\varepsilon}$ goes to zero. Comparable bounds on the realized open-loop cost in terms of the optimal cost of the data-driven optimal control problem are also provided in [28] and [63] using methods from distributionally robust optimization and a min-max formulation, respectively. Bounds such as (4.84) are useful since they not only allow us to guarantee bounds on the achieved performance a priori but they also reveal interesting insights, e.g., connections between the noise level, the data length, and the resulting performance [28].

We proceed with our analysis by deriving suboptimality estimates for the proposed approach, which are comparable to [43, 44, 149], i.e., bounds on the realized open-loop cost in terms of the optimal achievable one in the presence of noise-free data / exact model knowledge. Recall that Proposition 4.3 bounds the distance between the optimal input of the robust MPC problem (Problem 4.3) and the nominal one (Problem 4.4) in terms of the noise level. This result can be generalized to a bound on the optimal cost of Problem 4.3 (robust problem) in terms of the optimal cost of Problem 4.4 (nominal problem), i.e., under the same assumptions as in Proposition 4.3, it can be shown that

$$J_L^*(\tilde{\xi}_t) \leq \left(1+\hat{c}_1\bar{\varepsilon}^{\beta_\sigma}\right) \mathring{J}_L^*(\xi_t) + \hat{c}_2\bar{\varepsilon}^{2-\beta_\sigma} \tag{4.85}$$

for appropriate constants $\hat{c}_1, \hat{c}_2 > 0$. The precise statement along with a detailed proof of (4.85) is provided in Section 5.2 (Proposition 5.4), where we use the result to analyze data-driven MPC for nonlinear systems. Inserting (4.85) into (4.84), we infer

$$\hat{J}_L \leq \left(1 + \bar{\varepsilon}^{\frac{\beta_\sigma}{2}}\right)\left(\left(1 + \hat{c}_1\bar{\varepsilon}^{\beta_\sigma}\right)\check{J}_L^*(\xi_t) + \hat{c}_2\bar{\varepsilon}^{2-\beta_\sigma}\right) + \hat{\beta}_1(\bar{\varepsilon}). \tag{4.86}$$

Using $\check{J}_L^*(\xi_t) \leq \bar{J}$ by assumption, this implies

$$\hat{J}_L \leq \check{J}_L^*(\xi_t) + \hat{\beta}_2(\bar{\varepsilon}) \tag{4.87}$$

for some $\hat{\beta}_2 \in \mathcal{K}_\infty$. Recall that $\check{J}_L^*(\xi_t)$ corresponds to the nominal MPC Problem 4.4 with noise-free data and thus, due to persistence of excitation (Assumption 4.10), it is equivalent to an optimal control problem with exact model knowledge. Hence, (4.87) means that the cost realized by the optimal input of Problem 4.3 is bounded in terms of the optimal achievable cost for the considered control problem. In particular, the finite-horizon (open-loop) performance under the optimal input of our MPC problem (Problem 4.3) approaches the optimal performance if the noise level is sufficiently small. This insight is analogous to the bounds provided by [149] and [43, 44], which address data-driven optimal control based on system level synthesis [8] and the input-output parametrization [45], respectively. There are, however, a number of differences, e.g., [149] considers a state-feedback formulation and the bounds in [43, 44] depend on the assumed model error instead of the noise level. Further, we note that $\hat{\beta}_2$ is *sublinear* in $\bar{\varepsilon}$, i.e., the smallest exponent of $\bar{\varepsilon}$ appearing in $\hat{\beta}_2(\bar{\varepsilon})$ is $\frac{\beta_\sigma}{2}$. Due to $\beta_\alpha + 2\beta_\sigma < 2$, the rate $\frac{\beta_\sigma}{2}$ is smaller than (but can be made arbitrarily close to) $\frac{1}{2}$. On the other hand, [43, 44, 149] obtained bounds analogous to (4.87) with $\hat{\beta}_2$ depending linearly on the noise / model error. However, we conjecture that the bound (4.87) can be improved using a more sophisticated analysis, and deriving possibly tight suboptimality estimates is an interesting issue for future research. Finally, we note that performance bounds and suboptimality estimates in different but related problem setups have been considered, e.g., for linear-quadratic regulation based on sequential least-squares estimation and robust control for state-feedback [32, 33] or output-feedback [159], and for linear-quadratic regulation via direct data-driven control [31, 37].

We emphasize that, in contrast to the existing works mentioned above, our main theoretical results in this thesis and, in particular, in the present chapter concern the *closed-loop* behavior, e.g., we prove closed-loop practical stability in a receding horizon implementation. Further, we not only obtain suboptimality bounds for the open-loop, finite-horizon performance as in (4.87) but these results can be extended to bounds on the infinite-horizon performance of our MPC scheme in terms of the optimal achievable one. For model-based (i.e., nominal data-driven) MPC, suboptimality estimates depending on the prediction horizon can be found in [50, Theorem 5.22], [51, Theorems 6.2 and 6.4], [68, Appendix A]. Deriving possibly tight infinite-horizon performance bounds of robust data-driven MPC is an interesting problem for future research.

Summary

In this section, we presented a robust data-driven tracking MPC scheme for controlling unknown affine systems using noisy data. The scheme contains the same robustifying ingredients as the one in Section 4.1 (slack variable, regularization) as well as an artificial setpoint $(u^{\mathrm{s}}(t), y^{\mathrm{s}}(t))$ which is optimized online, similar to the nominal tracking MPC scheme in Section 3.3. We proved closed-loop practical exponential stability of the optimal reachable equilibrium based on a separation argument, and we derived performance bounds and suboptimality estimates for the open loop. As in Chapter 3, the inclusion of an artificial setpoint brings significant advantages such as a larger region of attraction, better robustness, and recursive feasibility despite online setpoint changes. Finally, the proposed scheme builds the basis for data-driven MPC of unknown *nonlinear* systems, which we address in Section 5.2.

4.4 Application: numerical examples

In this section, we apply the proposed robust data-driven MPC framework to numerical examples. First, in Section 4.4.1, we apply the scheme with terminal equality constraints from Section 4.1 to the linearized four-tank system also considered in Section 3.4, and we study the influence of different parameters on the closed-loop performance. Moreover, in Section 4.4.2, we apply the robust constraint tightening from Section 4.2 to a numerical example from the literature. For the tracking MPC scheme presented in Section 4.3, we postpone the numerical case study to Chapter 5, where a closely related scheme is used to control unknown nonlinear systems from data.

This section is based on and taken in parts literally from [JB13][7].

4.4.1 Robust data-driven MPC for the linearized four-tank system

In the following, we apply the robust data-driven MPC scheme with terminal equality constraints from Section 4.1 to the linearized four-tank system considered in Section 3.4. We refer the reader to Section 3.4 for a description of the system dynamics and parameters. Our goal is stabilization of the equilibrium

$$(u^{\mathrm{s}}, y^{\mathrm{s}}) = \left(\begin{bmatrix}1\\1\end{bmatrix}, \begin{bmatrix}0.65\\0.77\end{bmatrix}\right)$$

while satisfying the input constraints $u_t \in \mathbb{U} = [-2, 2]^2$ for all $t \in \mathbb{I}_{\geq 0}$. We assume that the order $n = 4$ of the system is known. Further, one noisy input-output trajectory $\{u_k^{\mathrm{d}}, y_k^{\mathrm{d}}\}_{k=0}^{N-1}$ of length $N = 400$ is available. The input generating the data is sampled uniformly from $[-1, 1]^2$ and the noise affecting the output data is sampled uniformly from $[-\bar{\varepsilon}, \bar{\varepsilon}]$, where the noise bound is equal to $\bar{\varepsilon} = 0.05$.

For the implementation, we choose the cost matrices $Q = I$, $R = 10^{-4}I$, the prediction horizon $L = 30$, as well as the regularization parameters $\lambda_\alpha \bar{\varepsilon} = 10^{-4}$, $\frac{\lambda_\sigma}{\bar{\varepsilon}} = 10^5$. Figure 4.2 shows the closed-loop input-output trajectory when applying the robust data-driven MPC scheme with terminal equality constraints (Algorithm 4.1). It can be seen that the MPC scheme approximately solves the task, i.e., the closed-loop trajectory converges close to the setpoint $(u^{\mathrm{s}}, y^{\mathrm{s}})$ while satisfying the input constraints $u_t \in \mathbb{U}$. Note that the trajectory does not exactly converge to the setpoint and has a non-zero stationary tracking error. This is in accordance with our theoretical analysis (Theorem 4.1) which showed *practical* exponential stability, i.e., the system converges to a neighborhood of the setpoint whose size depends on the noise level.

Influence of parameters

We now investigate the influence of design and data parameters on the closed-loop performance. First, we note that the influence of the cost matrices Q and R as well as the prediction horizon L is analogous to nominal data-driven / model-based

[7] J. Berberich, J. Köhler, M. A. Müller, and F. Allgöwer. "Robust constraint satisfaction in data-driven MPC." In: *Proc. 59th Conf. Decision and Control (CDC)*. 2020, pp. 1260–1267. ©2020 IEEE

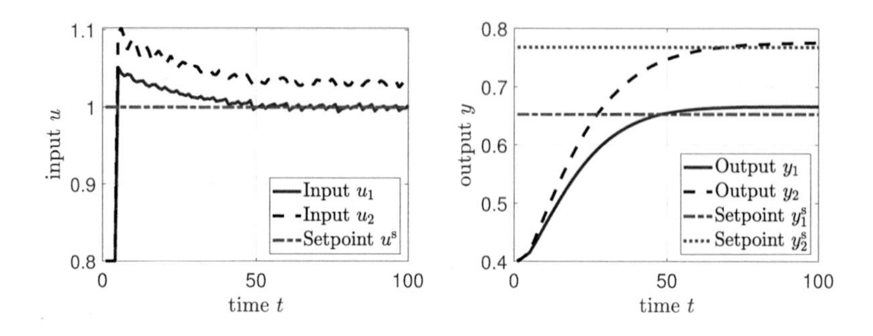

Figure 4.2. Closed-loop trajectory resulting from the application of the robust data-driven MPC scheme with terminal equality constraints (Algorithm 4.1) to the linearized four-tank system with output measurement noise.

MPC. In the following, we analyze the influence of varying the regularization parameters $\hat{\lambda}_\alpha := \bar{\varepsilon}\lambda_\alpha$ and $\hat{\lambda}_\sigma := \frac{\lambda_\sigma}{\bar{\varepsilon}}$ as well as the assumed upper bound on the system order n and the noise bound $\bar{\varepsilon}$. For each configuration, we apply the robust data-driven MPC scheme with terminal equality constraints (Algorithm 4.1) and compute the resulting closed-loop cost, i.e.,

$$\sum_{t=0}^{100} \|u_t - u^{\mathrm{s}}\|_R^2 + \|y_t - y^{\mathrm{s}}\|_Q^2. \tag{4.88}$$

Moreover, we normalize the cost by the corresponding cost of a nominal data-driven MPC scheme with terminal equality constraints, noise-free data, and parameters as above (compare Chapter 3), i.e., we multiply (4.88) by $\frac{1}{J_{\mathrm{mdl}}}$, where J_{mdl} is the closed-loop cost of the nominal MPC (Algorithm 3.1) for $\bar{\varepsilon} = 0$.

Influence of $\hat{\lambda}_\alpha$: First, we vary the parameter $\hat{\lambda}_\alpha = \bar{\varepsilon}\lambda_\alpha$ in the interval $[10^{-7}, 100]$ and keep all other parameters as above. The normalized cost is displayed in Figure 4.3. Note that the robust data-driven MPC (Algorithm 4.1) is consistently outperformed by the nominal one (Algorithm 3.1) due to the noisy data. Further, if $\hat{\lambda}_\alpha = \lambda_\alpha\bar{\varepsilon}$ is chosen too small, then the norm of $\alpha^*(t)$ is too large such that the noise is amplified too strongly, deteriorating the prediction accuracy. On the other hand, if $\hat{\lambda}_\alpha$ is too large, then the regularization of $\alpha(t)$ dominates in the cost and the stationary tracking error increases. This confirms our theoretical guarantees

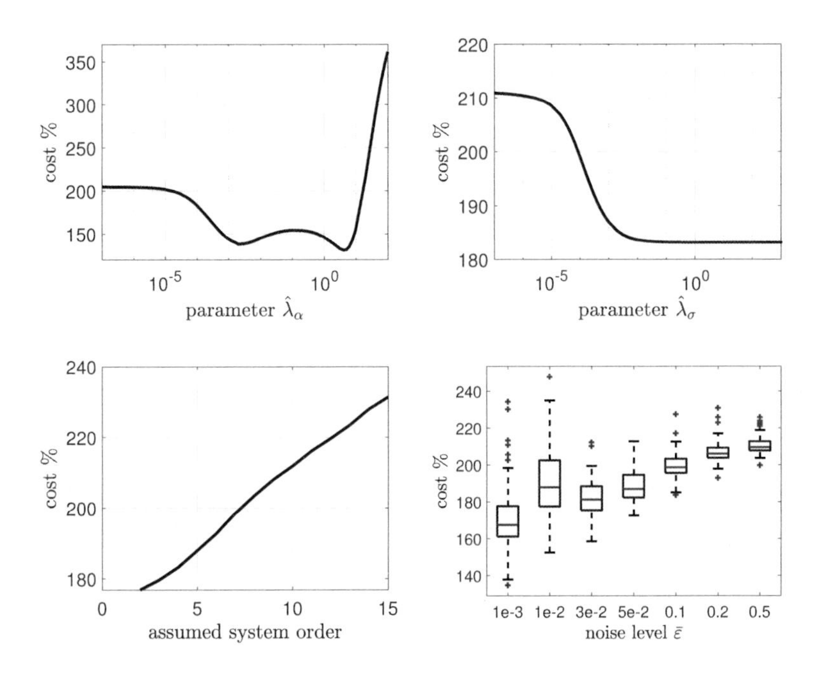

Figure 4.3. Normalized closed-loop cost when applying the robust data-driven MPC scheme with terminal equality constraints (Algorithm 3.1) to the linearized four-tank system with output measurement noise for different choices of the parameters $\hat{\lambda}_\alpha = \bar{\varepsilon}\lambda_\alpha$, $\hat{\lambda}_\sigma = \frac{\lambda_\sigma}{\bar{\varepsilon}}$, the assumed system order, and the noise level $\bar{\varepsilon}$.

in Theorem 4.1 which require that λ_α is neither too small nor too large. Finally, note that there is a wide corridor of values for $\hat{\lambda}_\alpha$ which provide satisfactory performance.

Influence of $\hat{\lambda}_\sigma$: Next, we perform the same analysis for $\hat{\lambda}_\sigma = \frac{\lambda_\sigma}{\bar{\varepsilon}}$ varying in $[10^{-7}, 10^3]$ with all other parameters as above. The resulting normalized closed-loop cost is also depicted in Figure 4.3. Note that, if $\hat{\lambda}_\sigma$ is too small, then the slack variable is too large, leading to unreliable predictions and thus poor closed-loop performance. On the other hand, for values $\hat{\lambda}_\sigma$ beyond a certain threshold, the closed-loop performance is good independently of the exact choice of $\hat{\lambda}_\sigma$.

Influence of assumed system order: We now investigate the influence of the assumed upper bound on the system order n, which enters the proposed MPC scheme via the initial conditions (4.1c), the terminal equality constraints (4.1d), and the number of consecutive open-loop applications in the multi-step scheme (Algorithm 4.1). Figure 4.3 shows the normalized closed-loop cost depending on the choice of n in the proposed data-driven MPC scheme. First, note that satisfactory performance can be achieved over a wide range of estimates for n, e.g., if $n = 8$ is assumed, i.e., twice as large as the actual system order. The best performance is achieved when assuming $n = 2$, whereas, for increasing estimates of n, the closed-loop performance slowly deteriorates. This can partly be explained by noting that larger values of n effectively shorten the prediction horizon in Problem 4.1 due to the terminal equality constraint (4.1d).

Influence of noise level $\bar{\varepsilon}$: Finally, we analyze the influence of the noise level on the closed-loop performance. To this end, we vary the noise level from which the output measurement noise is sampled as $\bar{\varepsilon} \in \{0.001, 0.01, 0.03, 0.05, 0.1, 0.2, 0.5\}$. We run 100 simulations with different random seeds for each noise level and plot the resulting closed-loop cost in Figure 4.3. Except for $\bar{\varepsilon} = 0.01$, the median closed-loop performance continuously deteriorates for increasing noise levels. This confirms our theoretical analysis in Theorem 4.1 which showed practical stability w.r.t. the noise level, i.e., the stationary tracking error increases with the noise level. We have also observed that, for huge noise levels (e.g., $\bar{\varepsilon} = 0.5$), the optimal input $\bar{u}^*_{[0,n-1]}(t)$ computed by Problem 4.1 is approximately equal to u^s. In this case, the closed loop converges to the setpoint since the system is open-loop stable, which explains why the closed-loop performance remains acceptable and its variance decreases when increasing $\bar{\varepsilon}$, compare Figure 4.3.

predictions at multiple time instants as well as the closed-loop output under the proposed MPC scheme. Due to the tightened constraints (4.43e), the output does not approach the boundary of the constraints but oscillates around $y = 7$. A similar behavior can be observed for the application of the MPC scheme in [133]. Notably, the noise level that can be handled in [133] for the present example is considerably larger than the one considered above. For larger noise levels, the proposed MPC scheme is initially infeasible since $a_{4,k} \geq y_{\max}$ already for small values of k. This is due to the fact that, in our proof, several conservative bounds are used. In particular, for bounding $\alpha'(t+n)$ in the proof of Proposition 4.2, we bound a product involving $H_{u\xi}^{\dagger}$ by c_{pe} and we replace $\|\xi_t\|_1$ by $\xi_{\max}$, cf. (4.48), which is the main source of conservatism. Nevertheless, the presented data-driven MPC scheme leads to end-to-end guarantees from noisy data of finite length to closed-loop output constraint satisfaction under mild assumptions. Compared to [133], our method is more general (i.e., it requires less assumptions) but more conservative, and it is very simple to apply since only a measured data trajectory as well as scalar estimates of the system constants Γ and ρ_k are required for its implementation.

4.5 Summary and discussion

4.5.1 Summary

In this chapter, we presented a framework for designing and analyzing robust data-driven MPC schemes, where, in contrast to Chapter 3, the available data were perturbed by noise. In Section 4.1, we showed that a data-driven MPC scheme with terminal equality constraints practically exponentially stabilizes the closed loop in such a scenario, provided that a slack variable as well as regularization terms in the cost are added. Next, in Section 4.2, we proposed a constraint tightening which was shown to be recursively feasible and ensures robust output constraint satisfaction in closed loop. Finally, in Section 4.3, we presented a robust data-driven tracking MPC scheme which contains the same robustifying modifications as the one in Section 4.1 as well as an artificial setpoint which is optimized online. For this scheme, we proposed a separation-type analysis, proving closed-loop practical exponential stability based on a continuity property of the underlying open-loop optimal control problem. Finally, we provided numerical results confirming our theoretical analysis in Section 4.4. We note that the recent paper [JB17] contains additional results on stability and robustness of data-driven MPC with noisy data and without any terminal ingredients, assuming a sufficiently long prediction horizon.

4.5.2 Discussion

In this chapter, we presented two different robust data-driven MPC schemes to ensure closed-loop practical exponential stability: one with terminal equality constraints (Section 4.1), enhanced by a robust constraint tightening in Section 4.2, and one with a tracking objective, online optimization of an artificial setpoint, and a data-driven prediction model for affine systems (Section 4.3). The benefits of including an artificial setpoint in this chapter are analogous to the nominal case in Chapter 3: the possibility to handle online setpoint changes, a significantly larger region of attraction, $(u^{\mathrm{r}}, y^{\mathrm{r}})$ need not be a feasible equilibrium, and improved robustness. This means that Algorithm 4.3 generally has superior theoretical properties and is thus preferable over Algorithm 4.1. The main advantage of Algorithm 4.1 is that it can be extended to guarantee robust output constraint satisfaction (see Section 4.2). Constructing an output constraint tightening based on Algorithm 4.3 is more challenging due to a different theoretical analysis which involves various constants that cannot easily be inferred without detailed model knowledge.

Let us discuss this point in more detail: The stability proof in Section 4.1 is an adaptation of arguments from model-based MPC, employing a standard candidate solution one would also use to prove stability in model-based robust MPC with terminal equality constraints. On the other hand, in Section 4.3, stability was proven by first bounding the difference between the optimal input of the proposed MPC scheme and a nominal version with noise-free data and, second, showing that the latter scheme is robust w.r.t. input disturbances. While the analysis in Section 4.1 is tailored to the specific MPC scheme under consideration, the separation principle of data-driven MPC employed in Section 4.3 is a novel result of independent interest and can be applied to a variety of different MPC formulations. For example, the same argument can also be used to provide a (shorter) stability proof of the approach in Section 4.1 or of the robust data-driven MPC approach without terminal ingredients in [JB17]. Further, as we show in Section 5.2, the same idea can be employed to prove stability in data-driven MPC for *nonlinear* systems. Finally, as we discuss next, (a variation of) Proposition 4.3 can also be used to derive robustness guarantees of a data-driven MPC scheme with general terminal ingredients as in Section 3.2.

Robust data-driven MPC with general terminal ingredients

Recall that we only considered noise-free data for the data-driven MPC scheme with terminal ingredients in Section 3.2. However, an extension of these results to noisy data is straightforward, where the noise may enter both the offline data used to build the Hankel matrices in (3.4b) and to compute the terminal ingredients via Proposition 3.1 as well as the online data used to specify initial conditions in (3.4c). In particular, based on a suitable modification of Proposition 4.3, it can be shown that the noise translates into an input disturbance for the corresponding nominal (noise-free) data-driven MPC scheme. Since model-based MPC with terminal ingredients possesses inherent robustness properties [154], we conjecture that the closed loop is practically exponentially stable w.r.t. the noise and disturbance level, provided that Problem 3.3 is slightly modified (by including a slack variable in (3.4b) and a regularization of $\alpha(t)$ in the cost). As a noteworthy advantage, this is possible based on a one-step MPC scheme as in Algorithm 3.2, whereas the robustness guarantees shown in Sections 4.1–4.3 require the application of a multi-step MPC scheme due to the presence of terminal equality constraints. Finally, the (offline) design by [JB14] which forms the basis for Proposition 3.1 also applies in the presence of noise and hence, our design of terminal ingredients can be carried out in this case. More precisely, replacing the definition of P_Δ^{w} in (3.8) by

$$P_\Delta^{\mathrm{w}} = \begin{bmatrix} -ZZ^\top & ZF^\top B_{\mathrm{w}} \\ B_{\mathrm{w}}^\top FZ^\top & \bar{d}I - B_{\mathrm{w}}^\top FF^\top B_{\mathrm{w}} \end{bmatrix}$$

for some $\bar{d} > 0$, Proposition 3.1 leads to terminal ingredients which satisfy Assumption 3.5 robustly for all systems (2.5) which are consistent with the measured data, when assuming that the data generated offline are perturbed by process noise $\{d_k\}_{k=0}^{N-1}$ and measurement noise $\{\varepsilon_k\}_{k=0}^{N}$ satisfying a bound $\sum_{k=0}^{N-1} \|d_k\|_2^2 + \|\varepsilon_{k+1} - \tilde{A}\varepsilon_k\|_2^2 \leq \bar{d}$ (compare [JB14] for details). To summarize, all results in Section 3.2 can be extended to the realistic scenario where the measured data are affected by noise and the system is subject to disturbances.

Computational complexity

Let us discuss the computational complexity of the robust data-driven MPC with terminal equality constraints in Section 4.1 (analogous considerations hold for the approaches in Sections 4.2 and 4.3). Similar to the nominal MPC scheme in Section 3.1, it is easy to see that the only free decision variables of Problem 4.1 are $\alpha(t)$ and $\sigma(t)$ with at least $m(L+2n)+n$ and $p(L+n)$ free parameters, respectively. On the contrary, to implement a model-based MPC scheme (with state measurements), mL parameters are required. The slack variable $\sigma(t)$ can be eliminated from Problem 4.1 by directly penalizing the norm of the model mismatch $\bar{y}(t) - H_{L+n}(\tilde{y}^{\mathrm{d}})\alpha(t)$ in the cost. Hence, considering the minimal amount of data required for persistence of excitation, Problem 4.1 has roughly the same number of decision variables as a model-based MPC problem. In contrast to the nominal case, however, Theorem 4.1 implies that larger data horizons N are beneficial for the theoretical properties of the proposed scheme as they typically decrease the constant c_{pe}. On the other hand, increasing values for N also lead to an increasing online complexity of Problem 4.1 since $\alpha(t) \in \mathbb{R}^{N-L-n+1}$, i.e., the presented MPC approach allows for a trade-off between computational complexity and desired closed-loop performance by appropriately selecting N. In Section 5.3, we compare the computational complexity of model-based and data-driven MPC with a numerical example.

Chapter 5

Linear tracking MPC for nonlinear systems

In this chapter, we design and analyze linear tracking MPC approaches to control nonlinear systems. First, in Section 5.1, we present a *model-based* MPC scheme which uses the linearized system dynamics at the current state as prediction model. By exploiting local linear approximations of the underlying nonlinear dynamics, we prove that this scheme exponentially stabilizes the optimal reachable steady-state for the underlying nonlinear system. In Section 5.2, we then propose a *data-driven* MPC scheme which, in contrast to the approaches in Chapter 3 and 4, updates the data used in the data-dependent Hankel matrices online at every time step. Extending the model-based analysis in Section 5.1, we prove that this scheme practically exponentially stabilizes the closed loop when controlling an unknown nonlinear system based only on input-output data. Finally, in Section 5.3, we illustrate the applicability of the MPC approaches with challenging nonlinear examples in simulation and in a real-world experiment.

The results presented in this chapter are based on Berberich et al. [JB8, JB10, JB11].

5.1 The model-based case

In this section, we propose a tracking MPC scheme for nonlinear systems using the linearized dynamics at the current state as a prediction model. We consider a tracking MPC formulation to steer the system to a desired target setpoint, which may potentially change online and can be unreachable by the system dynamics and

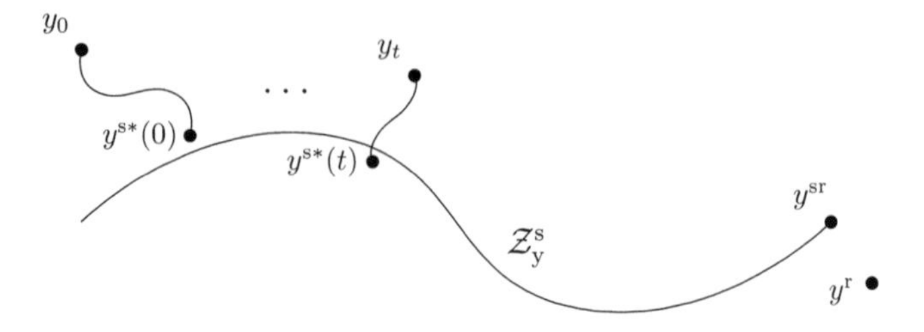

Figure 5.1. Scheme illustrating the basic idea of our model-based MPC approach. The figure displays the output equilibrium manifold $\mathcal{Z}^{\text{s}}_{\text{y}}$, the closed-loop output y_0 and artificial equilibrium $y^{\text{s}*}(0)$ at time 0, the closed-loop output y_t and artificial equilibrium $y^{\text{s}*}(t)$ at time t, the optimal reachable equilibrium y^{sr}, and the setpoint reference y^{r}. ©2021 IEEE

constraints, analogous to Sections 3.3 and 4.3. The cost function contains a tracking cost w.r.t. an artificial equilibrium as well as a penalty of the deviation between this artificial equilibrium and the desired target setpoint. We prove that, if the weight matrix of this penalty is chosen sufficiently small, then the linearization-based prediction model is sufficiently accurate such that the closed loop converges to the optimal reachable equilibrium. The latter is the target setpoint or the closest possible equilibrium, in case this setpoint is not an equilibrium of the system and / or does not satisfy the constraints.

Figure 5.1 illustrates the main idea: At time $t = 0$, the open-loop predictions reach an (artificial) equilibrium output $y^{\text{s}*}(0)$ for the linearized system dynamics lying in a neighborhood of the nonlinear equilibrium manifold $\mathcal{Z}^{\text{s}}_{\text{y}}$. At time t, after repeated application of the proposed MPC scheme, the artificial equilibrium $y^{\text{s}*}(t)$ is closer to y^{sr}, the optimal reachable equilibrium given the (potentially unreachable) setpoint reference y^{r}. The model used for prediction at time t relies on the linearized system dynamics at the current state x_t. By repeated application of the proposed MPC scheme, the artificial equilibrium $y^{\text{s}*}(t)$ slides along the manifold $\mathcal{Z}^{\text{s}}_{\text{y}}$ towards y^{sr} and eventually, the closed-loop output converges to y^{sr}.

The main contribution of this section is to prove that the proposed MPC scheme exponentially stabilizes y^{sr} under reasonable assumptions on the linearized dynamics. A key advantage of our scheme if compared to existing model-based nonlinear tracking MPC schemes [69, 83] lies in its computational efficiency, since only a convex QP needs to be solved online. Moreover, the implementation only requires an accurate linear model around the steady-state manifold, which is easier to obtain than a complex nonlinear model. Finally, this section provides the basis for the results in Section 5.2 where we merge the idea of linearization-based tracking MPC with the results in Chapters 3 and 4 to develop a data-driven MPC scheme for unknown nonlinear systems with closed-loop stability guarantees.

The section is structured as follows. In Section 5.1.1, we state the problem setting and the key assumptions required for our theoretical results. Next, we present our linearization-based MPC scheme for nonlinear systems in Section 5.1.2, and we prove theoretical properties such as recursive feasibility and closed-loop exponential stability in Section 5.1.3. Finally, in Section 5.1.4, we derive sufficient conditions for a technical assumption required in our theoretical results.

This section is based on and taken in parts literally from [JB10][1].

5.1.1 Problem setting

In this chapter, we consider discrete-time nonlinear systems of the form

$$x_{k+1} = f(x_k, u_k) = f_0(x_k) + Bu_k \tag{5.1}$$

with $f_0 : \mathbb{R}^n \to \mathbb{R}^n$, $B \in \mathbb{R}^{n \times m}$, and output

$$y_k = h(x_k, u_k) = h_0(x_k) + Du_k \tag{5.2}$$

for some $h_0 : \mathbb{R}^n \to \mathbb{R}^p$, $D \in \mathbb{R}^{p \times m}$. We impose pointwise-in-time constraints on the input, i.e., $u_t \in \mathbb{U} \subseteq \mathbb{R}^m$ for all $t \in \mathbb{I}_{\geq 0}$.

Assumption 5.1. *(Input constraints) The input constraint set* $\mathbb{U}$ *is compact.*

Due to the inexact prediction model, including state constraints into our framework would necessitate additional robust constraint tightening methods, which is an interesting issue for future research.

We propose a state-feedback MPC scheme to track a desired setpoint reference $y^{\mathrm{r}} \in \mathbb{R}^p$ with the nonlinear system (5.1)–(5.2). In contrast to existing works with

[1] J. Berberich, J. Köhler, M. A. Müller, and F. Allgöwer. "Linear tracking MPC for nonlinear systems part I: the model-based case." In: *IEEE Trans. Automat. Control* (2022). doi: 10.1109/TAC.2022.3166872. ©2021 IEEE

this goal such as [69, 83], the prediction model relies on the dynamics linearized at the current state. In the following, we state key assumptions required to prove that the closed loop under our MPC scheme asymptotically tracks the desired output setpoint.

Smoothness assumptions

We assume that both f and h are continuously differentiable and hence, for some $\tilde{x} \in \mathbb{R}^n$, we can define

$$A_{\tilde{x}} := \left.\frac{\partial f_0}{\partial x}\right|_{\tilde{x}}, \; e_{\tilde{x}} := f_0(\tilde{x}) - A_{\tilde{x}}\tilde{x}, \; C_{\tilde{x}} := \left.\frac{\partial h_0}{\partial x}\right|_{\tilde{x}}, \; r_{\tilde{x}} := h_0(\tilde{x}) - C_{\tilde{x}}\tilde{x}. \tag{5.3}$$

We write $f_{\tilde{x}}(x,u) := A_{\tilde{x}}x + Bu + e_{\tilde{x}}$ for the system dynamics linearized at $(x,u) = (\tilde{x},0)$, and $h_{\tilde{x}}(x,u) := C_{\tilde{x}}x + Du + r_{\tilde{x}}$ for the output linearized at $(x,u) = (\tilde{x},0)$. Note that, since System (5.1) is control-affine, linearizing the dynamics at time t only requires the state x_t and no knowledge of the input u_t, which is a crucial fact for the proposed MPC scheme and its theoretical analysis. If f or h are not affine in u, then it can be readily enforced by defining a new, incremental input $\Delta u_k := u_{k+1} - au_k$ for some $a \in \mathbb{R}$ with $|a| \leq 1$ (e.g., $a = 1$ corresponds to a standard incremental input) and considering the new state $\chi_k := \begin{bmatrix} x_k \\ u_k \end{bmatrix}$. In this case, the exact "prediction model" $u_{k+1} = au_k + \Delta u_k$ still allows us to enforce hard constraints on u as well as Δu while ensuring that the system is control-affine. The key idea of this section relies on the fact that a nonlinear system can be approximated locally by its linearization, given that f is suitably smooth. Indeed, by definition, it holds that

$$f_x(x,u) = f(x,u) \text{ and } h_x(x,u) = h(x,u). \tag{5.4}$$

That is, at the linearization point, the linearization is equal to the nonlinear function f. This insight is important for the theoretical analysis provided in the remainder of this section since it implies that the prediction error of the proposed MPC scheme, which uses a local linearization-based prediction model, is zero in the first time step. Throughout this chapter, we assume that all vector fields involved in the system dynamics are twice continuously differentiable.

Assumption 5.2. *(Smoothness) The vector fields f_0 and h_0 are twice continuously differentiable.*

Assumption 5.2 implies a useful quantitative bound on the difference between the nonlinear vector fields $f(x,u), h(x,u)$ and their linearizations $f_{\tilde{x}}(x,u), h_{\tilde{x}}(x,u)$ for $x \neq \tilde{x}$.

Proposition 5.1. *If Assumptions 5.1 and 5.2 hold, then for any compact set* $X \subset \mathbb{R}^n$*, there exist* $c_X, c_{Xh} > 0$ *such that for any* $x, \tilde{x} \in X, u \in \mathbb{U}$*, it holds that*

$$\|f(x,u) - f_{\tilde{x}}(x,u)\|_2 \leq c_X \|x - \tilde{x}\|_2^2, \tag{5.5}$$

$$\|h(x,u) - h_{\tilde{x}}(x,u)\|_2 \leq c_{Xh} \|x - \tilde{x}\|_2^2. \tag{5.6}$$

Proof. This follows directly from Taylor's Theorem in the multivariable case, together with compactness of X. ∎

For Proposition 5.1, it is crucial to consider a compact set X, since the error bound is in general not necessarily satisfied globally. We later show that a certain (compact) Lyapunov function sublevel set is positively invariant and can hence be used to define X. Although using uniform constants c_X, c_{Xh} in (5.5) and (5.6) over the set X may be conservative, the error bound still becomes arbitrarily small if x and $\tilde{x}$ are sufficiently close to each other. Furthermore, Assumption 5.2 implies that f is locally Lipschitz continuous in x, i.e., for any $c_{\text{Lip}} > 0$ and any compact set X there exists a constant $L_f \geq 0$ such that for any $x, \tilde{x} \in X, u \in \mathbb{U}$ satisfying $\|x - \tilde{x}\|_2 \leq c_{\text{Lip}}$ it holds that

$$\|f(x,u) - f(\tilde{x},u)\|_2 \leq L_f \|x - \tilde{x}\|_2, \tag{5.7}$$

and similarly for h. Using Assumption 5.2, we can derive the following bound between two linear models obtained by linearizations at two different points $x, \tilde{x}$ when evaluated at two further points $x_{\text{a}}, x_{\text{b}}$: For any constant $c_{\text{Lip}} > 0$ and any compact set X, there exist $L_f \geq 0$ and $c_X > 0$ such that

$$\begin{aligned} &\|f_x(x_{\text{a}},u) - f_{\tilde{x}}(x_{\text{b}},u)\|_2 \\ \leq &\|f(x_{\text{a}},u) - f_x(x_{\text{a}},u)\|_2 + \|f(x_{\text{b}},u) - f_{\tilde{x}}(x_{\text{b}},u)\|_2 + \|f(x_{\text{a}},u) - f(x_{\text{b}},u)\|_2 \\ \overset{(5.5),(5.7)}{\leq} & c_X \|x_{\text{a}} - x\|_2^2 + c_X \|x_{\text{b}} - \tilde{x}\|_2^2 + L_f \|x_{\text{a}} - x_{\text{b}}\|_2, \end{aligned} \tag{5.8}$$

for all $x, x_{\text{a}}, \tilde{x}, x_{\text{b}} \in X, u \in \mathbb{U}$ with $\|x_{\text{a}} - x_{\text{b}}\|_2 \leq c_{\text{Lip}}$.

Steady-state manifold

The control goal is to steer the nonlinear system (5.1) to a desired target setpoint, i.e., to track a user-specified output $y^{\mathrm{r}} \in \mathbb{R}^p$. Since the output y may depend on the input u, this also allows us to consider input setpoints. Let us now define the steady-state manifold as the set of all feasible steady-states

$$\mathcal{Z}^{\mathrm{s}} := \{(x^{\mathrm{s}}, u^{\mathrm{s}}) \in \mathbb{R}^n \times \mathbb{U}^{\mathrm{s}} \mid x^{\mathrm{s}} = f(x^{\mathrm{s}}, u^{\mathrm{s}})\} \tag{5.9}$$

with some (user-chosen) convex and compact set $\mathbb{U}^{\mathrm{s}} \subseteq \operatorname{int}(\mathbb{U})$, which is required for a local controllability argument in our proofs. Further, we denote the projection of $\mathcal{Z}^{\mathrm{s}}$ on the state component by $\mathcal{Z}^{\mathrm{s}}_x$ and the projection of $\mathcal{Z}^{\mathrm{s}}$ on the output as

$$\mathcal{Z}^{\mathrm{s}}_{\mathrm{y}} := \{y^{\mathrm{s}} \in \mathbb{R}^p \mid \exists (x^{\mathrm{s}}, u^{\mathrm{s}}) \in \mathcal{Z}^{\mathrm{s}} : y^{\mathrm{s}} = h(x^{\mathrm{s}}, u^{\mathrm{s}})\}.$$

The optimal equilibrium cost is defined as

$$J^*_{\mathrm{eq}} := \min_{y^{\mathrm{s}} \in \mathcal{Z}^{\mathrm{s}}_{\mathrm{y}}} \|y^{\mathrm{s}} - y^{\mathrm{r}}\|^2_S \tag{5.10}$$

for some $S \succ 0$. We denote a minimizer of (5.10) by y^{sr}. In Section 5.1.3, we provide sufficient conditions under which this minimizer is unique. Let us define the set of equilibria of the linearized system at some state $\tilde{x}$ as

$$\mathcal{Z}^{\mathrm{s}}_{\mathrm{Lin}}(\tilde{x}) := \{(x^{\mathrm{s}}, u^{\mathrm{s}}) \in \mathbb{R}^n \times \mathbb{U}^{\mathrm{s}} \mid x^{\mathrm{s}} = A_{\tilde{x}} x^{\mathrm{s}} + B u^{\mathrm{s}} + e_{\tilde{x}}\}$$

and the projection on the output as

$$\mathcal{Z}^{\mathrm{s}}_{\mathrm{y,Lin}}(\tilde{x}) := \{y^{\mathrm{s}} \in \mathbb{R}^p \mid \exists (x^{\mathrm{s}}, u^{\mathrm{s}}) \in \mathcal{Z}^{\mathrm{s}}_{\mathrm{Lin}}(\tilde{x}) : y^{\mathrm{s}} = C_{\tilde{x}} x^{\mathrm{s}} + D u^{\mathrm{s}} + r_{\tilde{x}}\}.$$

The optimal reachable equilibrium of the linearized system at $\tilde{x}$ is the minimizer of

$$J^*_{\mathrm{eq,Lin}}(\tilde{x}) := \min_{y^{\mathrm{s}} \in \mathcal{Z}^{\mathrm{s}}_{\mathrm{y,Lin}}(\tilde{x})} \|y^{\mathrm{s}} - y^{\mathrm{r}}\|^2_S. \tag{5.11}$$

We denote the minimizer of (5.11) by $y^{\mathrm{sr}}_{\mathrm{Lin}}(\tilde{x})$. It follows from $S \succ 0$ that (5.11) is strongly convex at any linearization point, i.e., for any $\tilde{x} \in \mathbb{R}^n$, $y^{\mathrm{s}} \in \mathcal{Z}^{\mathrm{s}}_{\mathrm{y,Lin}}(\tilde{x})$, it holds that

$$\|y^{\mathrm{s}} - y^{\mathrm{r}}\|^2_S - J^*_{\mathrm{eq,Lin}}(\tilde{x}) \geq \|y^{\mathrm{s}} - y^{\mathrm{sr}}_{\mathrm{Lin}}(\tilde{x})\|^2_S, \tag{5.12}$$

compare [69, Inequality (11)].

Assumption 5.3. *(Unique steady-state) There exists $\sigma_{\mathrm{s}} > 0$ such that, for any $\tilde{x} \in \mathbb{R}^n$,*

$$\sigma_{\min}\left(\begin{bmatrix} A_{\tilde{x}} - I & B \\ C_{\tilde{x}} & D \end{bmatrix}\right) \geq \sigma_{\mathrm{s}}. \tag{5.13}$$

Assumption 5.3 implies that $\begin{bmatrix} A_{\tilde{x}} - I & B \\ C_{\tilde{x}} & D \end{bmatrix}$ has full column rank, which is a standard condition in tracking, compare [118, Lemma 1.8], [83, Remark 1], or Assumption 4.9. This condition means that for any steady-state output of the linearized system, the corresponding input-state pair is unique. More precisely, for any $\tilde{x} \in \mathbb{R}^n$, there exists a linear map $\hat{g}_{\tilde{x}} : \mathcal{Z}^{\mathrm{s}}_{\mathrm{y,Lin}}(\tilde{x}) \to \mathcal{Z}^{\mathrm{s}}_{\mathrm{Lin}}(\tilde{x})$ such that

$$\hat{g}_{\tilde{x}}(y^{\mathrm{s}}) = (x^{\mathrm{s}}, u^{\mathrm{s}}), \tag{5.14}$$

where $(x^{\mathrm{s}}, u^{\mathrm{s}}) \in \mathcal{Z}^{\mathrm{s}}_{\mathrm{Lin}}(\tilde{x})$ is the steady-state corresponding to y^{s}, i.e., $y^{\mathrm{s}} = h_{\tilde{x}}(x^{\mathrm{s}}, u^{\mathrm{s}})$. Due to the uniform lower bound (5.13), the map $\hat{g}_{\tilde{x}}$ is uniformly Lipschitz continuous for all $\tilde{x} \in \mathbb{R}^n$. Further, the condition (5.13) implies that the linearized dynamics have no transmission zeros at 1 [90, Ass. 1] and that the number of outputs p is greater than or equal to the number of inputs m. Finally, by a global version of the inverse function theorem [115, Condition (1.1)], Assumption 5.3 implies the existence of a unique equilibrium $(x^{\mathrm{s}}, u^{\mathrm{s}}) \in \mathcal{Z}^{\mathrm{s}}$ for the *nonlinear* system for any given output equilibrium $y^{\mathrm{s}} \in \mathcal{Z}^{\mathrm{s}}_{\mathrm{y}}$, compare also [83, Remark 1]. Throughout this chapter, we write

$$(x^{\mathrm{sr}}_{\mathrm{Lin}}(\tilde{x}), u^{\mathrm{sr}}_{\mathrm{Lin}}(\tilde{x})) := \hat{g}_{\tilde{x}}(y^{\mathrm{sr}}_{\mathrm{Lin}}(\tilde{x})) \tag{5.15}$$

for the unique input-state pair corresponding to the minimizer $y^{\mathrm{sr}}_{\mathrm{Lin}}(\tilde{x})$ of (5.11), and we write $(x^{\mathrm{sr}}, u^{\mathrm{sr}})$ for the input-state pair corresponding to y^{sr}, i.e.,

$$(x^{\mathrm{sr}}, u^{\mathrm{sr}}) = \hat{g}_{x^{\mathrm{sr}}}(y^{\mathrm{sr}}). \tag{5.16}$$

Moreover, we require that the linearized dynamics are uniformly controllable.

Assumption 5.4. *(Controllability) The pair $(A_{\tilde{x}}, B)$ is uniformly controllable for all $\tilde{x} \in \mathbb{R}^n$, i.e., the minimum singular value of $\begin{bmatrix} B & \dots & A_{\tilde{x}}^{n-1}B \end{bmatrix}$ is uniformly lower bounded.*

Analogously to (2.10), it can be shown that Assumption 5.4 implies the existence of a constant $\Gamma > 0$ such that for any $\tilde{x} \in \mathbb{R}^n$, $(x^{\mathrm{s}}, u^{\mathrm{s}}) \in \mathcal{Z}^{\mathrm{s}}_{\mathrm{Lin}}(\tilde{x})$, and any initial condition x_0, there exists an input trajectory $\hat{u}_k \in \mathbb{R}^m$, $k \in \mathbb{I}_{[0,n-1]}$, steering the linearized system from x_0 to x^s, i.e, $\hat{x}_0 = x_0$, $\hat{x}_{k+1} = f_{\tilde{x}}(\hat{x}_k, \hat{u}_k)$, $\hat{x}_n = x^s$, while satisfying

$$\sum_{k=0}^{n-1} \|\hat{x}_k - x^s\|_2 + \|\hat{u}_k - u^s\|_2 \leq \Gamma \|x^s - x_0\|_2. \tag{5.17}$$

Moreover, the following assumption is required for the linearized system dynamics at any state.

Assumption 5.5. *(Non-singular dynamics) There exists $\underline{\sigma} > 0$ such that $\sigma_{\min}(I - A_{\tilde{x}}) \geq \underline{\sigma}$ for any $\tilde{x} \in \mathbb{R}^n$.*

Assumption 5.5 implies that, for any $\tilde{x} \in \mathbb{R}^n$, the matrix $I - A_{\tilde{x}}$ has full rank and hence, for any equilibrium input u^s there exists a unique equilibrium state x^s such that

$$x^s = A_{\tilde{x}} x^s + B u^s + e_{\tilde{x}}.$$

In case that $\mathbb{U} = \mathbb{R}^m$, it is straightforward to relax Assumption 5.5 by requiring that there exists a state-feedback gain K such that $A_K = A_{\tilde{x}} + BK$ satisfies the non-singularity condition $\sigma_{\min}(I - A_K) \geq \underline{\sigma}$ for some $\underline{\sigma}$ and for all $\tilde{x} \in \mathbb{R}^n$. We conjecture that it is possible to relax Assumption 5.5 further at the price of a more involved analysis.

Note that the conditions in Assumptions 5.3–5.5 are imposed for all $\tilde{x} \in \mathbb{R}^n$. This is mainly done for notational convenience. As will become clear in our theoretical results, it actually suffices if these assumptions hold for all $\tilde{x}$ in a suitably defined compact set depending on the (positively invariant) sublevel set of the Lyapunov function used to prove stability. Finally, we make an additional assumption on the steady-state manifold of the linearized dynamics.

Assumption 5.6. *(Compact steady-state manifold) There exists a compact set $\mathcal{B}$ such that $\mathcal{Z}^s_{\text{Lin}}(\tilde{x}) \subseteq \mathcal{B}$ for all $\tilde{x} \in \mathbb{R}^n$.*

Assumption 5.6 means that the union of all steady-state manifolds for the linearized dynamics at any point is contained in a compact set. If the input equilibrium constraints $\mathbb{U}^s$ are compact and Assumption 5.5 holds, then this is satisfied if the Jacobian is uniformly bounded. Assumption 5.6 is required for our theoretical results to obtain a uniform bound on the optimal equilibrium cost (5.11) and to conclude compactness of certain Lyapunov function sublevel sets. The assumption can be dropped if it is known that the closed-loop trajectories lie within a compact invariant subset of the state-space.

5.1.2 Proposed MPC scheme

In this section, we propose a linear tracking MPC scheme to steer the nonlinear system (5.1)–(5.2) to a desired target setpoint. The key idea is to use a local linearization-based model of the nonlinear system for prediction, and to update the linearization online using the current measurements.

Given the current state x_t at time t as well as the linearization of the nonlinear system at x_t according to (5.3), the following optimal control problem will be the basis for our MPC scheme:

Problem 5.1.

$$\underset{\substack{\bar{u}(t),\bar{x}(t)\\ u^{\mathrm{s}}(t),x^{\mathrm{s}}(t),y^{\mathrm{s}}(t)}}{\text{minimize}} \quad \sum_{k=0}^{L-1}\left(\|\bar{x}_k(t)-x^{\mathrm{s}}(t)\|_Q^2+\|\bar{u}_k(t)-u^{\mathrm{s}}(t)\|_R^2\right)+\|y^{\mathrm{s}}(t)-y^{\mathrm{r}}\|_S^2 \tag{5.18a}$$

subject to

$$\bar{x}_{k+1}(t)=A_{x_t}\bar{x}_k(t)+B\bar{u}_k(t)+e_{x_t}, \tag{5.18b}$$

$$\bar{x}_0(t)=x_t,\ \bar{x}_L(t)=x^{\mathrm{s}}(t), \tag{5.18c}$$

$$\bar{u}_k(t)\in\mathbb{U},\ k\in\mathbb{I}_{[0,L-1]}, \tag{5.18d}$$

$$(x^{\mathrm{s}}(t),u^{\mathrm{s}}(t))\in\mathcal{Z}^{\mathrm{s}}_{\mathrm{Lin}}(x_t), \tag{5.18e}$$

$$y^{\mathrm{s}}(t)=C_{x_t}x^{\mathrm{s}}(t)+Du^{\mathrm{s}}(t)+r_{x_t}. \tag{5.18f}$$

Compared to a standard linear MPC scheme with terminal equality constraints (compare Section 2.3), Problem 5.1 has two additional ingredients. First, the present scheme contains an artificial setpoint $(x^{\mathrm{s}}(t),u^{\mathrm{s}}(t))$ which is optimized online and whose distance w.r.t. the desired target setpoint y^{r} is penalized in the cost, similar to the tracking MPC formulations in Sections 3.3 and 4.3 or in [69, 83]. Further, the prediction of future trajectories of the nonlinear system is not based on the full nonlinear model (5.1), but instead on a local linearization around the current state x_t. Therefore, also the artificial setpoint $x^{\mathrm{s}}(t)$ is an equilibrium for the linearized dynamics according to (5.18e), but not necessarily for the nonlinear system, and the artificial output setpoint $y^s(t)$ satisfies the linearized output equation (5.18f).

We assume that the cost matrices in Problem 5.1 are positive definite, i.e., $Q, R, S \succ 0$, and we denote the optimal solution of Problem 5.1 by $\bar{x}^*(t)$, $\bar{u}^*(t)$, $x^{s*}(t)$, $u^{s*}(t)$, $y^{s*}(t)$ as well as the corresponding optimal cost by $J_L^*(x_t)$. Our results can be extended to positive semidefinite state weightings $Q \succeq 0$ by invoking an input-output-to-state stability argument as in Chapters 3 and 4. Moreover, we assume the following lower bound on the prediction horizon.

Assumption 5.7. *(Length of prediction horizon) The prediction horizon satisfies $L \geq n$.*

If $\mathbb{U}$, $\mathbb{U}^s$ are convex polytopes, then Problem 5.1 is a convex QP and can be solved efficiently. On the contrary, solving the non-convex problems associated with nonlinear MPC to optimality is, in general, computationally intractable. It is worth noting that the computational complexity of the proposed MPC approach is also smaller than that of alternative linearization-based approaches such as the real-time iteration scheme [35, 80, 157] since i) only the linearization w.r.t. x_t instead of the previously optimal solution $\bar{x}^*(t-1)$ is needed and ii) comparable stability guarantees of the real-time iteration scheme require a sufficiently small sampling time [157], i.e., solving the underlying optimization problem more frequently [80].

Algorithm 5.1. Model-based linear tracking MPC scheme for nonlinear systems
Offline: Choose prediction horizon L, cost matrices $Q, R, S \succ 0$, constraint sets $\mathbb{U}$, $\mathbb{U}^s$, setpoint y^r.
Online: At time $t = n \cdot i$, $i \in \mathbb{I}_{\geq 0}$, measure x_t, compute A_{x_t}, e_{x_t}, C_{x_t}, r_{x_t} (compare (5.3)) and solve Problem 5.1. Apply the input $u_{[t,t+n-1]} = \bar{u}^*_{[0,n-1]}(t)$ over the next n time steps.

In this section, we consider the n-step MPC scheme shown in Algorithm 5.1. As in Chapter 4, we employ a multi-step MPC scheme due to the joint occurrence of a model mismatch (induced by the linearized model) and terminal equality constraints.

Clearly, the prediction model in the proposed MPC scheme is not exact due to the linearization of the nonlinear system. As we will see in the remainder of this section, by suitably tuning the cost parameters (S needs to be small) and when starting close to the steady-state manifold $\mathcal{Z}^{\text{s}}$ of the nonlinear system, the artificial steady-state $x^{\text{s}}(t)$ is always close to the current state x_t such that the prediction error is sufficiently small. Then, $x^{\text{s}}(t)$ remains close to the nonlinear steady-state manifold and slowly drifts towards the optimal reachable equilibrium x^{sr} such that, asymptotically, the closed-loop state trajectory converges to x^{sr}, compare Figure 5.1.

5.1.3 Closed-loop guarantees

In this section, we provide a theoretical analysis of the closed loop under Algorithm 5.1. After deriving lower and upper bounds on the value function of Problem 5.1, we show a useful contraction property of the Lyapunov function. Thereafter, we present our main result on closed-loop exponential stability of the optimal reachable equilibrium.

Value function bound

In order to prove closed-loop exponential stability, we consider a Lyapunov function candidate of the form

$$V(x_t) := J_L^*(x_t) - J_{\text{eq,Lin}}^*(x_t)$$

with $J_{\text{eq,Lin}}^*(x_t)$ as in (5.11). The following result shows that V admits suitable quadratic lower and upper bounds, which will be required to prove desired stability properties.

Lemma 5.1. *Suppose Assumptions 5.1–5.4 and 5.7 hold. For any compact set $\mathbb{X} \subset \mathbb{R}^n$, there exist $c_{\text{l}}, c_{\text{u}}, \delta > 0$ such that*

(i) for all $x_t \in \mathbb{X}$, the function V is lower bounded as

$$V(x_t) \geq c_{\text{l}} \|x_t - x_{\text{Lin}}^{\text{sr}}(x_t)\|_2^2, \tag{5.19}$$

(ii) for all $x_t \in \mathbb{X}$ with $\|x_t - x_{\text{Lin}}^{\text{sr}}(x_t)\|_2 \leq \delta$, the function V is upper bounded as

$$V(x_t) \leq c_{\text{u}} \|x_t - x_{\text{Lin}}^{\text{sr}}(x_t)\|_2^2. \tag{5.20}$$

Proof. Throughout this proof and further proofs in this section, we will make repeated use of the inequalities

$$\|a+b\|_P^2 \leq 2\|a\|_P^2 + 2\|b\|_P^2, \tag{5.21}$$

$$\|a\|_P^2 - \|b\|_P^2 \leq \|a-b\|_P^2 + 2\|a-b\|_P \|b\|_P, \tag{5.22}$$

which hold for any vectors a, b and matrix $P = P^\top \succ 0$.

(i) Lower bound

Note that

$$\begin{aligned} V(x_t) =& J_L^*(x_t) - J_{\mathrm{eq,Lin}}^*(x_t) \\ \geq& \|x_t - x^{\mathrm{s}*}(t)\|_Q^2 + \|y^{\mathrm{s}*}(t) - y^{\mathrm{r}}\|_S^2 - J_{\mathrm{eq,Lin}}^*(x_t) \\ \overset{(5.12)}{\geq}& \|x_t - x^{\mathrm{s}*}(t)\|_Q^2 + \lambda_{\min}(S)\|y^{\mathrm{s}*}(t) - y_{\mathrm{Lin}}^{\mathrm{sr}}(x_t)\|_2^2. \end{aligned}$$

Assumption 5.3 (i.e., the maps in (5.14)) implies the existence of a constant $\bar{c}_{\mathrm{l}} > 0$, which can be chosen uniformly over x_t, such that

$$\|y^{\mathrm{s}*}(t) - y_{\mathrm{Lin}}^{\mathrm{sr}}(x_t)\|_2^2 \geq \bar{c}_{\mathrm{l}}\|x^{\mathrm{s}*}(t) - x_{\mathrm{Lin}}^{\mathrm{sr}}(x_t)\|_2^2.$$

Hence, we obtain

$$\begin{aligned} V(x_t) &\geq \|x_t - x^{\mathrm{s}*}(t)\|_Q^2 + \bar{c}_{\mathrm{l}}\lambda_{\min}(S)\|x^{\mathrm{s}*}(t) - x_{\mathrm{Lin}}^{\mathrm{sr}}(x_t)\|_2^2 \\ &\overset{(5.21)}{\geq} \underbrace{\frac{\min\{\lambda_{\min}(Q), \bar{c}_{\mathrm{l}}\lambda_{\min}(S)\}}{2}}_{c_{\mathrm{l}}:=}\|x_t - x_{\mathrm{Lin}}^{\mathrm{sr}}(x_t)\|_2^2. \end{aligned}$$

(ii) Upper bound

In the following, we construct a candidate solution to Problem 5.1 which will then be used to bound the optimal cost $J_L^*(x_t)$. We choose the candidate equilibrium as $x^{\mathrm{s}}(t) = x_{\mathrm{Lin}}^{\mathrm{sr}}(x_t)$, $u^{\mathrm{s}}(t) = u_{\mathrm{Lin}}^{\mathrm{sr}}(x_t)$, $y^{\mathrm{s}}(t) = y_{\mathrm{Lin}}^{\mathrm{sr}}(x_t)$, i.e., the optimal reachable equilibrium of the system linearized at x_t. Using $L \geq n$ and Assumption 5.4, there exists a trajectory of the linearized dynamics steering the state to $x^{\mathrm{s}}(t)$ within L steps while satisfying

$$\sum_{k=0}^{L-1} \|\bar{x}_k(t) - x^{\mathrm{s}}(t)\|_2 + \|\bar{u}_k(t) - u^{\mathrm{s}}(t)\|_2 \leq \Gamma\|x_t - x^{\mathrm{s}}(t)\|_2$$

for some $\Gamma > 0$ (compare (5.17)). If δ is sufficiently small, then $u_{\mathrm{Lin}}^{\mathrm{sr}}(x_t) \in \mathrm{int}(\mathbb{U})$ implies that the corresponding input satisfies the constraints, i.e., $\bar{u}_k(t) \in \mathbb{U}$ for all

$k \in \mathbb{I}_{[0,L-1]}$. Clearly, the above inequality implies

$$\begin{aligned}
&\sum_{k=0}^{L-1} \|\bar{x}_k(t) - x^{\mathrm{s}}(t)\|_Q^2 + \|\bar{u}_k(t) - u^{\mathrm{s}}(t)\|_R^2 \\
&\leq \lambda_{\max}(Q,R) \sum_{k=0}^{L-1} (\|\bar{x}_k(t) - x^{\mathrm{s}}(t)\|_2^2 + \|\bar{u}_k(t) - u^{\mathrm{s}}(t)\|_2^2) \\
&\leq \lambda_{\max}(Q,R)\Gamma^2 \|x_t - x^{\mathrm{s}}(t)\|_2^2.
\end{aligned}$$

Thus, the following holds for the Lyapunov function candidate

$$\begin{aligned}
V(x_t) \leq &\lambda_{\max}(Q,R)\Gamma^2 \|x_t - x^{\mathrm{s}}(t)\|_2^2 + \|y^{\mathrm{s}}(t) - y^{\mathrm{r}}\|_S^2 - J^*_{\mathrm{eq,Lin}}(x_t) \\
= &\underbrace{\lambda_{\max}(Q,R)\Gamma^2}_{c_{\mathrm{u}}:=} \|x_t - x^{\mathrm{sr}}_{\mathrm{Lin}}(x_t)\|_2^2.
\end{aligned}$$

■

Lemma 5.1 provides bounds on the Lyapunov function candidate $V(x_t)$ that will be employed to prove closed-loop exponential stability. As an alternative to $V(x_t)$, one could consider the candidate $J^*_L(x_t) - J^*_{\mathrm{eq}}$, which depends on the cost of the optimal reachable equilibrium for the *nonlinear* system instead of the cost $J^*_{\mathrm{eq,Lin}}(x_t)$ for the linearized system. This may simplify the subsequent stability analysis and allow us to drop a technical assumption (Assumption 5.8) required for the proof of our main result (Theorem 5.1). However, deriving a useful lower bound for $J^*_L(x_t) - J^*_{\mathrm{eq}}$ similar to (5.19) is difficult, which is why we consider the proposed Lyapunov function candidate $V(x_t)$ instead.

Contraction property

The following result shows that feasibility of Problem 5.1 at time t implies, under additional assumptions, feasibility at time $t+n$ and a certain contraction property for the Lyapunov function candidate V on suitable sublevel sets.

Proposition 5.2. *Suppose Assumptions 5.1–5.7 hold. Then, there exist* $V_{\max}, J^{\max}_{\mathrm{eq}} > 0$ *such that, if* $V(x_t) \leq V_{\max}$, $J^*_{\mathrm{eq,Lin}}(x_t) \leq J^{\max}_{\mathrm{eq}}$, $J^*_{\mathrm{eq,Lin}}(x_{t+n}) \leq J^{\max}_{\mathrm{eq}}$, *then Problem 5.1 is feasible at time* $t+n$ *and there exists a constant* $0 < c_{\mathrm{V}} < 1$ *such that* $V(x_{t+n}) \leq c_{\mathrm{V}} V(x_t)$.

The proof of Proposition 5.2 is provided in the appendix (Appendix A.3). It relies on a case distinction with two different candidate solutions: First, the case that the tracking cost w.r.t. the artificial steady-state is relatively large is considered, compare Inequality (A.63). In this case, the candidate solution is the previously optimal input appended by a local deadbeat controller compensating the model mismatch due to the linearization. If $V_{\max}$ and $J_{\mathrm{eq}}^{\max}$ are sufficiently small, then this model mismatch is also small such that a decrease of the optimal cost can be derived. Thereafter, the converse case is considered, where the current state is close to the artificial steady-state. In this case, the candidate solution results from shifting the artificial equilibrium towards the optimal reachable equilibrium for the linearized dynamics. Proposition 5.2 can also be seen as an extension of [69, 83], where similar properties are shown for MPC with a *nonlinear* prediction model, whereas our MPC scheme contains a linear prediction model.

Even for $V_{\max}$ arbitrarily small, the bound $V(x_t) \leq V_{\max}$ holds in a neighborhood of the steady-state manifold if S is chosen sufficiently small using compactness (cf. the proof of Theorem 5.1). In this case, $x^{s*}(t)$ is close to x_t and hence the stage cost $\sum_{k=0}^{L-1} \|\bar{x}_k^*(t) - x^{s*}(t)\|_Q^2 + \|\bar{u}_k^*(t) - u^{s*}(t)\|_R^2$ becomes small. Similarly, as we exploit in the next section, the bounds $J_{\mathrm{eq,Lin}}^*(x_t) \leq J_{\mathrm{eq}}^{\max}$ and $J_{\mathrm{eq,Lin}}^*(x_{t+n}) \leq J_{\mathrm{eq}}^{\max}$ hold for arbitrarily small $J_{\mathrm{eq}}^{\max}$ if S is chosen sufficiently small since the steady-state manifold is compact by Assumption 5.6. Note that Proposition 5.2 does not prove any recursive closed-loop properties since the assumed bounds are not necessarily satisfied recursively. We discuss these conditions in relation to the proof of closed-loop recursive feasibility and stability in the following.

Exponential stability

We now use Lemma 5.1 and Proposition 5.2 to prove closed-loop exponential stability of the optimal reachable steady-state x^{sr} of the nonlinear system. Proposition 5.2 implies that the function V satisfies $V(x_{t+n}) \leq c_V V(x_t)$. However, this does not yet prove the desired stability result since 1) it remains to show that the inequality is satisfied recursively and 2) V is only lower and upper bounded by the distance w.r.t. the optimal reachable steady-state of the *linearized* dynamics (compare Lemma 5.1). In the following, we make the additional assumption that the current state x_t is close to $x_{\mathrm{Lin}}^{\mathrm{sr}}(x_t)$ if and only if it is close to the optimal reachable steady-state x^{sr}.

Assumption 5.8. *For any compact set X with $\mathcal{B} \subseteq X \times \mathbb{U}$ (cf. Assumption 5.6), there exist constants $c_{\mathrm{eq},1}, c_{\mathrm{eq},2} > 0$ such that, for any $\hat{x} \in X$, it holds that*

$$c_{\mathrm{eq},1} \|\hat{x} - x_{\mathrm{Lin}}^{\mathrm{sr}}(\hat{x})\|_2^2 \leq \|\hat{x} - x^{\mathrm{sr}}\|_2^2 \leq c_{\mathrm{eq},2} \|\hat{x} - x_{\mathrm{Lin}}^{\mathrm{sr}}(\hat{x})\|_2^2. \tag{5.23}$$

Assumption 5.8 requires that the distance between some state $\hat{x}$ and x^{sr} is lower and upper bounded by the distance between $\hat{x}$ and the optimal reachable steady-state for the linearized dynamics $x^{\mathrm{sr}}_{\mathrm{Lin}}(\hat{x})$. In Section 5.1.4, we show that Assumption 5.8 holds if, in addition to the assumptions in Section 5.1.1, the setpoint y^{r} is reachable and $m = p$ holds. The following theorem shows that the optimal reachable equilibrium of the nonlinear system is exponentially stable under the proposed MPC scheme.

Theorem 5.1. *Suppose Assumptions 5.1–5.8 hold. Then, there exist $V_{\max}, \bar{S} > 0$ such that, if $\lambda_{\max}(S) \leq \bar{S}$ and $V(x_0) \leq V_{\max}$, then Problem 5.1 is feasible at any time $n \cdot i, i \in \mathbb{I}_{\geq 0}$, and x^{sr} in (5.16) is exponentially stable, i.e., there exist constants $C > 0, c_{\mathrm{V}} < 1$ such that for all $i \in \mathbb{I}_{\geq 0}$*

$$\|x_{ni} - x^{\mathrm{sr}}\|_2^2 \leq c_{\mathrm{V}}^i C \|x_0 - x^{\mathrm{sr}}\|_2^2. \tag{5.24}$$

Proof. Assumption 5.6 implies that the union of all output equilibrium manifolds $\mathcal{Z}^{\mathrm{s}}_{\mathrm{y,Lin}}(\tilde{x})$ is compact. Thus, there exists a uniform upper bound $J^{\max}_{\mathrm{eq}}$ on $J^*_{\mathrm{eq,Lin}}(\tilde{x})$, i.e.,

$$J^*_{\mathrm{eq,Lin}}(\tilde{x}) \leq J^{\max}_{\mathrm{eq}} \text{ for all } \tilde{x} \in \mathbb{R}^n. \tag{5.25}$$

Note that $J^{\max}_{\mathrm{eq}}$ can be chosen arbitrarily small when $\lambda_{\max}(S)$ is sufficiently small. Hence, choosing $\lambda_{\max}(S)$ sufficiently small and using (5.25), we can apply Proposition 5.2 to conclude $V(x_{t+n}) \leq c_{\mathrm{V}} V(x_t)$, which in turn implies $V(x_{t+n}) \leq V_{\max}$. Applying this argument inductively, we conclude $V(x_{t+n}) \leq c_{\mathrm{V}} V(x_t)$ for all $t = n \cdot i, i \in \mathbb{I}_{\geq 0}$, where $c_{\mathrm{V}} < 1$. Using Lemma 5.1 (the upper bound holds if $V_{\max}$ is sufficiently small), this implies

$$\|x_{t+n} - x^{\mathrm{sr}}_{\mathrm{Lin}}(x_{t+n})\|_2^2 \leq c_{\mathrm{V}}^i \frac{c_{\mathrm{u}}}{c_{\mathrm{l}}} \|x_t - x^{\mathrm{sr}}_{\mathrm{Lin}}(x_t)\|_2^2. \tag{5.26}$$

Finally, using (5.23), this leads to (5.24) with $C := \frac{c_{\mathrm{u}} c_{\mathrm{eq},2}}{c_{\mathrm{l}} c_{\mathrm{eq},1}}$. ■

Discussion

Theorem 5.1 is our main stability result in this section. It shows that, if $V_{\max}$ and S are sufficiently small, then the optimal reachable steady-state x^{sr} is exponentially stable under the proposed n-step MPC scheme, i.e., Inequality (5.24) holds, and hence also the output y_t exponentially converges towards the optimal reachable output equilibrium y^{sr}. Intuitively, Theorem 5.1 shows that the artificial steady-state $x^{\mathrm{s}*}(t)$ and thus also the state trajectory x_t slides along the steady-state manifold in closed loop towards the optimal reachable steady-state. Thus, the guaranteed region of attraction of x^{sr} is a neighborhood around the steady-state manifold, which increases if S is chosen smaller (for a given value of $V_{\max}$). Note that, although the prediction model of the proposed MPC scheme is not exact, the closed loop is nevertheless exponentially stable (i.e., it is not only practically stable) since the prediction accuracy improves as x_t gets closer to the steady-state manifold $\mathcal{Z}^{\mathrm{s}}$.

Remark 5.1. *(Qualitative guarantees) Theorem 5.1 should be interpreted as a* qualitative *result since it does not provide explicit values of $V_{\max}$ and $\bar{S}$ leading to closed-loop stability. Loosely speaking, Theorem 5.1 guarantees stability if the initial state is* sufficiently close *to $\mathcal{Z}^{\mathrm{s}}_x$ and $\lambda_{\max}(S)$ is* sufficiently small. *This is due to the fact that $V(x_t) \leq V_{\max}$ can be ensured for an arbitrarily small $V_{\max}$ if x_t lies in a neighborhood of $\mathcal{Z}^{\mathrm{s}}_x$ and $\bar{S}$ is sufficiently small. In [42], it is shown for linear systems that a tracking MPC formulation recovers optimality properties of a standard MPC scheme if the weight on the distance between the artificial setpoint and the reference setpoint (i.e., the matrix S in our setting) is suitably large. This indicates a trade-off when designing the matrix S: It needs to be suitably small such that the linearization error is small and stability can be guaranteed, but the performance deteriorates if S is chosen too small.*

Remark 5.2. *(Related work) Let us discuss the relation of the linear model-based MPC proposed in Section 5.1 to other linearization-based MPC approaches for nonlinear systems. Note that Theorem 5.1 guarantees closed-loop stability of Algorithm 5.1 which only requires solving one convex QP online. Alternative nonlinear MPC approaches based on convex optimization typically employ an LTI prediction model by linearizing at the setpoint or, as in the real-time iteration scheme [35, 80, 157], a linear time-varying (LTV) prediction model by linearizing along the candidate solution. Closed-loop guarantees of such approaches require either an additional bounding of the linearization error [23] or a sufficiently small sampling time [157], i.e., solving the underlying optimization problem more frequently [80]. The latter ensures that the linearized dynamics do not change too rapidly, which is analogous to our condition on $\lambda_{\max}(S)$ being sufficiently small. The proposed approach has a large region of attraction due to the online optimization of the artificial steady-state. In particular, while a standard (linearization-based) MPC only guarantees stability when starting in a region around the setpoint, our MPC scheme ensures stability for initial conditions far* *away from the setpoint as long as they are close to the steady-state manifold. This fact will be illustrated with a numerical example in Section 5.3.1. Furthermore, as is standard in MPC, our guarantees remain true if the QP is not solved up to optimality by using a warm start and results on suboptimality in MPC [118, Section 2.7].*

Further related approaches are [76, 112] which estimate (approximate) LTV models of nonlinear systems from data and employ these in MPC schemes. While these approaches lead to convex optimization problems which can be solved efficiently, they do not provide guarantees on closed-loop stability. Another line of research to control nonlinear systems via MPC with linear prediction models relies on Koopman operator theory [72]. The key differences to our approach are that we exploit local linearity properties whereas the (infinite-dimensional) Koopman operator provides a globally valid linear model, and that we provide desirable closed-loop stability guarantees.

Remark 5.3. *(Strength of assumptions) Theorem 5.1 requires a number of assumptions, most of which are not too restrictive when proving closed-loop stability of a linearization-based MPC scheme utilizing the benefits of an artificial setpoint: Assumption 5.1 (compact input constraints) is standard. Assumption 5.2 (smoothness) is clearly required for a linearization-based MPC scheme. Further, Assumption 5.3 is a common condition in the literature on tracking MPC, compare [69, 83]. If Assumption 5.3 does not hold, we can still guarantee asymptotic stability of some steady-state, but not necessarily convergence to* x^{sr}. *Assumptions on controllability (Assumption 5.4) are also standard in the presence of terminal equality constraints, see [118]. In order to ensure that the employed bounds hold uniformly, it is crucial to make assumptions on compactness of the steady-state manifold (Assumption 5.6). Moreover, as we show in Section 5.1.4, Assumption 5.8 holds, in fact, as long as* $m = p$ *(equal number of inputs and outputs) and* y^{r} *is reachable. In the absence of Assumption 5.8, we can still guarantee closed-loop stability of the optimal reachable steady-state of* some *linearization, compare Inequality* (5.26), *which may not necessarily be optimal for the nonlinear dynamics. On the other hand, Assumption 5.5 (non-singular dynamics) might possibly be relaxed, compare the discussion after Assumption 5.5. Finally, we note that these assumptions hold in many practical applications, e.g., for the CSTR example we consider in Section 5.3.1.*

To conclude, the proposed MPC scheme based on repeatedly solving Problem 5.1 leads to desirable closed-loop guarantees when applied to a nonlinear system. Moreover, the scheme can be tuned based on a *single* design parameter S, which allows for a trade-off between the size of the region of attraction and the convergence speed. Compared to nonlinear tracking MPC schemes such as [69, 83], our approach has the drawback that convergence may be slower since S needs to be chosen sufficiently small. On the other hand, Problem 5.1 is a convex QP which can be solved up to global optimality very efficiently. Further, the prediction model only requires an accurate description of the system close to the steady-state manifold, which may be simpler to obtain than a globally accurate model which is required to obtain superior performance with existing nonlinear MPC approaches. Finally, as we show in Section 5.2, the presented idea can be extended in order to develop a data-driven MPC scheme to control unknown nonlinear systems by continuously updating the measured data used for prediction.

5.1.4 Sufficient conditions for Assumption 5.8

In the following, we present sufficient conditions for Assumption 5.8. To this end, we assume that $m = p$, i.e., the numbers of inputs and outputs coincide. Further, we assume that the target setpoint y^{r} is reachable, as captured in the following assumption.

Assumption 5.9. *(Reachability) For any $\tilde{x} \in \mathbb{R}^n$, the target setpoint y^{r} is reachable under the linearized dynamics, i.e., $y^{\mathrm{sr}}_{\mathrm{Lin}}(\tilde{x}) = y^{\mathrm{r}}$.*

Assumption 5.9 means that the optimal reachable output equilibrium $y^{\mathrm{sr}}_{\mathrm{Lin}}(\tilde{x})$ at any linearization point $\tilde{x} \in \mathbb{R}^n$ is equal to the target setpoint y^{r} and hence, inserting $\tilde{x} = x^{\mathrm{sr}}$, the same holds true for the nonlinear optimal reachable output equilibrium, i.e, $y^{\mathrm{sr}} = y^{\mathrm{r}}$. Assumption 5.9 is only restrictive if the equilibrium input leading to the target setpoint y^{r} does not satisfy the input constraints. In particular, Assumption 5.9 always holds in case of no input constraints, i.e., if $\mathbb{U} = \mathbb{R}^m$, due to Assumption 5.3 and $m = p$. Before verifying Assumption 5.8, we first show a technical intermediate result.

Lemma 5.2. *Suppose Assumptions 5.1–5.3, 5.5, and 5.6 hold and $m = p$. Then, for any compact set X with $\mathcal{B} \subseteq X \times \mathbb{U}$ (cf. Assumption 5.6), there exists a constant $c_{\mathrm{s1}} > 0$ such that for any steady-state $(x^{\mathrm{s}}, u^{\mathrm{s}}) \in \mathcal{Z}^{\mathrm{s}}$ of the nonlinear system, any state $\tilde{x} \in X$, and any input $\tilde{u} \in \mathbb{U}$, it holds that*

$$\|(x^{\mathrm{s}}, u^{\mathrm{s}}) - (\tilde{x}, \tilde{u})\|_2 \leq c_{\mathrm{s1}} \big(\|h(x^{\mathrm{s}}, u^{\mathrm{s}}) - h(\tilde{x}, \tilde{u})\|_2 + \|\tilde{x} - f(\tilde{x}, \tilde{u})\|_2\big). \tag{5.27}$$

Proof. Define the map $s : X \times \mathbb{U} \to X \times h(X, \mathbb{U})$ with

$$s(x,u) = \begin{bmatrix} x - f(x,u) \\ h(x,u) \end{bmatrix}. \tag{5.28}$$

First, recall that, by Assumption 5.3, $s(x,u)$ is invertible with smooth and hence (on the compact set $h(X, \mathbb{U})$) Lipschitz continuous inverse, cf. [115, Condition (1.1)]. Thus, there exists $c_{s1} > 0$ such that, for any $x_1, x_2 \in X$, $u_1, u_2 \in \mathbb{U}$, it holds that

$$\|(x_1,u_1) - (x_2,u_2)\|_2 \leq c_{s1} \|s(x_1,u_1) - s(x_2,u_2)\|_2. \tag{5.29}$$

Choosing $(x^s, u^s) \in \mathcal{Z}^s \subseteq \mathcal{B} \times \mathbb{U}$, $\tilde{x} \in X$, $\tilde{u} \in \mathbb{U}$, (5.29) implies

$$\begin{aligned} \|(x^s,u^s) - (\tilde{x},\tilde{u})\|_2 &\leq c_{s1} \|s(x^s,u^s) - s(\tilde{x},\tilde{u})\|_2 \\ &\leq c_{s1} \big(\|x^s - f(x^s,u^s) - \tilde{x} + f(\tilde{x},\tilde{u})\|_2 + \|h(x^s,u^s) - h(\tilde{x},\tilde{u})\|_2 \big). \end{aligned} \tag{5.30}$$

Using $x^s = f(x^s, u^s)$, we infer (5.27). ■

Let us now prove (5.23) based on Lemma 5.2 and the given assumptions.

Proposition 5.3. *If Assumptions 5.1–5.3, 5.5, 5.6, and 5.9 hold and $m = p$, then Assumption 5.8 holds.*

Proof. **Proof of $c_{\mathrm{eq},1} \|\hat{x} - x^{\mathrm{sr}}_{\mathrm{Lin}}(\hat{x})\|_2^2 \leq \|\hat{x} - x^{\mathrm{sr}}\|_2^2$**
The linear map in (5.14) can be written explicitly as

$$\hat{g}_{\hat{x}}(y^s) = \underbrace{\begin{bmatrix} A_{\hat{x}} - I & B \\ C_{\hat{x}} & D \end{bmatrix}^{-1}}_{M_{\hat{x}}^{-1} :=} \begin{bmatrix} -e_{\hat{x}} \\ y^s - r_{\hat{x}} \end{bmatrix} \tag{5.31}$$

for any $y^s \in \mathcal{Z}^s_{\mathrm{y,Lin}}(\hat{x})$. In the following, we derive a bound on the difference $M_{\hat{x}}^{-1} - M_{x^{\mathrm{sr}}}^{-1}$ which we then use to obtain the desired statement. To this end, it is readily derived that

$$\begin{aligned} \|M_{\hat{x}}^{-1} - M_{x^{\mathrm{sr}}}^{-1}\|_2 &\leq \|M_{\hat{x}}^{-1}\|_2 \|I - M_{\hat{x}} M_{x^{\mathrm{sr}}}^{-1}\|_2, \\ \|I - M_{\hat{x}} M_{x^{\mathrm{sr}}}^{-1}\|_2 &\leq \|M_{x^{\mathrm{sr}}} - M_{\hat{x}}\|_2 \|M_{x^{\mathrm{sr}}}^{-1}\|_2. \end{aligned}$$

Combining these inequalities and using that $\|M_{\hat{x}}^{-1}\|_2 \leq \frac{1}{\sigma_s}$ and $\|M_{x^{\mathrm{sr}}}^{-1}\|_2 \leq \frac{1}{\sigma_s}$ due to Assumption 5.3, we obtain

$$\|M_{\hat{x}}^{-1} - M_{x^{\mathrm{sr}}}^{-1}\|_2 \leq \frac{1}{{\sigma_s}^2} \|M_{x^{\mathrm{sr}}} - M_{\hat{x}}\|_2. \tag{5.32}$$

Similar to (5.8), it holds that

$$\|M_{x^{\mathrm{sr}}} - M_{\hat{x}}\|_2 \leq \tilde{c}_{\mathrm{X}} \|x^{\mathrm{sr}} - \hat{x}\|_2^2 \tag{5.33}$$

for some $\tilde{c}_{\mathrm{X}} > 0$. Here, we use that $\hat{x} \in \mathrm{X}$ by assumption and $x^{\mathrm{sr}} \in \mathcal{B}_{\mathrm{x}} \subseteq \mathrm{X}$ due to Assumption 5.6, where $\mathcal{B}_{\mathrm{x}}$ denotes the projection of $\mathcal{B}$ on the state component. To summarize, combining (5.32) and (5.33) and using that $y^{\mathrm{sr}} = y^{\mathrm{sr}}_{\mathrm{Lin}}(\hat{x}) = y^{\mathrm{r}}$ by Assumption 5.9, we have

$$\begin{aligned}
&\|x^{\mathrm{sr}}_{\mathrm{Lin}}(\hat{x}) - x^{\mathrm{sr}}\|_2 \\
\overset{(5.31)}{\leq} &\left\| M_{\hat{x}}^{-1} \begin{bmatrix} -e_{\hat{x}} \\ y^{\mathrm{sr}} - r_{\hat{x}} \end{bmatrix} - M_{x^{\mathrm{sr}}}^{-1} \begin{bmatrix} -e_{x^{\mathrm{sr}}} \\ y^{\mathrm{sr}} - r_{x^{\mathrm{sr}}} \end{bmatrix} \right\|_2 \\
\leq &\left\| (M_{\hat{x}}^{-1} - M_{x^{\mathrm{sr}}}^{-1}) \begin{bmatrix} -e_{x^{\mathrm{sr}}} \\ y^{\mathrm{sr}} - r_{x^{\mathrm{sr}}} \end{bmatrix} \right\|_2 + \left\| M_{\hat{x}}^{-1} \begin{bmatrix} e_{x^{\mathrm{sr}}} - e_{\hat{x}} \\ r_{x^{\mathrm{sr}}} - r_{\hat{x}} \end{bmatrix} \right\|_2 \\
\leq &\frac{\tilde{c}_{\mathrm{X}}}{\sigma_{\mathrm{s}}^2} \left\| \begin{bmatrix} -e_{x^{\mathrm{sr}}} \\ y^{\mathrm{sr}} - r_{x^{\mathrm{sr}}} \end{bmatrix} \right\|_2 \|x^{\mathrm{sr}} - \hat{x}\|_2^2 + \|M_{\hat{x}}^{-1}\|_2 \left\| \begin{bmatrix} e_{\hat{x}} - e_{x^{\mathrm{sr}}} \\ r_{\hat{x}} - r_{x^{\mathrm{sr}}} \end{bmatrix} \right\|_2 .
\end{aligned} \tag{5.34}$$

Further, using a similar argument as in Proposition 5.1 for the vector fields f_0, h_0, there exists a constant $c_{\mathrm{X}0} > 0$ such that

$$\|M_{\hat{x}}^{-1}\|_2 \left\| \begin{bmatrix} e_{\hat{x}} - e_{x^{\mathrm{sr}}} \\ r_{\hat{x}} - r_{x^{\mathrm{sr}}} \end{bmatrix} \right\|_2 \leq \frac{1}{\sigma_{\mathrm{s}}} c_{\mathrm{X}0} \|\hat{x} - x^{\mathrm{sr}}\|_2^2 .$$

Together with (5.34), this implies

$$\|x^{\mathrm{sr}}_{\mathrm{Lin}}(\hat{x}) - x^{\mathrm{sr}}\|_2 \leq \bar{c}_{\mathrm{eq},1} \|x^{\mathrm{sr}} - \hat{x}\|_2$$

with

$$\bar{c}_{\mathrm{eq},1} := \left(\frac{\tilde{c}_{\mathrm{X}}}{\sigma_{\mathrm{s}}^2} \left\| \begin{bmatrix} -e_{x^{\mathrm{sr}}} \\ y^{\mathrm{sr}} - r_{x^{\mathrm{sr}}} \end{bmatrix} \right\|_2 + \frac{c_{\mathrm{X}0}}{\sigma_{\mathrm{s}}} \right) \max_{\hat{x} \in \mathrm{X}, x^{\mathrm{s}} \in \mathcal{B}_{\mathrm{x}}} \|x^{\mathrm{s}} - \hat{x}\|_2 .$$

Finally, using

$$\|\hat{x} - x^{\mathrm{sr}}_{\mathrm{Lin}}(\hat{x})\|_2 \leq \|\hat{x} - x^{\mathrm{sr}}\|_2 + \|x^{\mathrm{sr}} - x^{\mathrm{sr}}_{\mathrm{Lin}}(\hat{x})\|_2,$$

we obtain the left inequality in (5.23) for $c_{\mathrm{eq},1} := \frac{1}{(1+\bar{c}_{\mathrm{eq},1})^2} > 0$.

Proof of $\|\hat{x} - x^{\mathrm{sr}}\|_2^2 \leq c_{\mathrm{eq},2} \|\hat{x} - x^{\mathrm{sr}}_{\mathrm{Lin}}(\hat{x})\|_2^2$

Applying Lemma 5.2 with $(x^{\mathrm{s}}, u^{\mathrm{s}}) = (x^{\mathrm{sr}}, u^{\mathrm{sr}})$, $(\tilde{x}, \tilde{u}) = (x^{\mathrm{sr}}_{\mathrm{Lin}}(\hat{x}), u^{\mathrm{sr}}_{\mathrm{Lin}}(\hat{x}))$, we obtain

$$\begin{aligned}
&\|x^{\mathrm{sr}} - x^{\mathrm{sr}}_{\mathrm{Lin}}(\hat{x})\|_2 \\
\leq &c_{\mathrm{s}1} \left(\|y^{\mathrm{sr}} - h(x^{\mathrm{sr}}_{\mathrm{Lin}}(\hat{x}), u^{\mathrm{sr}}_{\mathrm{Lin}}(\hat{x}))\|_2 + \|x^{\mathrm{sr}}_{\mathrm{Lin}}(\hat{x}) - f(x^{\mathrm{sr}}_{\mathrm{Lin}}(\hat{x}), u^{\mathrm{sr}}_{\mathrm{Lin}}(\hat{x}))\|_2 \right).
\end{aligned} \tag{5.35}$$

Using $y^{\mathrm{sr}} = h_{\hat{x}}(x^{\mathrm{sr}}_{\mathrm{Lin}}(\hat{x}), u^{\mathrm{sr}}_{\mathrm{Lin}}(\hat{x}))$ due to Assumption 5.9 together with a bound of the form (5.8) for the vector field h, the first term is bounded as

$$\|h_{\hat{x}}(x^{\mathrm{sr}}_{\mathrm{Lin}}(\hat{x}), u^{\mathrm{sr}}_{\mathrm{Lin}}(\hat{x})) - h_{x^{\mathrm{sr}}_{\mathrm{Lin}}(\hat{x})}(x^{\mathrm{sr}}_{\mathrm{Lin}}(\hat{x}), u^{\mathrm{sr}}_{\mathrm{Lin}}(\hat{x}))\|_2 \leq c_{Xh}\|\hat{x} - x^{\mathrm{sr}}_{\mathrm{Lin}}(\hat{x})\|_2^2$$

for some $c_{Xh} > 0$. Moreover, the second term on the right-hand side of (5.35) is bounded as

$$\begin{aligned}\|x^{\mathrm{sr}}_{\mathrm{Lin}}(\hat{x}) - f(x^{\mathrm{sr}}_{\mathrm{Lin}}(\hat{x}), u^{\mathrm{sr}}_{\mathrm{Lin}}(\hat{x}))\|_2 =& \|f_{\hat{x}}(x^{\mathrm{sr}}_{\mathrm{Lin}}(\hat{x}), u^{\mathrm{sr}}_{\mathrm{Lin}}(\hat{x})) - f_{x^{\mathrm{sr}}_{\mathrm{Lin}}(\hat{x})}(x^{\mathrm{sr}}_{\mathrm{Lin}}(\hat{x}), u^{\mathrm{sr}}_{\mathrm{Lin}}(\hat{x}))\|_2 \\ \overset{(5.8)}{\leq}& c_X\|\hat{x} - x^{\mathrm{sr}}_{\mathrm{Lin}}(\hat{x})\|_2^2.\end{aligned}$$

Combining the above inequalities, we obtain

$$\begin{aligned}\|\hat{x} - x^{\mathrm{sr}}\|_2 \leq& \|\hat{x} - x^{\mathrm{sr}}_{\mathrm{Lin}}(\hat{x})\|_2 + \|x^{\mathrm{sr}}_{\mathrm{Lin}}(\hat{x}) - x^{\mathrm{sr}}\|_2 \\ \leq& \|\hat{x} - x^{\mathrm{sr}}_{\mathrm{Lin}}(\hat{x})\|_2 + c_{\mathrm{s}1}(c_{Xh} + c_X)\|\hat{x} - x^{\mathrm{sr}}_{\mathrm{Lin}}(\hat{x})\|_2^2,\end{aligned}$$

leading to the right inequality in (5.23) for

$$c_{\mathrm{eq},2} := (1 + c_{\mathrm{s}1}(c_{Xh} + c_X)\max_{\hat{x}\in X}\|\hat{x} - x^{\mathrm{sr}}_{\mathrm{Lin}}(\hat{x})\|_2)^2.$$

∎

Summary

In this section, we presented a tracking MPC scheme (Algorithm 5.1) which uses the linearized dynamics at the current state x_t for prediction at time t. In contrast to the other (data-driven) MPC schemes developed in this thesis, Algorithm 5.1 relies on state measurements as well as explicit knowledge of a model, i.e., the Jacobians of the underlying nonlinear system. We proved that the optimal reachable steady-state is exponentially stable in closed loop under the proposed MPC scheme. The results in this section serve two purposes: First, they are an alternative to existing nonlinear MPC approaches which either require solving a non-convex optimization problem or use LTV models as in the real-time iteration scheme. In Section 5.3, we showcase the practicality of our approach and compare its performance and computational complexity to existing approaches. Second, the model-based analysis in this section provides the basis for the design and analysis of data-driven MPC schemes for nonlinear systems with closed-loop guarantees, which we pursue in the following section.

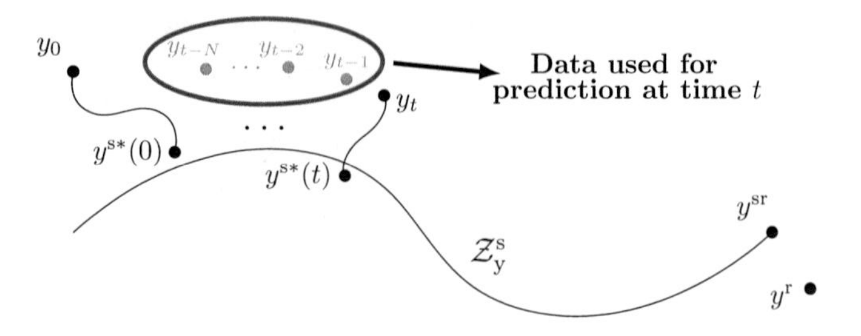

Figure 5.2. Graphical scheme illustrating the basic idea of our data-driven MPC approach. The figure displays the output equilibrium manifold $\mathcal{Z}^{\mathrm{s}}_{\mathrm{y}}$, the closed-loop output y_0 and artificial equilibrium $y^{\mathrm{s}*}(0)$ at time 0, the closed-loop output y_t and artificial equilibrium $y^{\mathrm{s}*}(t)$ at time t, the past N measurements $\{y_k\}_{k=t-N}^{t-1}$ used for prediction at time t, the optimal reachable equilibrium y^{sr}, and the setpoint y^{r}. ©2021 IEEE

5.2 The data-driven case

In this section, we propose an MPC approach to control unknown nonlinear systems with closed-loop stability guarantees by updating the data used in the Fundamental Lemma online, thereby exploiting that nonlinear systems can be approximated locally via linearization. The basic idea is depicted in Figure 5.2.

As in Section 5.1, the goal is to stabilize the optimal reachable output equilibrium y^{sr} corresponding to a given setpoint y^{r}, which may not lie on the output equilibrium manifold $\mathcal{Z}^{\mathrm{s}}_{\mathrm{y}}$. To this end, we employ a tracking MPC formulation with an artificial equilibrium $y^{\mathrm{s}}(t)$ which is optimized online, similar to [69, 83]. In Section 5.1, we showed that, if the current linearization is used for prediction, then, under suitable assumptions on the design parameters and for initial conditions close to $\mathcal{Z}^{\mathrm{s}}_{\mathrm{y}}$, $y^{\mathrm{s}}(t)$ and thus the closed-loop output y_t slide along $\mathcal{Z}^{\mathrm{s}}_{\mathrm{y}}$ towards y^{sr}. In the present section, the key difference to Section 5.1 is that no model of the nonlinear system or its linearization is available and we use past N input-output measurements to predict future trajectories based on the Fundamental Lemma. Since these measurements originate from the nonlinear system, they do not provide an exact description of the linearized dynamics which poses additional challenges if compared to the model-based MPC in Section 5.1. In this section, we show that, if the system does not evolve too rapidly during the initial data collection, then the predictions are sufficiently accurate during the closed-loop operation such that practical stability of y^{sr} can be guaranteed. Our MPC scheme relies on solving strictly convex QPs and the only prior knowledge about the nonlinear system required for the implementation is a (potentially rough) upper bound on its order. For our theoretical analysis, we assume that the closed-loop input generated by the MPC scheme is persistently exciting, but we discuss multiple practical approaches for ensuring this property and we plan to investigate this assumption in more detail in future research.

We first describe the problem setup including the required assumptions in Section 5.2.1. In Section 5.2.2, we present our data-driven MPC scheme for nonlinear systems. Based on a continuity property of the underlying optimization problem (Section 5.2.3), we prove practical exponential stability of the closed loop in Section 5.2.4.

This section is based on and taken in parts literally from [JB11][2].

5.2.1 Nonlinear dynamics and linearization

Problem setting

The control problem considered in this section is analogous to that in Section 5.1 and it is detailed in the following for completeness. We consider nonlinear systems of the form

$$\begin{aligned} x_{k+1} &= f(x_k, u_k) = f_0(x_k) + Bu_k, \\ y_k &= h(x_k, u_k) = h_0(x_k) + Du_k \end{aligned} \tag{5.36}$$

with state $x_k \in \mathbb{R}^n$, input $u_k \in \mathbb{R}^m$, and output $y_k \in \mathbb{R}^p$, all at time $k \in \mathbb{I}_{\geq 0}$, and $f_0 : \mathbb{R}^n \to \mathbb{R}^n$, $B \in \mathbb{R}^{n \times m}$, $h_0 : \mathbb{R}^n \to \mathbb{R}^p$, $D \in \mathbb{R}^{p \times m}$. Note that we assume control-affine system dynamics, which can be enforced via an input transformation, see Section 5.1.1 for details. System (5.36) is subject to pointwise-in-time input constraints $u_t \in \mathbb{U}$ for $t \in \mathbb{I}_{\geq 0}$ with some set $\mathbb{U} \subseteq \mathbb{R}^m$.

[2] J. Berberich, J. Köhler, M. A. Müller, and F. Allgöwer. "Linear tracking MPC for nonlinear systems part II: the data-driven case." In: *IEEE Trans. Automat. Control* (2022). doi: 10.1109/TAC.2022.3166851. ©2021 IEEE

Assumption 5.10. *(Input constraint set) The input constraint set* $\mathbb{U}$ *is a convex, compact polytope.*

For some convex and compact polytope $\mathbb{U}^{\mathrm{s}} \subseteq \operatorname{int}(\mathbb{U})$, which is required for a local controllability argument, we define the steady-state manifold and its projection on the output by

$$\begin{aligned} \mathcal{Z}^{\mathrm{s}} &:= \{(x^{\mathrm{s}}, u^{\mathrm{s}}) \in \mathbb{R}^n \times \mathbb{U}^{\mathrm{s}} \mid x^{\mathrm{s}} = f(x^{\mathrm{s}}, u^{\mathrm{s}})\}, \\ \mathcal{Z}^{\mathrm{s}}_{\mathrm{y}} &:= \{y^{\mathrm{s}} \in \mathbb{R}^p \mid y^{\mathrm{s}} = h(x^{\mathrm{s}}, u^{\mathrm{s}}), (x^{\mathrm{s}}, u^{\mathrm{s}}) \in \mathcal{Z}^{\mathrm{s}}\}. \end{aligned}$$

Further, the projection of $\mathcal{Z}^{\mathrm{s}}$ on the state is denoted by $\mathcal{Z}^{\mathrm{s}}_{\mathrm{x}}$. In this section, we propose a data-driven MPC scheme to track a desired setpoint reference $y^{\mathrm{r}} \in \mathbb{R}^p$. In contrast to Section 5.1, our MPC scheme relies only on input-output data of (5.36), whereas the vector fields f and h are *unknown*. More precisely, we assume that, at time t, we only have access to the last $N \in \mathbb{I}_{\geq 0}$ input-output measurements $\{u_k, y_k\}_{k=t-N}^{t-1}$ to compute the next control input. While we assume noise-free data for our main theoretical results, it is straightforward to extend these results to noisy data, compare Remark 5.7. Similar to the data-driven MPC approaches in Chapters 3 and 4, the scheme does not involve state measurements and therefore, the analysis relies on the *extended* state vector

$$\xi_t := \begin{bmatrix} u_{[t-n,t-1]} \\ y_{[t-n,t-1]} \end{bmatrix} \tag{5.37}$$

for $t \in \mathbb{I}_{\geq n}$. In general, $y^{\mathrm{r}} \notin \mathcal{Z}^{\mathrm{s}}_{\mathrm{y}}$ such that our scheme will guarantee stability of the optimal reachable equilibrium y^{sr}, which is the minimizer of

$$J^*_{\mathrm{eq}} := \min_{y^{\mathrm{s}} \in \mathcal{Z}^{\mathrm{s}}_{\mathrm{y}}} \|y^{\mathrm{s}} - y^{\mathrm{r}}\|_S^2 \tag{5.38}$$

with some $S \succ 0$. Assumptions made later will imply that this minimizer and the corresponding input-state pair $(x^{\mathrm{sr}}, u^{\mathrm{sr}})$ are unique, and we denote by ξ^{sr} the corresponding extended state. We assume that all vector fields in (5.36) are continuously differentiable and we define, for a linearization point $\tilde{x} \in \mathbb{R}^n$,

$$\begin{aligned} A_{\tilde{x}} &:= \left.\frac{\partial f_0}{\partial x}\right|_{\tilde{x}}, \quad e_{\tilde{x}} := f_0(\tilde{x}) - A_{\tilde{x}}\tilde{x}, \\ C_{\tilde{x}} &:= \left.\frac{\partial h_0}{\partial x}\right|_{\tilde{x}}, \quad r_{\tilde{x}} := h_0(\tilde{x}) - C_{\tilde{x}}\tilde{x}. \end{aligned} \tag{5.39}$$

Moreover, we define the affine dynamics resulting from the linearization of (5.36) at $(x,u) = (\tilde{x},0)$ by $f_{\tilde{x}}(x,u) := A_{\tilde{x}}x + Bu + e_{\tilde{x}}$ and $h_{\tilde{x}}(x,u) := C_{\tilde{x}}x + Du + r_{\tilde{x}}$. Let now $\mathcal{D} = \{u'_k, y'_k\}_{k=0}^{N-1}$ be a trajectory of the *linearized* dynamics at some $\tilde{x} \in \mathbb{R}^n$ (i.e., a trajectory of (5.36), replacing f and h by $f_{\tilde{x}}$ and $h_{\tilde{x}}$, respectively), whose input-state component is persistently exciting in the sense that

$$\operatorname{rank}\left(\begin{bmatrix} H_{L+n+1}(u') \\ H_1(x'_{[0,N-L-n-1]}) \\ \mathbb{1}_{N-L-n}^\top \end{bmatrix}\right) = m(L+n+1)+n+1, \tag{5.40}$$

compare (3.17). For our theoretical analysis, we define the optimal steady-state problem for the linearized dynamics by

$$J^*_{\text{eq,Lin}}(\tilde{x}) := \min_{u^s, y^s, \alpha^s} \|y^s - y^r\|_S^2 \tag{5.41}$$

$$\text{s.t.} \begin{bmatrix} H_{L+n+1}(u') \\ H_{L+n+1}(y') \\ \mathbb{1}_{N-L-n}^\top \end{bmatrix} \alpha^s = \begin{bmatrix} \mathbb{1}_{L+n+1} \otimes u^s \\ \mathbb{1}_{L+n+1} \otimes y^s \\ 1 \end{bmatrix}, \ u^s \in \mathbb{U}^s.$$

Note that this problem is equivalent to (5.11) since the constraint provides an equivalent parametrization of all steady-states of the linearization, compare Theorem 3.3. We write $u^{sr}_{\text{Lin}}(\tilde{x})$, $y^{sr}_{\text{Lin}}(\tilde{x})$ for the optimal solution and $x^{sr}_{\text{Lin}}(\tilde{x})$ ($\xi^{sr}_{\text{Lin}}(\tilde{x})$) for the (extended) steady-state of the linearized dynamics, which are unique due to assumptions made in the following.

Further, $\alpha^{sr}_{\text{Lin}}(\mathcal{D})$ denotes the optimal solution with minimum 2-norm. Note that, in contrast to the optimal input, state, and output, the optimal value of α^s depends not only on $\tilde{x}$ but also on the data set $\mathcal{D}$ in (5.41). Throughout this section, we assume that $\alpha^{sr}_{\text{Lin}}(\mathcal{D})$ is uniformly bounded. [3]

Assumptions

Since our analysis relies on similar arguments as Section 5.1, we require the following assumption, compare Section 5.1 for details.

[3] This is always fulfilled in closed loop if the data generated by the proposed MPC scheme are uniformly persistently exciting (cf. Assumption 5.16), using compactness of the steady-state manifold (Assumption 5.11). In case this is not guaranteed a priori, a uniform bound on $\alpha^{sr}_{\text{Lin}}(\mathcal{D})$ can be ensured by using incremental data $\Delta u_k = u_{k+1} - u_k$, $\Delta y_k = y_{k+1} - y_k$ in the proposed data-driven MPC scheme and its analysis. In this case, $(u^{sr}_{\text{Lin}}(\tilde{x}), y^{sr}_{\text{Lin}}(\tilde{x})) = (0,0)$ for any $\tilde{x} \in \mathbb{R}^n$ and thus, $\alpha^{sr}_{\text{Lin}}(\mathcal{D}) = 0$, i.e., $\alpha^{sr}_{\text{Lin}}(\mathcal{D})$ is uniformly bounded.

Assumption 5.11. *System (5.36) satisfies Assumptions 5.2–5.6.*

Moreover, we make the following assumption on the extended state vector ξ_t in (5.37).

Assumption 5.12. *(Observability) There exists a locally Lipschitz continuous map $T_L : \mathbb{R}^{n(m+p)} \to \mathbb{R}^n$ such that*

$$x_t = T_{\mathrm{L}}(\xi_t). \tag{5.42}$$

Further, for any $x_t \in \mathbb{R}^n$, there exists an extended state $\xi_t' = \begin{bmatrix} u'^{\top}_{[t-n,t-1]} & y'^{\top}_{[t-n,t-1]} \end{bmatrix}^\top$ corresponding to x_t in the dynamics linearized at x_t, i.e., there exists an affine map $T_{x_t} : \mathbb{R}^{(m+p)n} \to \mathbb{R}^n$ such that

$$x_t = T_{x_t}(\xi_t'). \tag{5.43}$$

Assumption 5.12 corresponds to (final state) observability (compare, e.g., [118, Definition 4.29]) of the linearized and nonlinear dynamics, see (2.4) for an analogous condition in the linear case which was required for our linear data-driven MPC results in Chapters 3 and 4. In the proof of Theorem 5.2, we provide a precise definition of the particular choice of ξ_t' used for our theoretical analysis.

Assumption 5.13. *For any compact set Ξ, there exist constants $c_{\mathrm{eq},1}, c_{\mathrm{eq},2} > 0$ such that, for any extended state $\hat{\xi} \in \Xi$ it holds that*

$$c_{\mathrm{eq},1}\|\hat{\xi} - \xi^{\mathrm{sr}}_{\mathrm{Lin}}(\hat{x})\|_2^2 \leq \|\hat{\xi} - \xi^{\mathrm{sr}}\|_2^2 \leq c_{\mathrm{eq},2}\|\hat{\xi} - \xi^{\mathrm{sr}}_{\mathrm{Lin}}(\hat{x})\|_2^2, \tag{5.44}$$

where $\hat{x} = T_{\mathrm{L}}(\hat{\xi})$, compare (5.42).

Assumption 5.13 is analogous to Assumption 5.8, where Inequality (5.44) is assumed for the state $\hat{x}$ and the optimal reachable steady-states x^{sr}, $x^{\mathrm{sr}}_{\mathrm{Lin}}(\hat{x})$ (see Section 5.1 for details). In the data-driven framework, we require Assumption 5.13 since the proposed data-driven MPC scheme only involves input-output values and hence, our theoretical analysis relies on the extended state ξ. Using similar arguments as in Section 5.1, it can be shown that Assumption 5.13 holds if i) Assumptions 5.2, 5.3, 5.5, and 5.6 hold for the extended state-space system with state[4] ξ, ii) the setpoint y^{r} is reachable, i.e., $y^{\mathrm{sr}} = y^{\mathrm{r}}$, and iii) $m = p$.

Finally, we assume the following lower bound on the prediction horizon.

Assumption 5.14. *(Length of prediction horizon) The prediction horizon satisfies $L \geq 2n$.*

[4]It is straightforward to verify that, if Assumptions 5.2, 5.3, 5.5, and 5.6 hold for the state x, then they also hold for the state ξ with different constants.

Bounding the influence of the nonlinearity on the data

Our goal is to control the unknown nonlinear system (5.36) via MPC using the last N input-output measurements $\{u_k, y_k\}_{k=t-N}^{t-1}$ to predict future trajectories at time t. By Theorem 3.3, data corresponding to the (affine) linearization of (5.36) provide an exact parametrization of all trajectories of the linearized dynamics. However, we only have access to data of the *nonlinear* dynamics. In the following, we bound the difference between the (artificial) data of the linearization at x_t and the actually available data of the nonlinear system.

Let $\{u_k, x_k, y_k\}_{k\in\mathbb{I}_{\geq 0}}$ be an arbitrary trajectory of the nonlinear system (5.36). For some $t \in \mathbb{I}_{\geq N}$, we write $\{x'_k(t)\}_{k=-N}^{n}$ and $\{y'_k(t)\}_{k=-N}^{n-1}$ for the state and output corresponding to the dynamics linearized at x_t resulting from an application of the input $\{u_k\}_{k=t-N}^{t+n-1}$ with the initial state of the nonlinear system x_{t-N} at time $t-N$, i.e., $x'_{-N}(t) = x_{t-N}$ and

$$x'_{k+1}(t) = f_{x_t}(x'_k(t), u_{t+k}),\ \ y'_k(t) = h_{x_t}(x'_k(t), u_{t+k}) \tag{5.45}$$

for $k \in \mathbb{I}_{[-N,n-1]}$. We denote this "artificial" input-output data trajectory by $\mathcal{D}_t := \{u_{t+k}, y'_k(t)\}_{k=-N}^{-1}$ and we define

$$H_{ux,t} := \begin{bmatrix} H_{L+n+1}(u_{[t-N,t-1]}) \\ H_1(x'_{[-N,-L-n-1]}(t)) \\ \mathbb{1}_{N-L-n}^\top \end{bmatrix}. \tag{5.46}$$

The following result provides a bound on the difference between the (known) output of the nonlinear system y and the (unknown) output of the affine dynamics y'.

Lemma 5.3. *Suppose Assumption 5.11 holds. For any compact set $X \subset \mathbb{R}^n$, there exists $c_\Delta > 0$ such that for any $t \in \mathbb{I}_{\geq N}$, $k \in \mathbb{I}_{[-N,n-1]}$, and $x_k, x_t \in X$, $\Delta_{t,k} := y_{t+k} - y'_k(t)$ satisfies*

$$\|\Delta_{t,k}\|_2 \leq c_\Delta \sum_{j=t-N}^{t+k} \|x_t - x_j\|_2^2. \tag{5.47}$$

Proof. Using $f(x,u) = f_x(x,u)$, $h(x,u) = h_x(x,u)$, we have

$$\begin{aligned} x_{k+1} &= A_{x_k}x_k + Bu_k + e_{x_k} \\ &= A_{x_t}x_k + Bu_k + e_{x_t} + \underbrace{(A_{x_k} - A_{x_t})x_k + e_{x_k} - e_{x_t}}_{\Delta_{x,k}:=}, \end{aligned} \tag{5.48}$$

$$\begin{aligned} y_k &= C_{x_k}x_k + Du_k + r_{x_k} \\ &= C_{x_t}x_k + Du_k + r_{x_t} + \underbrace{(C_{x_k} - C_{x_t})x_k + r_{x_k} - r_{x_t}}_{\Delta_{y,k}:=}, \end{aligned} \tag{5.49}$$

for any $k \in \mathbb{I}_{[t-N,t+n-1]}$. By repeatedly applying (5.48), we obtain, for $k \in \mathbb{I}_{[t-N+1,t+n-1]}$, $x_k = x'_{k-t}(t) + \Delta_{x,k-1} + \sum_{j=t-N}^{k-2} A_{x_{k-1}} \cdot \ldots \cdot A_{x_{j+1}}\Delta_{x,j}$. In combination with (5.49), this yields

$$y_k = y'_{k-t}(t) + \Delta_{y,k} + C_{x_t}\Delta_{x,k-1} + C_{x_t}\sum_{j=t-N}^{k-2} A_{x_{k-1}} \cdot \ldots \cdot A_{x_{j+1}}\Delta_{x,j} = y'_{k-t}(t) + \Delta_{t,k-t}.$$

This implies

$$\Delta_{t,k} = \Delta_{y,t+k} + C_{x_{t+k}}\Delta_{x,t+k-1} + C_{x_{t+k}}\sum_{j=t-N}^{t+k-2} A_{x_{t+k-1}} \cdot \ldots \cdot A_{x_{j+1}}\Delta_{x,j} \tag{5.50}$$

for $k \in \mathbb{I}_{[-N,n-1]}$. Using smoothness of the dynamics in (5.36), we can apply Inequalities (5.5) and (5.6) to derive

$$\begin{aligned} \|\Delta_{x,k}\|_2 &= \|f_{x_k}(x_k,u_k) - f_{x_t}(x_k,u_k)\|_2 \le c_X\|x_t - x_k\|_2^2, \\ \|\Delta_{y,k}\|_2 &= \|h_{x_k}(x_k,u_k) - h_{x_t}(x_k,u_k)\|_2 \le c_{Xh}\|x_t - x_k\|_2^2 \end{aligned}$$

for $k \in \mathbb{I}_{[t-N,t+n-1]}$ and with suitable $c_X, c_{Xh} > 0$. Using these inequalities to bound $\Delta_{t,k}$ in (5.50) and exploiting that, on the compact set X, the Jacobians A_x and C_x are uniformly bounded, there exists $c_\Delta > 0$ such that (5.47) holds. ∎

Lemma 5.3 shows that the difference between y_{t+k} and $y'_k(t)$, i.e., the "output measurement noise" affecting the ideal data of the linearized dynamics, is bounded by the squared distance of x_t to the past states. This means that, if the state trajectory does not move too rapidly, then the data collected from the nonlinear system (5.36) are close to those of the linearized dynamics and can therefore be used to (approximately) parametrize trajectories of the linearized dynamics via Theorem 3.3.

5.2.2 Proposed MPC scheme

At time $t \in \mathbb{I}_{\geq N}$, given past N input-output measurements $\{u_k, y_k\}_{k=t-N}^{t-1}$ of the nonlinear system (5.36), we define the following open-loop optimal control problem:

Problem 5.2.

$$\begin{aligned}\underset{\substack{\alpha(t),\sigma(t),\bar{u}(t)\\ \bar{y}(t),u^{\mathrm{s}}(t),y^{\mathrm{s}}(t)}}{\text{minimize}} \quad & \sum_{k=-n}^{L}\left(\|\bar{u}_k(t)-u^{\mathrm{s}}(t)\|_R^2+\|\bar{y}_k(t)-y^{\mathrm{s}}(t)\|_Q^2\right)+\|y^{\mathrm{s}}(t)-y^{\mathrm{r}}\|_S^2 \qquad (5.51\mathrm{a})\\ & +\lambda_\alpha\|\alpha(t)-\alpha_{\mathrm{Lin}}^{\mathrm{sr}}(\mathcal{D}_t)\|_2^2+\lambda_\sigma\|\sigma(t)\|_2^2\end{aligned}$$

subject to

$$\begin{bmatrix}\bar{u}(t)\\ \bar{y}(t)+\sigma(t)\\ 1\end{bmatrix} = \begin{bmatrix}H_{L+n+1}\left(u_{[t-N,t-1]}\right)\\ H_{L+n+1}\left(y_{[t-N,t-1]}\right)\\ \mathbb{1}_{N-L-n}^\top\end{bmatrix}\alpha(t), \qquad (5.51\mathrm{b})$$

$$\begin{bmatrix}\bar{u}_{[-n,-1]}(t)\\ \bar{y}_{[-n,-1]}(t)\end{bmatrix} = \begin{bmatrix}u_{[t-n,t-1]}\\ y_{[t-n,t-1]}\end{bmatrix}, \qquad (5.51\mathrm{c})$$

$$\begin{bmatrix}\bar{u}_{[L-n,L]}(t)\\ \bar{y}_{[L-n,L]}(t)\end{bmatrix} = \begin{bmatrix}\mathbb{1}_{n+1}\otimes u^{\mathrm{s}}(t)\\ \mathbb{1}_{n+1}\otimes y^{\mathrm{s}}(t)\end{bmatrix}, \qquad (5.51\mathrm{d})$$

$$\bar{u}_k(t)\in\mathbb{U},\ k\in\mathbb{I}_{[0,L]},\ u^{\mathrm{s}}(t)\in\mathbb{U}^{\mathrm{s}}. \qquad (5.51\mathrm{e})$$

The cost (5.51a) contains a tracking term with weights $Q, R \succ 0$ w.r.t. an artificial setpoint $(u^{\mathrm{s}}(t), y^{\mathrm{s}}(t))$ which is optimized online and whose distance to the setpoint y^{r} is penalized in the cost by the weight $S \succ 0$, compare Problem 5.1. Similar to the linear data-driven MPC schemes in Chapters 3 and 4, Problem 5.2 uses a prediction model based on Hankel matrices (5.51b). In contrast to the previous data-driven MPC approaches, the data used for prediction in Problem 5.2 are updated at any time step using the last N measurements of the *nonlinear* system. Hence, the prediction model of Problem 5.2 can be seen as an approximation of the affine dynamics resulting from the linearization at x_t, cf. Lemma 5.3, which in turn provides a local approximation of the nonlinear dynamics (5.36). In order to account for the model mismatch due to the nonlinearity, the slack variable $\sigma(t)$ is introduced and the cost (5.51a) additionally contains regularization terms of $\alpha(t)$ and $\sigma(t)$ with parameters $\lambda_\alpha, \lambda_\sigma > 0$. We note that analogous ingredients are employed to handle noise in linear data-driven MPC, see [27, 28] or Chapter 4.

In (5.51c), the first n components of the predictions are set to the past n input-output measurements in order to specify initial conditions, compare [94]. As in Chapters 3 and 4, n can be replaced by any upper bound on the system order (or, more specifically, on the system's lag), i.e., the application of the proposed MPC does *not* require accurate knowledge of the system order. Moreover, (5.51d) represents the terminal equality constraint w.r.t. the artificial equilibrium $(u^{\mathrm{s}}(t), y^{\mathrm{s}}(t))$. It is defined over $n+1$ steps since this ensures that $(u^{\mathrm{s}}(t), y^{\mathrm{s}}(t))$ is an (approximate) equilibrium of the dynamics linearized at x_t, compare Definition 2.2. In order to parametrize (approximate) trajectories of the *affine* dynamics corresponding to the linearization at x_t, the last line of (5.51b) implies that $\alpha(t)$ sums up to 1, compare Theorem 3.3. Further, the scheme contains constraints on the input equilibrium and on the input trajectory in (5.51e). Note that Problem 5.2 is a strictly convex QP and, hence, it can be solved efficiently.

Remark 5.4. *(Computation of $\alpha^{\mathrm{sr}}_{\mathrm{Lin}}(\mathcal{D}_t)$) Problem 5.2 is analogous to Problem 4.3, which is the basis for robust data-driven tracking MPC from noisy data (cf. Section 4.3), with the main difference that the data in* (5.51b) *are updated online. In particular, the regularization of $\alpha(t)$ is not w.r.t. zero but depends on $\alpha^{\mathrm{sr}}_{\mathrm{Lin}}(\mathcal{D}_t)$, which can be (approximately) computed as the least-squares solution of* (5.41) *by inserting the past N input-output measurements $\{u_k, y_k\}_{k=t-N}^{t-1}$ of the nonlinear system* (5.36)*. Since these measurements are not a trajectory of the linearized dynamics, one can resort to robustifying modifications, compare Remark 4.8 for details. As in Section 4.3, our qualitative theoretical results remain true if an approximation $\alpha^{\mathrm{s}\prime}$ of $\alpha^{\mathrm{sr}}_{\mathrm{Lin}}(\mathcal{D}_t)$ with $\|\alpha^{\mathrm{s}\prime} - \alpha^{\mathrm{sr}}_{\mathrm{Lin}}(\mathcal{D}_t)\|_2 \leq c$ for some $c > 0$ is known, at the cost of additional conservatism for increasing values of c. Since $\alpha^{\mathrm{sr}}_{\mathrm{Lin}}(\mathcal{D}_t)$ is uniformly bounded, our guarantees hold for any uniformly bounded $\alpha^{\mathrm{s}\prime}$ (e.g., $\alpha^{\mathrm{s}\prime} = 0$).*

We denote the optimal solution of Problem 5.2 at time t by $\bar{u}^*(t)$, $\bar{y}^*(t)$, $\alpha^*(t)$, $\sigma^*(t)$, $u^{\mathrm{s}*}(t)$, $y^{\mathrm{s}*}(t)$, and the closed-loop input, state, and output at time t by u_t, x_t, and y_t, respectively. Further, we write $J_L^*(\xi_t)$ for the corresponding optimal cost. Problem 5.2 is applied in a multi-step fashion, see Algorithm 5.2.

Considering a multi-step MPC scheme instead of a standard (one-step) MPC scheme simplifies the theoretical analysis with terminal equality constraints due to a local controllability argument in the proof, similar to the model-based MPC in Section 5.1.

Note that Algorithm 5.2 allows to control unknown nonlinear systems based only on measured data and without any model knowledge except for a (potentially

Algorithm 5.2. Nonlinear data-driven MPC scheme
Offline: Choose upper bound on system order n, prediction horizon L, cost matrices $Q, R, S \succ 0$, regularization parameters $\lambda_\alpha, \lambda_\sigma > 0$, constraint sets $\mathbb{U}$, $\mathbb{U}^{\mathrm{s}}$, setpoint y^{r}, and generate initial data $\{u_k, y_k\}_{k=0}^{N-1}$.
Online: At time $t = N + n \cdot i$, $i \in \mathbb{I}_{\geq 0}$, take the past N measurements $\{u_k, y_k\}_{k=t-N}^{t-1}$ and compute an approximation of $\alpha_{\mathrm{Lin}}^{\mathrm{sr}}(\mathcal{D}_t)$ by solving (5.41) or (4.72), compare Remark 5.4. Solve Problem 5.2 and apply the input $u_{[t,t+n-1]} = \bar{u}^*_{[0,n-1]}(t)$ over the next n time steps.

rough) upper bound on the system order. Moreover, it only requires solving one (or two, if $\alpha_{\mathrm{Lin}}^{\mathrm{sr}}(\mathcal{D}_t)$ is computed online) strictly convex QPs. In the remainder of this section, we prove that, under suitable assumptions, the closed loop under Algorithm 5.2 is practically exponentially stable. Our analysis builds on the MPC approach based on linearized models in Section 5.1. Similar to Section 5.1, the key idea is that, if $\lambda_{\max}(S)$ is sufficiently small and the initial state is sufficiently close to $\mathcal{Z}_{\mathrm{x}}^{\mathrm{s}}$, then the data-driven prediction model in (5.51b) provides a good approximation of the nonlinear dynamics in closed loop, thus ensuring closed-loop stability.

5.2.3 Continuity of the optimal input and cost

In this section, we present a key technical result for our closed-loop analysis bounding the distance of the optimal input/cost of Problem 5.2 to the corresponding optimal input/cost with "ideal" data of the linearized dynamics. At time $t \in \mathbb{I}_{\geq N}$, given a data set $\mathcal{D}_t = \{u_{t+k}, y'_k(t)\}_{k=-N}^{-1}$ of the dynamics linearized at x_t (compare (5.45)), an extended state $\xi'_t = \begin{bmatrix} u'_{[t-n,t-1]} \\ y'_{[t-n,t-1]} \end{bmatrix}$ corresponding to x_t in the dynamics linearized at x_t (compare Assumption 5.12), and a vector $\tilde{\sigma} = \begin{bmatrix} \tilde{\sigma}_{\mathrm{init}} \\ \tilde{\sigma}_{\mathrm{dyn}} \end{bmatrix} \in \mathbb{R}^{p(L+2n+1)+mn}$, we define the following optimization problem:

Problem 5.3.

$$\underset{\substack{\alpha(t),\bar{u}(t),\bar{y}(t)\\u^{s}(t),y^{s}(t)}}{\text{minimize}} \quad \sum_{k=-n}^{L}\left(\|\bar{u}_k(t)-u^{s}(t)\|_R^2+\|\bar{y}_k(t)-y^{s}(t)\|_Q^2\right)+\|y^{s}(t)-y^{r}\|_S^2 \tag{5.52a}$$

$$+\lambda_\alpha\|\alpha(t)-\alpha^{\text{sr}}_{\text{Lin}}(\mathcal{D}_t)\|_2^2$$

subject to

$$\begin{bmatrix}\bar{u}(t)\\ \bar{y}(t)+\tilde{\sigma}_{\text{dyn}}\\ 1\end{bmatrix}=\begin{bmatrix}H_{L+n+1}\left(u_{[t-N,t-1]}\right)\\ H_{L+n+1}\left(y'(t)\right)\\ \mathbb{1}_{N-L-n}^{\top}\end{bmatrix}\alpha(t), \tag{5.52b}$$

$$\begin{bmatrix}\bar{u}_{[-n,-1]}(t)\\ \bar{y}_{[-n,-1]}(t)\end{bmatrix}=\begin{bmatrix}u'_{[t-n,t-1]}\\ y'_{[t-n,t-1]}\end{bmatrix}+\tilde{\sigma}_{\text{init}}, \tag{5.52c}$$

$$\begin{bmatrix}\bar{u}_{[L-n,L]}(t)\\ \bar{y}_{[L-n,L]}(t)\end{bmatrix}=\begin{bmatrix}\mathbb{1}_{n+1}\otimes u^{s}(t)\\ \mathbb{1}_{n+1}\otimes y^{s}(t)\end{bmatrix}, \tag{5.52d}$$

$$\bar{u}_k(t)\in\mathbb{U},\ k\in\mathbb{I}_{[0,L]},\ u^{s}(t)\in\mathbb{U}^{s}. \tag{5.52e}$$

We denote the optimal solution of Problem 5.3 with[5] $\tilde{\sigma}=0$ by $\check{\alpha}^*(t)$, $\check{u}^{s*}(t)$, $\check{y}^{s*}(t)$, $\check{u}^*(t)$, $\check{y}^*(t)$. Moreover, we write $\check{J}_L^*(\xi_t',\mathcal{D}_t)$ for the optimal cost to emphasize its dependence on the initial condition ξ_t' as well as the data set $\mathcal{D}_t=\{u_{t+k},y_k'(t)\}_{k=-N}^{-1}$. For $\tilde{\sigma}=0$, the constraints of Problem 5.3 correspond to those of Problem 5.2 with $\sigma(t)=0$ and replacing the data $\{u_k,y_k\}_{k=t-N}^{t-1}$ and initial condition ξ_t of the nonlinear system by that of the dynamics linearized at x_t, i.e., $\{u_{t+k},y_k'(t)\}_{k=-N}^{-1}$ and ξ_t'. Thus, according to Theorem 3.3, any $\bar{u}(t)$, $\bar{y}(t)$ satisfying the constraints of Problem 5.3 with $\tilde{\sigma}=0$ are a trajectory of the dynamics linearized at x_t. Note that the initial conditions $\{u_{t+k}',y_{t+k}'\}_{k=-n}^{-1}$ in (5.52c) are in general different from $\{u_{t+k},y_k'(t)\}_{k=-n}^{-1}$ in (5.52b), although both correspond to the dynamics linearized at x_t and both are close to the input-output trajectory $\{u_{t+k},y_{t+k}\}_{k=-n}^{-1}$ of the nonlinear system.

Assumption 5.15. *(LICQ) For any $t\in\mathbb{I}_{\geq N}$, Problem 5.3 satisfies a linear independence constraint qualification (LICQ), i.e., the row entries of the equality and active inequality constraints are linearly independent.*

[5] We require the flexibility of choosing $\tilde{\sigma}\neq 0$ for a technical argument in the proof of Proposition 5.4 below.

Similar to Section 4.3, we require Assumption 5.15 for a technical step in the following result. Such an LICQ assumption is common in linear MPC, compare [13], and may possibly be relaxed at the price of a more involved analysis. We now derive a bound on the difference between the *nominal* optimal cost $\check{J}_L^*(\xi_t', \mathcal{D}_t)$ and input $\check{u}^*(t)$ corresponding to Problem 5.3 with $\bar{\sigma} = 0$ and the *perturbed* optimal cost $J_L^*(\xi_t)$ and input $\bar{u}^*(t)$ corresponding to Problem 5.2.

Proposition 5.4. *Suppose Assumptions 5.10, 5.11, and 5.15 hold and the matrix $H_{ux,t}$ in (5.46) has full row rank, i.e.,* $\text{rank}(H_{ux,t}) = m(L+n+1)+n+1$*. Moreover, for some $\bar{\varepsilon} > 0$, consider*

$$\|\xi_t - \xi_t'\|_2 \leq \bar{\varepsilon}, \ \|y_{t+k} - y_k'(t)\|_2 \leq \bar{\varepsilon} \ \ \forall k \in \mathbb{I}_{[-N,-1]}, \tag{5.53}$$

$$\lambda_\alpha = \bar{\lambda}_\alpha \bar{\varepsilon}^{\beta_\alpha}, \ \lambda_\sigma = \frac{\bar{\lambda}_\sigma}{\bar{\varepsilon}^{\beta_\sigma}} \tag{5.54}$$

for some $\bar{\lambda}_\alpha, \bar{\lambda}_\sigma, \beta_\alpha, \beta_\sigma > 0$ with $\beta_\alpha + 2\beta_\sigma < 2$, where ξ_t' satisfies (5.43) and $y'(t)$ is the output of the dynamics linearized at x_t with input $u_{[t-N,t-1]}$ and initial condition x_{t-N}, compare (5.45).

i) There exist $\bar{\varepsilon}_{\max}, c_{J,a}, c_{J,b} > 0$ such that, if $\bar{\varepsilon} \leq \bar{\varepsilon}_{\max}$, (5.53), and (5.54) hold, then

$$J_L^*(\xi_t) \leq \left(1 + c_{J,a}\bar{\varepsilon}^{\beta_\sigma}\right) \check{J}_L^*(\xi_t', \mathcal{D}_t) + c_{J,b}\bar{\varepsilon}^{2-\beta_\sigma}\left(1 + \|H_{ux,t}^\dagger\|_2^2 \bar{\varepsilon}^{\beta_\alpha}\right) \tag{5.55}$$

with $H_{ux,t}$ as in (5.46).

ii) For any $\bar{J} > 0$, there exist $\bar{\varepsilon}_{\max} > 0$, $\beta_u \in \mathcal{K}_\infty$ such that, if $\check{J}_L^*(\xi_t', \mathcal{D}_t) \leq \bar{J}$, $\bar{\varepsilon} \leq \bar{\varepsilon}_{\max}$, (5.53), and (5.54) hold, then

$$\|\bar{u}^*(t) - \check{u}^*(t)\|_2 \leq \beta_u(\bar{\varepsilon}). \tag{5.56}$$

The proof of Proposition 5.4 is provided in the appendix (Appendix A.4). The result bounds the difference between the optimal input/cost for the robust MPC problem (Problem 5.2) with data of the nonlinear system and the optimal input/cost for the nominal MPC problem (Problem 5.3) with data of the linearized dynamics, i.e., with $\bar{\sigma} = 0$. In Part (i) of the proof, we bound the optimal cost $J_L^*(\xi_t)$ in terms of $\check{J}_L^*(\xi_t', \mathcal{D}_t)$ using a simple candidate solution, i.e., we show that Problem 5.2 is feasible whenever Problem 5.3 is feasible. Part (ii), i.e., the proof of (5.56), relies on a combination of strong convexity of Problem 5.3, another candidate solution for Problem 5.2, and properties of multi-parametric QPs.

For this technical result, we assume that the difference between the data and initial condition of the nonlinear system and that of the linearized dynamics is bounded, compare (5.53). This condition can be interpreted as a bound on the "noise" affecting the data of the linearized dynamics and it is only required for the open-loop analysis of Problem 5.2 in Proposition 5.4. In Section 5.2.4, we prove that a bound as in (5.53) is implicitly enforced by our MPC scheme in closed loop. Since the bounds (5.55) and (5.56) depend on the linearization-induced error $\bar{\varepsilon}$, Proposition 5.4 can be interpreted as a continuity property of the optimal solution of Problem 5.2 w.r.t. additive and multiplicative perturbations.

Proposition 5.4 plays a crucial role in the theoretical analysis of the proposed data-driven MPC scheme for nonlinear systems. Nevertheless, the result is of independent interest beyond its application for the stability proof in Section 5.2.4. In particular, it enables a novel separation-type proof of stability and robustness of linear data-driven MPC, compare Section 4.3. Moreover, as discussed in more detail in Section 4.5, Proposition 5.4 can be used to derive additional and more general robustness results for linear data-driven MPC.

5.2.4 Closed-loop guarantees

Before presenting our main theoretical result of this section, we make an additional assumption on closed-loop persistence of excitation.

Assumption 5.16. *(Closed-loop persistence of excitation) There exists $c_{\mathrm{H}} > 0$ such that, for all $t \in \mathbb{I}_{\geq N}$, the matrix $H_{ux,t}$ from* (5.46) *has full row rank and $\|H_{ux,t}^{\dagger}\|_2 \leq c_{\mathrm{H}}$.*

Assumption 5.16 ensures that the persistence of excitation condition in (5.40) holds uniformly in closed loop. To be precise, in addition to $H_{ux,t}$ having full row rank, we also require a uniform upper bound on $\|H_{ux,t}^{\dagger}\|_2$ which holds if the singular values of $H_{ux,t}$ are uniformly lower and upper bounded. Assumption 5.16 is crucial for our theoretical results since, if the data used for prediction are updated online, they may in general not be persistently exciting and can thus lead to inaccurate predictions for the nonlinear system. We note that Assumption 5.16 can be restrictive and may not be satisfied in practice, in particular upon convergence of the closed loop. Nevertheless, there are various pragmatic approaches to ensure the availability of persistently exciting data for our purposes such as (i) stopping the data updates as soon as a neighborhood of the setpoint is reached, (ii) adding suitable excitation signals to the closed-loop input applied to the system, or (iii) incentivizing persistently exciting inputs similar to adaptive MPC approaches [88].

In the following, we present our main theoretical result on closed-loop stability. Our analysis relies on the Lyapunov function candidate $V(\xi_t', \mathcal{D}_t) := \check{J}_L^*(\xi_t', \mathcal{D}_t) - J_{\text{eq,Lin}}^*(x_t)$.

Theorem 5.2. *Suppose Assumptions 5.10–5.15 hold. There exists $\bar{\theta} > 0$ such that for any $\theta \in (0, \bar{\theta}]$, if Assumption 5.16 holds with $c_{\text{H}} = \frac{\bar{c}_{\text{H}}}{\theta}$ for some $\bar{c}_{\text{H}} > 0$, then there exist $V_{\max}, \bar{S}, C > 0$, regularization parameters $\lambda_\alpha, \lambda_\sigma > 0$, and $0 < c_{\text{V}} < 1$, $\beta_\theta \in \mathcal{K}_\infty$, such that, if*

$$V(\xi_N', \mathcal{D}_N) \leq V_{\max}, \ \lambda_{\max}(S) \leq \bar{S} \tag{5.57}$$

$$\|\xi_N - \xi_i\|_2 \leq \theta, \ i \in \mathbb{I}_{[0,N-1]}, \tag{5.58}$$

then, for any $t = N + ni$, $i \in \mathbb{I}_{\geq 0}$, Problem 5.2 is feasible and the closed loop under Algorithm 5.2 satisfies

$$\|\xi_t - \xi^{\text{sr}}\|_2^2 \leq c_{\text{V}}^i C \|\xi_N - \xi^{\text{sr}}\|_2^2 + \beta_\theta(\theta). \tag{5.59}$$

The proof of Theorem 5.2 is provided in the appendix (Appendix A.5).

Discussion

Theorem 5.2 shows that the closed-loop state trajectory converges close to the optimal reachable equilibrium (cf. (5.59)) if the cost matrix S is sufficiently small, the regularization parameters are chosen suitably (roughly speaking, λ_α is small and λ_σ is large), and if the initial data satisfy $\|\xi_N - \xi_i\|_2 \leq \theta$ for $i \in \mathbb{I}_{[0,N-1]}$ with a sufficiently small θ. Inequality (5.59) implies that the optimal reachable setpoint ξ^{sr} is practically exponentially stable, i.e., the closed-loop trajectory exponentially converges to a region around ξ^{sr} whose size increases with θ. In particular, under suitable assumptions, the proposed MPC scheme steers the closed loop arbitrarily close to ξ^{sr} if the initial state evolution, i.e., the parameter θ, is sufficiently small. The proof relies on a separation argument, combining the continuity of data-driven MPC in Proposition 5.4 with the model-based MPC analysis from Section 5.1.

In the following, we provide an interpretation for the above stability result. As shown in Lemma 5.3, the perturbation of the available data and thus the prediction accuracy of the implicit prediction model in Problem 5.2 depends on the distance of past state values to the current state. The parameter θ provides a bound on this distance at initial time $t = N$, which is shown to hold recursively for all $t \in \mathbb{I}_{\geq N}$ in the proof. Thus, if θ is sufficiently small at $t = N$, then the closed loop does not move too rapidly such that the prediction model remains accurate and stability can be shown. The conditions $\lambda_{\max}(S) \leq \bar{S}$ and $V(\xi_0'(N), \mathcal{D}_N) \leq V_{\max}$ for sufficiently small $\bar{S}$ and $V_{\max}$ are similar to the results for model-based MPC in Section 5.1, and they guarantee that the closed-loop trajectory remains close to the steady-state manifold and that the bound $\|\xi_t - \xi_k\|_2 \leq c_{\theta,1}\theta$, $k \in \mathbb{I}_{[t-N,t-1]}$, holds recursively for some $c_{\theta,1} > 0$, compare (A.130) in the proof of Theorem 5.2. Hence, in comparison to Section 5.1, $\bar{S}$ as well as $V_{\max}$ need to be potentially smaller such that the bound $\|\xi_t - \xi_k\|_2 \leq c_{\theta,1}\theta$, $k \in \mathbb{I}_{[t-N,t-1]}$ holds and, therefore, closed-loop stability can be guaranteed. Further, the proof of Theorem 5.2 shows that the choice of the regularization parameters λ_α and λ_σ leading to closed-loop practical stability depends on θ. In particular, we have $\lambda_\alpha = \bar{\lambda}_\alpha \theta^{2\beta_\alpha}$ and $\lambda_\sigma = \frac{\bar{\lambda}_\sigma}{\theta^{2\beta_\sigma}}$ for some $\bar{\lambda}_\alpha, \bar{\lambda}_\sigma, \beta_\alpha, \beta_\sigma > 0$, i.e., the regularization of $\alpha(t)$ / $\sigma(t)$ increases / decreases with the parameter θ, which quantifies the prediction error.

Remark 5.5. *(Verification of assumptions) The theoretical stability result (Theorem 5.2) uses Assumptions 5.11–5.16. If the nonlinear dynamics* (5.36) *are known, then they can be verified analogously to Section 5.1 (see Section 5.3.1 for a numerical example). If the dynamics are unknown but only input-output data are available, then performing this verification is an interesting issue for future research with preliminary results for linear systems in Section 4.2.3.*

Remark 5.6. *(Persistence of excitation) Note that Assumption 5.16 requires a minimum amount of persistence of excitation, i.e., a uniform bound $\|H_{ux,t}^{\dagger}\|_2 \leq c_{\mathrm{H}}$. On the other hand, Theorem 5.2 also requires an (initial) upper bound on $\|\xi_t - \xi_i\| \leq \theta$, $i \in \mathbb{I}_{[t-N,t-1]}$, compare* (5.58)*, which in turn limits the amount of persistence of excitation. By allowing c_{H} to depend on θ as $c_{\mathrm{H}} = \frac{\bar{c}_{\mathrm{H}}}{\theta}$, our analysis takes into account the fact that θ cannot be arbitrarily small for a fixed c_{H}, i.e., for a fixed lower bound on the amount of persistence of excitation. Assumption 5.16 is the main reason why Theorem 5.2 only provides a* practical *stability result, i.e., asymptotic stability can in general not be proven if we assume uniform closed-loop persistence of excitation. This is in contrast to the results in Section 5.1 which showed closed-loop* exponential *stability assuming that an exact model of the linearization is available, i.e., without any requirements on persistence of excitation for the closed loop.*

For the numerical examples in Section 5.3, we observe that the closed loop indeed converges very closely to the optimal reachable equilibrium ξ^{sr} *and, moreover, persistence of excitation can be ensured by stopping the data updates. Finally, the condition* $\|\xi_t - \xi_i\|_2 \leq \theta$, $i \in \mathbb{I}_{[t-N,t-1]}$, *is generally easier to satisfy for smaller values of the data length N and the prediction model in (5.51b) is less accurate for too large values of N. This is different from linear data-driven MPC as in Chapter 4, where larger values of N usually improve the closed-loop performance in case of noisy data. For nonlinear systems, on the other hand, "older" data points correspond to an "older" approximate linear model of the system which typically yields a worse approximation at the current state.*

Remark 5.7. *(Noisy data) While we assume that the data* $\{u_k, y_k\}_{k=t-N}^{t-1}$ *used in Problem 5.2 are noise-free, our results can be extended to output measurement noise. To be more precise, suppose output measurements* $\tilde{y}_k := y_k + \epsilon_k$ *with* $\|\epsilon_k\|_2 \leq \bar{\epsilon}$ *are available for some sufficiently small* $\bar{\epsilon} > 0$*. Then, under the conditions in Theorem 5.2, it can be shown analogously to the proof of Theorem 5.2 that there exist* $\beta_\epsilon \in \mathcal{K}_\infty$ *as well as (possibly different)* $0 < c_{\mathrm{V}} < 1$, $C > 0$, $\beta_\theta \in \mathcal{K}_\infty$, *such that*

$$\|\xi_t - \xi^{\mathrm{sr}}\|_2^2 \leq c_{\mathrm{V}}^i C \|\xi_N - \xi^{\mathrm{sr}}\|_2^2 + \beta_\theta(\theta) + \beta_\epsilon(\bar{\epsilon}) \tag{5.60}$$

for any $t = N + ni$, $i \in \mathbb{I}_{\geq 0}$*. Inequality (5.60) means that the closed loop converges to a region around the optimal reachable steady-state whose size increases with the parameter* θ *and the noise bound* $\bar{\epsilon}$*. This extension is possible since the perturbation of Problem 5.2 caused by the linearization error is, in fact, handled as output measurement noise affecting the input-output data, compare Lemma 5.3. Thus, small levels of noise will not affect the qualitative theoretical guarantees, but may only deteriorate the quantitative properties, e.g., increase the tracking error and decrease the region of attraction.*

Related work

We now address the relation of the proposed data-driven MPC scheme and its theoretical analysis to existing alternatives. The works [76, 112] are related to our approach since they estimate linear time-varying models of nonlinear systems from data online, but they do not provide closed-loop stability guarantees. Further, controlling nonlinear systems using linear models via Koopman operator arguments has received increasing attention in recent years [72], also in connection to the Fundamental Lemma [77, 79], but typically no closed-loop guarantees can be given. Finally, data-driven control methods based on machine learning, cf. [56, 130], have been successfully applied but, also, they often do not provide closed-loop guarantees.

An obvious alternative to our results in the present section is provided by sequential adaptive system identification (e.g., online LTI system identification or recursive least squares estimation [85]) and model-based MPC. As we show in Section 5.3.2 with a numerical example, such an identification-based approach indeed has comparable performance to the proposed data-driven one. However, to the best of our knowledge, the existing literature does not contain any results on closed-loop stability based on linearization arguments under similar assumptions as we consider for either identification-based or data-driven MPC. Alternative system identification approaches for nonlinear systems combine Lipschitz continuity-like properties with set membership estimation [21, 105], possibly leading to increasingly complex models, or they require appropriately chosen basis functions, see, e.g., [128] or approaches from nonlinear adaptive MPC [53]. We note that set membership techniques can also be employed for direct nonlinear data-driven control [131]. In contrast to the above approaches, the proposed data-driven MPC is simple, direct, applicable to a broad class of nonlinear systems, and it admits stability guarantees, thereby indicating potential advantages over the classical identification-based approaches. This is possible since we implicitly encourage the closed-loop trajectory to remain in vicinity of the steady-state manifold, where our data-driven prediction model is a good approximation of the underlying nonlinear dynamics. Moreover, the presented results are also related to nonlinear offset-free MPC [109], which deals with setpoint tracking based on an uncertain model, whereas our approach achieves asymptotic convergence to the setpoint using only input-output data.

Finally, we discuss the relation of our results to existing applications and extensions of the Fundamental Lemma (Theorem 2.1) to nonlinear systems. The recent literature contains extensions of the Fundamental Lemma to Hammerstein and Wiener systems [JB2], second-order Volterra systems [122], linear parameter-varying systems [142], flat systems [JB1], nonlinear autoregressive exogenous systems [107], or bilinear systems [156]. Further results on nonlinear data-driven control inspired by the Fundamental Lemma include explicit analysis and controller design methods for bilinear systems [16], polynomial systems [54, 96, 98], systems with rational or more general non-polynomial dynamics [JB28], systems with nonlinear uncertainties in the loop [JB14, 89, 136], or LTV systems [110]. We also mention [121], which employs a linear data-driven controller based on online data updates to control unknown switched systems. However, most of these works assume that the system is linearly parametrized in known basis functions, which restricts their practical applicability. The approach presented in this section takes an entirely different route, resorting to local linear approximations of the nonlinear dynamics by updating the data used in (an affine version of) the Fundamental Lemma online. This enables us to prove closed-loop stability guarantees based on a very simple implementation and without any restrictive assumptions on the availability of basis functions. Further, each of the approaches listed above is tailored to a specific class of nonlinear systems, mainly due to a choice of basis functions, and no guarantees are provided for a receding-horizon implementation of data-driven optimal control. In contrast, our results in this section provide a generic approach to controlling a broad class of nonlinear systems based only on input-output data. The main drawback of online data updates as in the proposed approach is the requirement on closed-loop persistence of excitation (Assumption 5.16), cf. the discussion below Assumptions 5.16 and the numerical and experimental results in Section 5.3.

Summary

In this section, we presented a linear data-driven MPC scheme to control unknown nonlinear systems based only on input-output data. In contrast to using the Jacobians for prediction as in Section 5.1, we employed (the affine version of) the Fundamental Lemma with online data updates, which posed additional technical challenges. Under suitable assumptions on the system and closed-loop persistence of excitation, we proved that our data-driven MPC scheme practically exponentially stabilizes the optimal reachable equilibrium for the given setpoint. As in Section 4.3, the analysis relied on a separation argument, combining continuity of data-driven MPC (Proposition 5.4) with the model-based analysis in Section 5.1. The implementation of the proposed approach is simple, only requiring the solution of strictly convex QPs, but admits strong theoretical guarantees under reasonable assumptions such as smooth system dynamics, controllability, observability, as well as compactness of the steady-state manifold. In Section 5.3, we illustrate the practical applicability of the presented data-driven MPC scheme for challenging nonlinear systems in both simulation and a real-world experiment.

5.3 Application: simulation and experiment

In this section, we apply the MPC schemes developed in Sections 5.1 and 5.2 to two nonlinear systems. First, in Section 5.3.1, we apply the model-based MPC from Section 5.1 to the well-known CSTR from [103] in simulation. Next, in Section 5.3.2, we illustrate that the data-driven MPC from Section 5.2 can be used to control the same CSTR, using only input-output data and no model knowledge. Finally, the data-driven MPC from Section 5.2 is applied to a nonlinear four-tank system in simulation (Section 5.3.3) and a real-world experiment (Section 5.3.4). Throughout this section, we put a particular focus on comparing the proposed MPC approaches to alternative model-based or data-driven MPC schemes and studying the influence of design parameters on the closed-loop performance.

This section is based on and taken in parts literally from [JB8][6], [JB10][7], and [JB11][8].

5.3.1 Model-based MPC of a continuous-stirred tank reactor

We apply the model-based MPC scheme proposed in Section 5.1 (Algorithm 5.1) to the CSTR from [103] with the nonlinear system dynamics

$$f(x,u) = \begin{bmatrix} x_1 + \frac{T_s}{\theta}(1-x_1) - T_s\bar{k}x_1 e^{-\frac{M}{x_2}} \\ x_2 + \frac{T_s}{\theta}(x_f - x_2) + T_s\bar{k}x_1 e^{-\frac{M}{x_2}} - T_s\alpha u(x_2 - x_c) \end{bmatrix}.$$

The states x_1 and x_2 are the temperature and the concentration, respectively, and the control input u is the coolant flow rate. These dynamics are obtained from the continuous-time dynamics in [103] via a simple Euler discretization with sampling

[6] J. Berberich, J. Köhler, M. A. Müller, and F. Allgöwer. "Data-driven model predictive control: closed-loop guarantees and experimental results." In: *at-Automatisierungstechnik*, 69.7 (2021), pp. 608–618.

[7] J. Berberich, J. Köhler, M. A. Müller, and F. Allgöwer. "Linear tracking MPC for nonlinear systems part I: the model-based case." In: *IEEE Trans. Automat. Control* (2022). doi: 10.1109/TAC.2022.3166872. ©2021 IEEE

[8] J. Berberich, J. Köhler, M. A. Müller, and F. Allgöwer. "Linear tracking MPC for nonlinear systems part II: the data-driven case." In: *IEEE Trans. Automat. Control* (2022). doi: 10.1109/TAC.2022.3166851. ©2021 IEEE

time $T_s = 0.2$. The other parameters appearing in the vector field are $\theta = 20$, $\bar{k} = 300$, $M = 5$, $x_f = 0.3947$, $x_c = 0.3816$, $\alpha = 0.117$. Our control goal is tracking of the output setpoint $y^r = 0.6519$ for the concentration, i.e., $h(x,u) = x_2$, while satisfying the input constraints $u_t \in \mathbb{U} = [0.1, 2]$ for $t \in \mathbb{I}_{\geq 0}$. In order to set up the MPC, we consider the cost matrices $Q = I$, $R = 0.05$, $S = 100$, the prediction horizon $L = 40$, and the input equilibrium constraints $\mathbb{U}^s = [0.11, 1.99]$. Since the dynamics are not of the control-affine form (5.1), we implement the MPC scheme with an incremental input formulation $\Delta u_k := u_{k+1} - u_k$ and include an additional penalty $\|\Delta u_k\|_2^2$ in the cost.

Satisfaction of assumptions

We first investigate whether our assumptions are met by the CSTR. For this verification, we only consider linearization points in the relevant operating range, i.e., $\tilde{x} \in (0,1]^2$. It is simple to verify that the considered system satisfies Assumption 5.2 (smoothness). Assumption 5.3 holds in $(0,1]^2$ except for a neighborhood of $x_2 = x_c$. Similarly, the linearized dynamics are controllable (Assumption 5.4) on $(0,1]^2$ except in a neighborhood of $x_2 = x_c$ or $x_2 = 0$. While Assumption 5.5 (non-singular dynamics) does not hold due to the integrator dynamics $u_{k+1} = u_k + \Delta u_k$, our theoretical results still apply since Assumption 5.5 holds for the original system (without Δu) and thus, for any given $(u^s, \Delta u^s) = (u^s, 0)$ there still exists a unique steady-state x^s. Further, Assumption 5.6 (compact steady-state manifold of the linearization) clearly holds on the set $\tilde{x} \in (0,1]^2$ due to Assumption 5.5. Finally, Assumption 5.8 holds since $m = p$ and the setpoint y^r is reachable, i.e., Assumption 5.9 holds, compare Section 5.1.4.

Results

Figure 5.3 shows the closed-loop trajectory under Algorithm 5.1 when starting at the initial state $x_0 = \begin{bmatrix} 0.9492 & 0.43 \end{bmatrix}^\top$. During the full closed-loop operation, the trajectory remains close to $\mathcal{Z}^s$ such that the prediction error induced by the linearization is small and y_t asymptotically converges to y^r. For comparison, we also apply the following MPC schemes, each with terminal equality constraints, online optimization of an artificial equilibrium, an incremental input penalty, and the same design parameters as above:

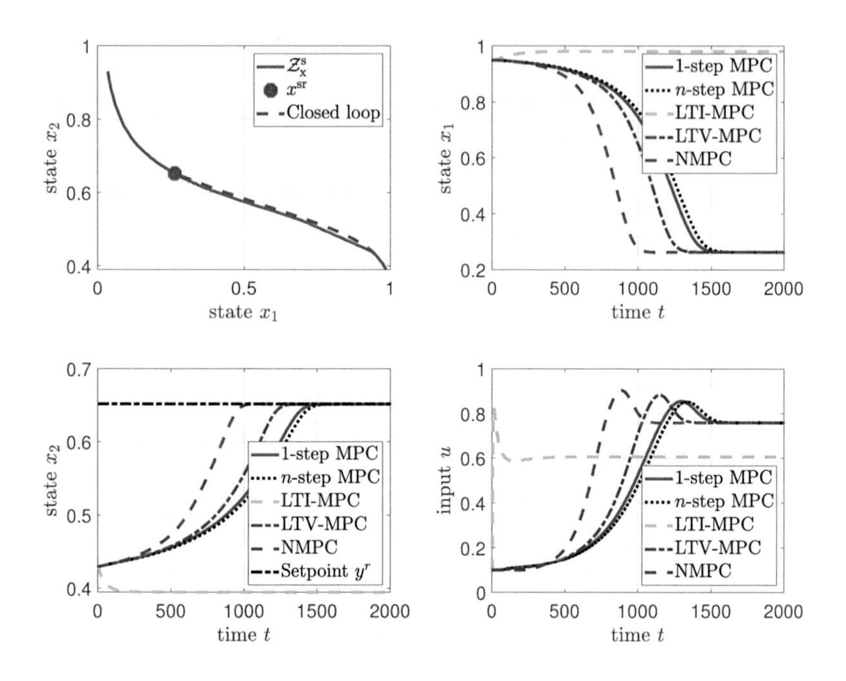

Figure 5.3. Closed-loop trajectory resulting from the application of the model-based MPC scheme (Algorithm 5.1) to the CSTR. Left upper plot: state trajectory in phase plane along with the steady-state manifold $\mathcal{Z}_x^s$ and the optimal reachable steady-state x^{sr}. Remaining plots: input-state trajectory in time domain for linearization-based 1-step MPC (solid), linearization-based n-step MPC (dotted), LTI-MPC (dashed), LTV-MPC (dash-dotted), and nonlinear MPC (dashed). ©2021 IEEE

	Setup QP	Optimization	Sum
Nonlinear MPC	–	21.6	21.6
LTV-MPC	2.2	6.8	9
Proposed 1- or n-step MPC	0.6	6.8	7.4

Table 5.1. Average computation times in milliseconds of the MPC schemes in the numerical example in Section 5.3.1.

1. the proposed MPC scheme in a one-step fashion (i.e., Algorithm 5.1 with $n = 1$),
2. a one-step MPC scheme using an LTI prediction model obtained by linearizing the nonlinear dynamics (5.1) at x^{sr} (called "LTI-MPC"),
3. a one-step MPC scheme using an LTV prediction model obtained by linearizing the nonlinear dynamics (5.1) at time t along the candidate solution $\bar{x}_1^*(t-1), \bar{x}_2^*(t-1), \ldots, \bar{x}_N^*(t-1)$ (called "LTV-MPC"), analogously to [23] and comparable to the real-time iteration scheme [35], and
4. the nonlinear tracking MPC scheme from [69].

All closed-loop state- and input-trajectories can be seen in Figure 5.3. First, note that, except for the LTI-MPC, all MPC schemes achieve asymptotic tracking of the desired setpoint. The nonlinear tracking MPC from [69] performs better than the LTV-MPC, which in turn outperforms the proposed n-step MPC scheme (Algorithm 5.1) as well as the corresponding 1-step MPC scheme. Table 5.1 lists the times required for setting up the QPs, including the computation of the Jacobians, and solving the optimization problems arising in the considered MPC schemes (using 'quadprog' for the QPs and CasADi [9] with solver 'IPOPT' for the nonlinear optimization problem in [69]). Note that the LTV-MPC has slightly larger computation times since, at each time step, L linearized dynamics need to be computed, whereas the proposed MPC scheme only requires the linearization at x_t. While the LTV-MPC provides a good trade-off between computational complexity and closed-loop performance, theoretical results in the literature require either an additional bounding of the linearization error [23] or sufficiently many iterations [80, 157]. Without online optimization of an artificial setpoint, all MPC schemes considered above are initially infeasible. Furthermore, note that the performance of the 1-step MPC scheme is superior if compared to that of the 2-step MPC scheme (note that $n = 2$) since the model is updated more frequently (twice as often) for the 1-step scheme and hence, the influence of the prediction error is smaller. Finally, the LTI-MPC based on the linearization at x^{sr} fails to track the desired setpoint. To summarize, in application to a CSTR, the presented tracking MPC scheme using a linearized prediction model leads to a closed-loop performance which is comparable to that of nonlinear tracking MPC while being significantly more computationally efficient.

5.3.2 Data-driven MPC of a continuous-stirred tank reactor

Next, we apply the data-driven MPC approach from Section 5.2 (Algorithm 5.2) to the nonlinear continuous stirred tank reactor from [103]. We consider the same discretized system as in Section 5.3.1 (see above for a detailed description of the system dynamics, parameters, and satisfaction of Assumption 5.11 as well as the counterpart of Assumption 5.13). In contrast to Section 5.3.1, the system model is unknown and we only have access to input-output data. For the following simulation study, we assume that the output can be measured exactly without noise since this allows us to better investigate and illustrate the relation to alternative schemes and the interplay between the nonlinear system dynamics and suitable design parameters of Problem 5.2 leading to a good closed-loop operation. As we discuss in Remark 5.7, our data-driven MPC scheme retains its qualitative theoretical guarantees (Theorem 5.2) in the presence of output measurement noise. Moreover, in Section 5.3.4, we show that the proposed MPC is also applicable in a real-world experiment with noisy measurements.

To implement the MPC, we choose the parameters

$$Q = 1, R = 5 \cdot 10^{-2}, S = 10, \lambda_\alpha = 3 \cdot 10^{-6}, \lambda_\sigma = 10^7,$$

and the prediction horizon $L = 40$. Further, the constraints are $\mathbb{U} = [0.1, 2]$, $\mathbb{U}^{\mathrm{s}} = [0.11, 1.99]$. In comparison to Section 5.3.1, where we chose $S = 100$, we use the smaller choice $S = 10$ for the data-driven MPC scheme to ensure that the closed loop does not change too rapidly, i.e., the bound $\|\xi_t - \xi_i\|_2 \leq c_{\theta,1}\theta$, $i \in \mathbb{I}_{[t-N,t-1]}$, holds in closed loop for a sufficiently small θ. This is required since, in contrast to Problem 5.1, Problem 5.2 only contains an *approximation* of the linearized dynamics whose accuracy deteriorates for larger values of θ, compare Lemma 5.3.

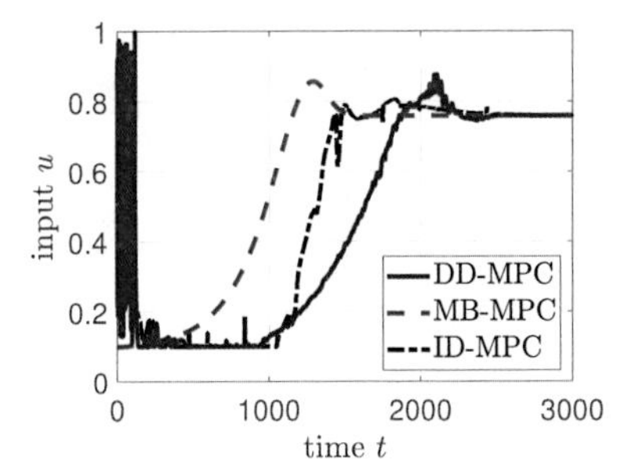

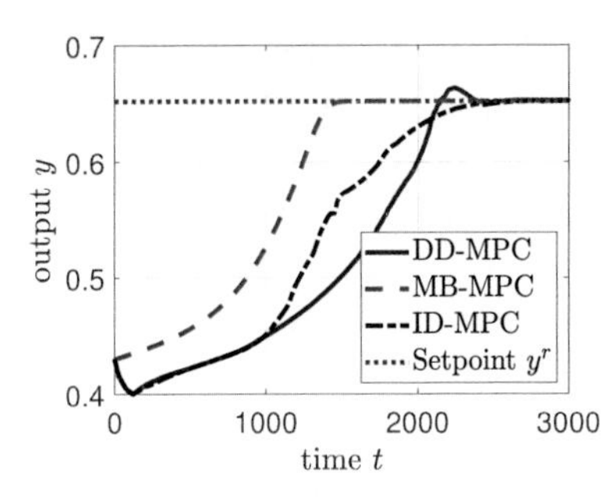

Figure 5.4. Closed-loop trajectory resulting from the application of the data-driven MPC scheme (Algorithm 5.2) to the CSTR (DD-MPC, solid) along with closed-loop trajectories for model-based MPC (Algorithm 5.1, MB-MPC, dashed) and identification-based MPC (ID-MPC, dash-dotted). ©2021 IEEE

In order to ensure that the dynamics are control-affine as in (5.36), we introduce an incremental input $\Delta u_k = u_{k+1} - u_k$ playing the role of the input u_k in Section 5.2, similar to the model-based MPC in Section 5.3.1. To guarantee satisfaction of the input constraints $u_t \in \mathbb{U}$, we consider an extended output vector $\hat{y}_t = \begin{bmatrix} y_t \\ u_t \end{bmatrix}$ and add output constraints of the form $\hat{y}_k(t) \in \mathbb{R} \times \mathbb{U}$, $k \in \mathbb{I}_{[0,L]}$, as well as $\hat{y}^s(t) \in \mathbb{R} \times \mathbb{U}^s$ in Problem 5.2. With these modifications, the theoretical guarantees of our MPC scheme remain true and the output constraints are satisfied in closed loop since the available data provide an exact "prediction model" of $\Delta u \mapsto u$.

Further, we replace $\alpha^{\mathrm{sr}}_{\mathrm{Lin}}(x_t)$ in Problem 5.2 by the "approximation" $\alpha^*(t-n)$ to regularize w.r.t. the previously optimal solution and thus to encourage stationary behavior. The data length is chosen as $N = 120$ and we generate initial data samples for $t \in \mathbb{I}_{[0,N-1]}$ by sampling the input uniformly from $u_t \in [0.1, 1]$. Finally, to ensure that the data used for prediction are persistently exciting, we stop updating the data in (5.51b) as soon as the tracking cost is sufficiently small, i.e.,

$$\sum_{k=-n}^{L} \|\Delta \bar{u}_k^*(t) - \overbrace{\Delta u^{s*}(t)}^{=0}\|_2^2 + \|\bar{u}_k^*(t) - u^{s*}(t)\|_R^2 + \|\bar{y}_k^*(t) - y^{s*}(t)\|_Q^2 \leq 10^{-5}.$$

Results

We now apply the multi-step MPC scheme in Algorithm 5.2 with the above parameters and modifications. First, we note that updating the data in Problem 5.2 online is a crucial ingredient of our approach. In particular, a data-driven MPC as in Algorithm 5.2 but using the initial data $\{u_k, y_k\}_{k=0}^{N-1}$ for prediction at all times leads to a relatively large tracking error, i.e., the output converges to $1.14 \neq y^{\mathrm{r}} = 0.6519$. The closed-loop input-output trajectory under our MPC approach is depicted in Figure 5.4, along with the trajectory resulting from the multi-step MPC based on a linearized model from Section 5.1 (using the same parameters as above, except for $S = 100$ instead of $S = 10$). In the data-driven MPC, the input is more unsteady which can be explained by the combination of the less accurate prediction model and terminal equality constraints. Moreover, since the matrix S is chosen smaller in comparison to the model-based MPC ($S = 10$ instead of $S = 100$), the convergence towards the setpoint y^{r} is slower. Solving Problem 5.2 online takes on average 13 milliseconds, i.e., longer than the model-based MPC which takes 7.4 milliseconds.

For comparison, we apply a model-based MPC scheme based on online least-squares identification[9] of an affine input-output model

$$y_{k+2} = a_1 y_{k+1} + a_0 y_k + b_2 u_{k+2} + b_1 u_{k+1} + b_0 u_k + c,$$

using the last N data points $\{u_k, y_k\}_{k=t-N}^{t-1}$ for identification at time t. The MPC implementation is analogous to that in Section 5.1 with an extended state-space model (2.5), and the parameters Q, R, horizon length L, and data length N are as above, including the incremental input penalty $\|\Delta u_k\|_2^2$. The matrix S is chosen as $S = 30$, i.e., larger than for the data-driven MPC, since the slack variable in Problem 5.2 relaxes the terminal equality constraint and, thereby, speeds up the closed-loop convergence. While this identification-based MPC scheme exhibits comparable performance to the proposed data-driven MPC scheme (cf. Figure 5.4), the existing literature does not contain any theoretical results on closed-loop stability under similar assumptions as in Section 5.2.

Regarding the condition (5.58) in Theorem 5.2, we note that the quantity $\bar{\theta}(t) := \max_{i \in \mathbb{I}_{[t-N,t-1]}} \|\xi_t - \xi_i\|_2$ takes its maximum at $t = 141$, i.e., directly after the initial

[9] For the identification, we implemented an incremental parameter update based on a regularization w.r.t. the previous parameter estimate. In particular, a model-based MPC using a simple least-squares parameter estimation does not suffice to successfully steer the system to the setpoint.

excitation phase. This yields a uniform bound on the distance between closed-loop state trajectories, i.e., $\|\xi_t - \xi_i\|_2 \leq c_{\theta,1}\theta$ for all $t \in \mathbb{I}_{[N,3000]}$, $i \in \mathbb{I}_{[t-N,t-1]}$, where $\theta := \max_t \bar{\theta}(t)$, compare (A.130) in the proof of Theorem 5.2. Let us now investigate whether this uniform upper bound conflicts with the requirements on closed-loop persistence of excitation in Assumption 5.16. For the present example, we observe that the norm of $H_{ux,t}^{\dagger}$ increases in closed loop under the data-driven MPC when the trajectory approaches the setpoint. However, the product $\|H_{ux,t}^{\dagger}\|_2\bar{\theta}(t)$ is uniformly bounded even when x_t is close to x^{sr} and, hence, there exists a constant $\bar{c}_{\mathrm{H}} > 0$ such that $\|H_{ux,t}^{\dagger}\|_2 \leq \frac{\bar{c}_{\mathrm{H}}}{\bar{\theta}(t)}$ holds, compare the conditions in Theorem 5.2. Thus, while $\|H_{ux,t}^{\dagger}\|_2$ increases for smaller bounds $\bar{\theta}(t)$, the order of increase is not larger than $\frac{1}{\bar{\theta}(t)}$ such that the persistence of excitation assumptions required for Theorem 5.2 are fulfilled.

Influence of design parameters

In Figure 5.5, we show the closed-loop cost $\sum_{t=0}^{3000}\|y_t - y^{\mathrm{r}}\|_2^2$ when varying the parameters S and λ_α, keeping all other parameters as above. The displayed cost is normalized, i.e., it is multiplied by $\frac{1}{J_{\mathrm{mdl}}}$, where J_{mdl} denotes the closed-loop cost $\sum_{t=0}^{3000}\|y_t - y^{\mathrm{r}}\|_2^2$ of the model-based MPC from Section 5.1. First, the closed-loop cost of the data-driven MPC is generally worse than that of the model-based MPC due to the initial excitation phase as well as the smaller choice of S and the resulting slower convergence. Further, recall that smaller values of S decrease the bound $\bar{\theta}(t)$ above and thus improve the prediction accuracy. If S is chosen too large, the closed-loop trajectory moves too rapidly such that the predictions are inaccurate and the performance is more sensitive w.r.t. variations of λ_α. On the other hand, too small values of S lead to a large cost since the convergence is slower and the term $\lambda_\alpha\|\alpha(t) - \alpha_{\mathrm{Lin}}^{\mathrm{sr}}(\mathcal{D}_t)\|_2^2$ dominates in the optimization of Problem 5.2, thus increasing the stationary tracking error (recall that we consider $\alpha_{\mathrm{Lin}}^{\mathrm{sr}}(\mathcal{D}_t) \approx \alpha^*(t-n)$). However, there exists a corridor $S \in [7, 30]$ for which the closed-loop performance is acceptable over a wide range of λ_α.

Finally, we note that the proof of Theorem 5.2 relies on multiple conservative estimates and should therefore be interpreted as a qualitative stability result, compare the discussion below Theorem 5.2. Correspondingly, the magnitude of the (initial) excitation bound θ and the choices of the tuning variables S, λ_α, λ_σ considered in the example are not necessarily chosen small (for θ, S, λ_α) or large

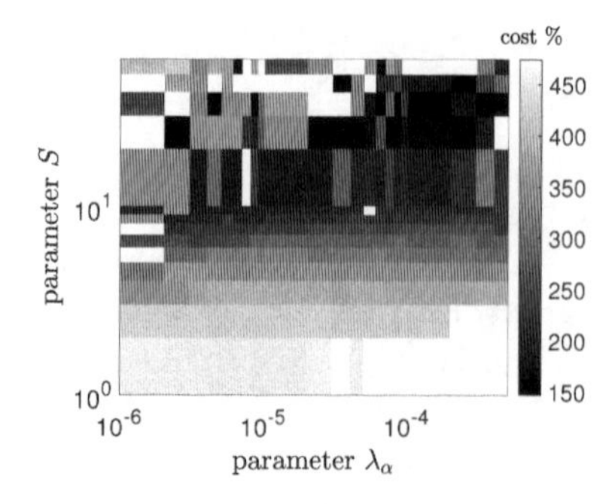

Figure 5.5. Normalized closed-loop cost depending on the parameters S and λ_α when applying the data-driven MPC scheme (Algorithm 5.2) to the CSTR. ©2021 IEEE

(for λ_σ) enough to satisfy the corresponding bounds in the proof.

5.3.3 Data-driven MPC of a four-tank system (simulation)

In this section, we apply the data-driven MPC scheme for nonlinear systems developed in Section 5.2 to a simulation model of the four-tank system originally considered in [116]. In contrast to the numerical case studies in Chapters 3 and 4, we now consider the *nonlinear* system dynamics which are given in continuous time by

$$\begin{aligned}
\dot{x}_1 &= -\frac{a_1}{A_1}\sqrt{2gx_1} + \frac{a_3}{A_1}\sqrt{2gx_3} + \frac{\gamma_1}{A_1}u_1, \\
\dot{x}_2 &= -\frac{a_2}{A_2}\sqrt{2gx_2} + \frac{a_4}{A_2}\sqrt{2gx_4} + \frac{\gamma_2}{A_2}u_2, \\
\dot{x}_3 &= -\frac{a_3}{A_3}\sqrt{2gx_3} + \frac{1-\gamma_2}{A_3}u_2, \\
\dot{x}_4 &= -\frac{a_4}{A_4}\sqrt{2gx_4} + \frac{1-\gamma_1}{A_4}u_1,
\end{aligned} \tag{5.61}$$

where x_i is the water level of tank i in cm, u_i the flow rate of pump i in cm^3/s, and the other terms are system parameters, whose values are taken from [116] and summarized in Table 5.2. The output of the system is given by $y = \begin{bmatrix} x_1 \\ x_2 \end{bmatrix}$. As in

$A_1 = A_2$:	$A_3 = A_4$:	a_1:	a_2:	$a_3 = a_4$:	$\gamma_1 = \gamma_2$:	g:
50.27cm^2	28.27cm^2	0.233cm^2	0.242cm^2	0.127cm^2	0.4	$981\text{cm}^2/\text{s}$

Table 5.2. Parameter values of the simulation model (5.61).

Section 5.3.2, we assume that exact (noise-free) output measurements are available.

We now apply the nonlinear data-driven MPC scheme from Section 5.2 (compare Algorithm 5.2) in a one-step fashion to the discrete-time nonlinear system obtained via Euler discretization with sampling time $T_s = 1.5$ seconds of (5.61). Our goal is to track the setpoint $y^{\mathrm{r}} = \begin{bmatrix} 15 \\ 15 \end{bmatrix}$ while satisfying the input constraints $u_t \in \mathbb{U} = [0, 60]^2$. To this end, we apply an input sequence sampled uniformly from[10] $u_k \in [20, 30]^2$ over the first N time steps to collect initial data, where the system is initialized at $x_0 = 0$. Thereafter, for each $t \in \mathbb{I}_{\geq N}$, we solve Problem 5.2, apply the first component of the optimal predicted input, and update the data $\{u_k, y_k\}_{k=t-N}^{t-1}$ used for prediction in (5.51b) in the next time step based on the current measurements. We use the parameters

$$\begin{aligned} &N = 150,\ L = 35,\ Q = I,\ R = 2I, \\ &S = 20I,\ \lambda_\alpha = 5 \cdot 10^{-5},\ \lambda_\sigma = 2 \cdot 10^5, \end{aligned} \tag{5.62}$$

and we choose the equilibrium input constraints as $\mathbb{U}^s = [0.6, 59.4]^2$. Further, the value of n used in Problem 5.2 (i.e., our estimate of the system order) is chosen as 3, which is an upper bound on the lag $\underline{l} = 2$ of (the linearization of) the above system. We note that comparable performance is achieved for estimates of the system order ranging from 2 to 4, see Table 5.3 and the discussion below. We choose the "approximation" $\alpha^{\mathrm{sr}}_{\mathrm{Lin}}(\mathcal{D}_t) \approx 0$ when solving Problem 5.2, for which we retain our theoretical guarantees, compare Remark 5.4.

Results

The closed-loop input and output trajectories under the MPC scheme with the above parameters can be seen in Fig. 5.6. After the initial excitation phase $t \in \mathbb{I}_{[0,N-1]}$,

[10] This interval is chosen sufficiently large and does not contain zero due to the fact that too small inputs imply that the outputs are also small and thus lie in a region where the sensors of the experimental setup in Section 5.3.4 are less accurate.

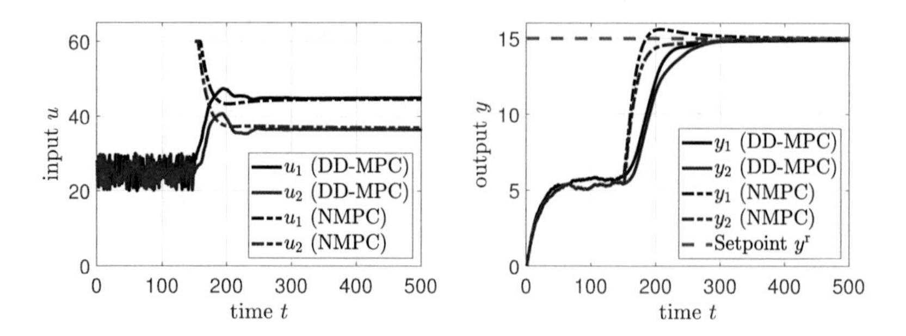

Figure 5.6. Closed-loop trajectory resulting from the application of the data-driven MPC scheme (DD-MPC, Algorithm 5.2) to the four-tank system in simulation (solid) along with closed-loop trajectory for model-based nonlinear MPC from [69] (NMPC, dash-dotted).

the MPC successfully steers the output to the desired target setpoint. First, we note that updating the data used for prediction in (5.51b) is a crucial ingredient of our MPC approach for nonlinear systems. In particular, if we do not update the data online but only use the first N input-output measurements $\{u_k, y_k\}_{k=0}^{N-1}$ for prediction, then the closed loop does not converge to the desired output y^r and instead yields a significant permanent offset due to the model mismatch. For comparison, Fig. 5.6 also shows the closed-loop trajectory starting at time $t = N$ resulting from a nonlinear tracking MPC scheme with full model knowledge and state measurements from [69], where the parameters are as above except for $S = 200I$ and $R = 0.1I$. The two MPC schemes exhibit similar convergence speed although the data-driven MPC uses "less aggressive" parameters due to the slack variable $\sigma(t)$ which implicitly relaxes the terminal equality constraint (5.51d).

Influence of design parameters

As observed before in this thesis as well as the related literature, e.g., [39], the choice of the regularization parameter λ_α has an essential impact on the closed-loop performance of data-driven MPC. In the following, we investigate in more detail how the specific choice of λ_α influences the closed-loop performance. To this end,

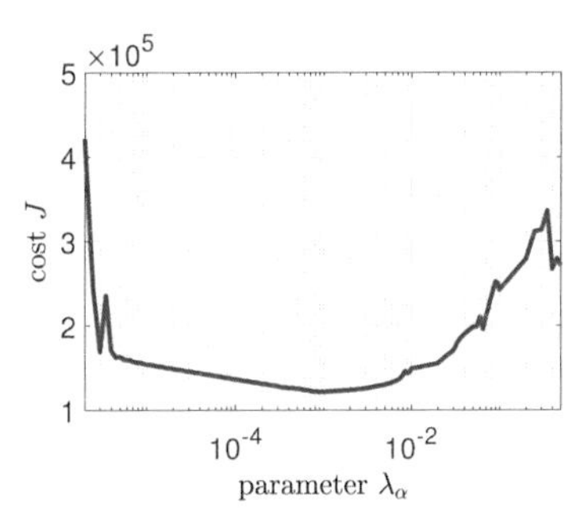

Figure 5.7. Closed-loop cost depending on the parameter λ_α when applying the data-driven MPC scheme (Algorithm 5.2) to the four-tank system in simulation.

we perform closed-loop simulations for a range of values λ_α and, for each of these simulations, we compute the corresponding cost as the deviation of the closed-loop output from the target setpoint y^{r}, i.e., $J = \sum_{t=N}^{500} \|y_t - y^{\mathrm{r}}\|_S^2$. For comparison, we note that the parameters in (5.62) lead to a closed-loop cost of $J = 1.42 \cdot 10^5$, whereas the model-based nonlinear MPC shown in Fig. 5.6 leads to $J = 3.1 \cdot 10^4$. Fig. 5.7 shows the closed-loop cost depending on the parameter λ_α with all other parameters as in (5.62). Although the cost strongly depends on λ_α, it can be seen that a wide range of values $\lambda_\alpha \in [2 \cdot 10^{-5}, 0.01]$ leads to a good performance, i.e., $J \leq 1.5 \cdot 10^5$. If λ_α is chosen too small, then the robustness w.r.t. the nonlinearity deteriorates and the influence of numerical inaccuracies increases, which leads to a cost increase. This is in accordance with Theorem 5.2 which requires that λ_α is suitably chosen (in particular, it cannot be arbitrarily small). On the other hand, if λ_α is chosen too large then the closed-loop cost increases significantly since too small choices of the vector $\alpha(t)$ shift the input and output to which the closed loop converges towards zero, i.e., large values of λ_α increase the asymptotic tracking error. To summarize, since a wide range of values λ_α leads (approximately) to the minimum achievable cost, tuning the parameter λ_α is easy for the present example.

Next, we analyze how different choices of other design parameters influence the closed-loop cost. Table 5.3 displays ranges for various parameters for which the cost J is less than $1.5 \cdot 10^5$, when keeping all other parameters as in (5.62). The data length N needs to be sufficiently large such that the input is persistently

N:	L:	assumed system order:
$\mathbb{I}_{[130,159]}$	$\mathbb{I}_{[32,41]}$	$\mathbb{I}_{[2,4]}$
$\bar{s}$:	λ_α:	λ_σ:
$[16, 3 \cdot 10^2]$	$[2 \cdot 10^{-5}, 0.01]$	$[4 \cdot 10^2, 10^6]$

Table 5.3. Parameter ranges leading to a closed-loop cost $J \leq 1.5 \cdot 10^5$ when applying the data-driven MPC scheme (Algorithm 5.2) to the four-tank system in simulation.

exciting, but choosing it too large deteriorates the performance since then the data used for prediction in (5.51b) cover a larger region of the state-space and the implicit linearization-based "model" is a less accurate approximation of the nonlinear dynamics (5.61). This is in contrast to the results on robust data-driven MPC for linear systems in Chapter 4, where larger data lengths typically improve the closed-loop performance. Similarly, too large values for the prediction horizon L are detrimental since they imply that the predicted trajectories are further away from the initial state, where the prediction accuracy deteriorates. On the other hand, too short horizons L lead to worse robustness due to the terminal equality constraints (5.51d). The assumed system order n cannot be larger than 4 due to the dependence of the required persistence of excitation on n and since larger values of n effectively shorten the prediction horizon due to the terminal equality constraints (5.51d), which are specified over $n+1$ time steps. If N and L are increased to $N = 190$ and $L = 40$, then the closed-loop output still converges to y^{r}, e.g., for the upper bound 10 on the system order.

Further, Table 5.3 displays values of $\bar{s}$ leading to a good closed-loop performance if the matrix S is chosen as $S = \bar{s}I$. The value $\bar{s}$ cannot be arbitrarily large since it needs to be small enough such that the artificial setpoint $(u^{\mathrm{s}}(t), y^{\mathrm{s}}(t))$ and hence the predicted trajectories remain close to the initial state, where the prediction accuracy of the data-dependent model (5.51b) is acceptable (compare the discussion in Section 5.2). On the other hand, for too small values of $\bar{s}$, the asymptotic tracking error increases since the artificial steady-state is close to the initial condition and thus, the regularization of α w.r.t. zero dominates the cost of Problem 5.2. Moreover, the parameter λ_σ can be chosen in a relatively large range. To summarize, the

MPC scheme shown in Section 5.2 can successfully control the nonlinear four-tank system from [116] in simulation, and the influence of system and design parameters on the closed-loop performance confirms our theoretical findings.

5.3.4 Data-driven MPC of a four-tank system (experiment)

In the following, we apply the data-driven MPC scheme presented in Section 5.2 in an experimental setup to the four-tank system by Quanser. This system possesses qualitatively the same dynamics as (5.61), but the parameter values differ (compare [114] for details). Nevertheless, as we show in the following, the presented nonlinear data-driven MPC scheme can successfully control the system using the same design parameters as in Section 5.3.3 due to its ability to adapt to changing operating conditions, in particular by updating the data used for prediction online.

As for the simulation results in Section 5.3.3, we apply the proposed data-driven MPC scheme in a one-step fashion. We use the same sampling time $T_s = 1.5$ seconds as in Section 5.3.3 and we first apply an open-loop input sampled uniformly from $u_k \in [20, 30]^2$ in order to generate data of length $N = 150$. Thereafter, we compute the input applied to the plant via the MPC scheme based on Problem 5.2, where the design parameters are chosen exactly as in Section 5.3.3, i.e., as in (5.62). In addition to only tracking the setpoint $y^{\mathrm{r}} = \begin{bmatrix} 15 \\ 15 \end{bmatrix}$ in the time interval $t \in \mathbb{I}_{[0,600]}$, we include an online setpoint change for the time interval $t \in \mathbb{I}_{[601,1200]}$ to $y^{\mathrm{r}} = \begin{bmatrix} 11 \\ 11 \end{bmatrix}$. We note that the computation time for solving the strictly convex QP Problem 5.2 is negligible compared to the sampling time of 1.5 seconds.

Results

The resulting closed-loop input-output trajectory is displayed in Fig. 5.8. After the initial exploration phase of length N, the closed-loop output first converges towards the setpoint $\begin{bmatrix} 15 \\ 15 \end{bmatrix}$ and after time $t = 600$, the output converges towards the second setpoint $\begin{bmatrix} 11 \\ 11 \end{bmatrix}$, i.e., the MPC approximately solves our control problem. Note that the system does not exactly track the setpoints which can be explained by i) the

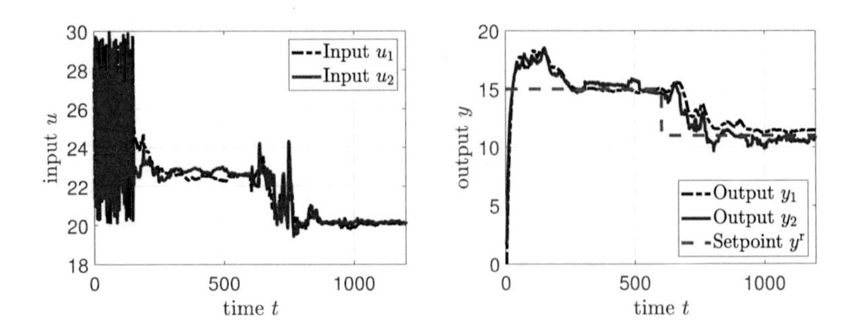

Figure 5.8. Closed-loop trajectory resulting from the application of the data-driven MPC scheme (Algorithm 5.2) to the four-tank system in an experiment.

practical stability guarantees in Theorem 5.2, i.e., convergence to a neighborhood of the setpoint, ii) measurement noise which is unavoidable in a real-world experiment and increases the neighborhood to which the closed loop converges, compare Remark 5.7, and iii) potential loss of persistence of excitation upon convergence since we did not implement any measures ensuring a persistently exciting closed loop (see the discussion below Assumption 5.16). The latter also explains the non-ideal transient behavior after initiating the setpoint change at time $t = 600$ since the past $N = 150$ input-output data points contain only little information about the system. Finally, similar to the simulation results in Section 5.3.3, the closed loop has a large steady-state tracking error if at all times only the first $N = 150$ data points are used for prediction, underpinning the importance of updating the measured data in (5.51b) online when controlling nonlinear systems.

Comparing Figures 5.6 and 5.8, we observe an important advantage of the presented MPC framework. Clearly, the two four-tank systems [116] and [114] have different parameters, e.g., the steady-state inputs leading to the output y^r differ significantly. In particular, the model (5.61) does not accurately describe the four-tank system [114], e.g., due to differing pump flow rates, differing tube diameters, manufacturing inaccuracies, aging, and since the model (5.61) is not even an *exact* representation of the physical reality for the four-tank system considered in [116]. In order to implement a (nonlinear) model-based MPC as in [116], all of the mentioned quantities need to be carefully modeled which can be a challenging and time-consuming task. On the other hand, estimating an accurate model based on an open-loop experiment is also difficult due to the nonlinear nature of (5.61) and since only input-output measurements are available, see, e.g., [21]. In contrast, the proposed MPC leads to an acceptable closed-loop performance without any modifications compared to the simulation in Section 5.3.3 due to the fact that it naturally adapts to the operating conditions. This makes our MPC framework both very simple to apply, since no modeling or nonlinear identification tasks need to be carried out, and reliable, since it allows for rigorous theoretical guarantees.

5.4 Summary and discussion

5.4.1 Summary

In this chapter, we presented a linear tracking MPC framework for nonlinear systems. First, in Section 5.1, we considered the model-based case, i.e., the proposed MPC scheme relied on the linearized dynamics at the current state of the nonlinear system. We proved that our MPC scheme exponentially stabilizes the optimal reachable equilibrium for the nonlinear system. The proof exploited the fact that, if the closed loop does not move too rapidly, then the linearization is a good approximation of the nonlinear dynamics.

In Section 5.2, we then addressed the case where only input-output data of the nonlinear system are available. In this case, the prediction model of the proposed MPC scheme relies on updating the data used in the Fundamental Lemma online at every time step, inducing a similar effect of local linear approximations as in Section 5.1. By combining the results in Section 5.1 with a novel continuity result for data-driven MPC, we proved that the data-driven MPC scheme practically exponentially stabilizes the optimal reachable equilibrium for the unknown nonlinear system. Finally, in Section 5.3, we demonstrated the practicality of the developed MPC approaches by means of nonlinear examples from the literature in simulation and in a real-world experiment.

5.4.2 Discussion

Controlling unknown nonlinear systems without explicit model knowledge is a very challenging problem, in particular if only input-output data are available. In this chapter, we proposed a framework for solving this problem by combining linear data-driven control techniques with online data updates. In addition to the provided theoretical guarantees, our results have the advantage of a simple implementation, only relying on the solution of strictly convex QPs, and not requiring any knowledge of basis functions for the unknown nonlinear dynamics. As discussed in more detail in Section 5.2, the existing literature does not contain any results on nonlinear data-driven control that provide similar guarantees under comparable assumptions.

The main limitation of the presented data-driven MPC approach lies in Assumption 5.16 (persistence of excitation), i.e., our main theoretical result (Theorem 5.2) requires that the data generated by our MPC scheme in closed loop are persistently exciting. However, without any modifications of the proposed algorithm, this need not be the case such that our guarantees may be lost. As discussed in Section 5.2 and illustrated empirically in Section 5.3, the data-driven MPC still provides good practical performance even if persistence of excitation is only ensured heuristically or not enforced at all. Nevertheless, tackling this issue by developing appropriate modifications of our MPC algorithm is a highly interesting and relevant future research direction, potentially leading to both superior theoretical guarantees and more reliable performance. To this end, we mention recent results on experiment design in the context of the Fundamental Lemma [64, 65, 135] and approaches from adaptive MPC [19, 88, 91], both of which may be useful to systematically handle closed-loop persistence of excitation in the presence of online data updates.

Finally, we note that multiple results presented in this chapter that were mainly motivated as technical intermediate steps for the stability proof of data-driven MPC (Theorem 5.2) are in fact interesting in their own right. First, the model-based MPC from Section 5.1 provides an alternative to existing linearization-based approaches for nonlinear MPC such as the real-time iteration scheme, leading to desirable closed-loop guarantees and satisfactory practical performance. Furthermore, Proposition 5.4 (continuity of data-driven MPC) suggests a novel separation-type approach for proving stability and robustness in data-driven MPC by analyzing nominal data-driven (i.e., model-based) MPC schemes and relating them to their robust counterparts. For example, the same result is employed in Section 4.3 to prove stability of robust data-driven tracking MPC with noisy data. It can also be used to derive performance bounds and suboptimality estimates (Section 4.3.4) which apply similarly for the nonlinear data-driven MPC approach in Section 5.2.

Chapter 6

Conclusions

6.1 Summary

In this thesis, we developed a framework for data-driven MPC with closed-loop guarantees on stability and robustness. Our results were structured into three different chapters, corresponding to the three problem settings we investigated:

- data-driven MPC for linear systems with noise-free data (Chapter 3),
- data-driven MPC for linear systems with noisy data (Chapter 4), and
- data-driven MPC for nonlinear systems (Chapter 5).

Let us now summarize our findings for each of these chapters.

Data-driven MPC for linear systems with noise-free data

If noise-free, persistently exciting input-output data of an LTI system are available, then Willems' Fundamental Lemma (Theorem 2.1) allows us to parametrize all trajectories of the system without any explicit model knowledge. Based on this fact, we designed and analyzed different data-driven MPC schemes in Chapter 3 with a focus on closed-loop guarantees. Inspired by model-based MPC, we started with a simple formulation based on terminal equality constraints (Section 3.1) for which we proved closed-loop stability. Since terminal equality constraints have various drawbacks such as poor robustness and a potentially small region of attraction, we then considered more general terminal ingredients, i.e., a terminal region and a terminal cost (Section 3.2). We not only proved desirable closed-loop properties for the resulting MPC scheme but also developed a procedure to design terminal ingredients based on input-output data. In Section 3.3, we then addressed a tracking MPC formulation employing an artificial setpoint which is optimized online and provided closed-loop stability guarantees when applying the resulting algorithm to systems with affine dynamics. In comparison to the previous approaches, this tracking formulation has several advantages. It leads to a larger region of attraction than the one with standard terminal equality constraints, it allows for online setpoint changes, and it does not require the setpoint to be a feasible equilibrium for the unknown system. Finally, we validated and compared the developed MPC schemes with a numerical example (Section 3.4).

Data-driven MPC for linear systems with noisy data

In practical applications, it is rarely the case that noise-free data are available. Therefore, in Chapter 4, we investigated data-driven MPC for linear systems based on one noisy input-output trajectory.

First, in Section 4.1, we proposed a robust modification of the nominal data-driven MPC scheme with terminal equality constraints from Section 3.1. This modification contained a slack variable to relax the conditions of the Fundamental Lemma as well as regularization terms in the cost. We proved that the resulting robust data-driven MPC scheme practically exponentially stabilizes the closed loop, i.e., the system converges to a neighborhood of the setpoint whose size depends on the noise level. Further, our analysis revealed various insights into the influence of the data quality on the resulting closed-loop performance, e.g., the tracking error decreases and the region of attraction increases if the data length increases, the norm of the input generating the data increases, or the noise level decreases.

While we only considered input constraints in Section 4.1, we extended our robust data-driven MPC scheme by adding a constraint tightening to ensure output constraint satisfaction in Section 4.2. We proved that this constraint tightening is recursively feasible and guarantees that the output satisfies the constraints in closed loop. Further, we developed approaches to estimate system constants related to controllability and observability, which are required to construct the constraint tightening.

In Section 4.3, we proposed a robust data-driven tracking MPC for affine systems which combines the schemes from Sections 3.3 and 4.1 regarding nominal tracking and robustness w.r.t. noise, respectively. We proved that the MPC scheme practically exponentially stabilizes the optimal reachable equilibrium in closed loop based on a novel separation argument of nominal and robust data-driven MPC.

Finally, in Section 4.4, we successfully applied the developed MPC schemes to numerical examples and investigated the influence of different parameters on the closed-loop performance.

Data-driven MPC for nonlinear systems

While Chapters 3 and 4 only addressed linear systems, we presented a data-driven MPC approach for nonlinear systems based on local linear approximations in Chapter 5.

First, we developed a model-based tracking MPC scheme which uses the linearized dynamics at the current state for prediction (Section 5.1). The main idea was that, if the closed loop moves slowly, then these linearized dynamics provide an accurate approximation of the nonlinear dynamics such that the predictions are accurate and closed-loop guarantees can be given. We proved that the MPC scheme exponentially stabilizes the optimal reachable steady-state for the nonlinear system if a design parameter in the cost is sufficiently small and the initial state is close to the steady-state manifold.

In Section 5.2, we proposed a data-driven MPC scheme to control unknown nonlinear systems based on input-output data. In contrast to all previous data-driven MPC approaches, this MPC scheme contains online data updates, i.e., at every time step the last N input-output measurements are used to predict trajectories via the Fundamental Lemma. If the distance between these data points is not too large, i.e., the closed loop does not move too rapidly, then the Fundamental Lemma provides a good approximation of the linearized dynamics at the current state, thus allowing us to employ results from our model-based analysis in Section 5.1. Based on this idea, we proved that the data-driven MPC scheme practically exponentially stabilizes the optimal reachable equilibrium for the unknown nonlinear system in closed loop. In addition to the conditions in Section 5.1, the stability result in Section 5.2 requires that the data generated in closed loop by the proposed MPC are persistently exciting, which is challenging to guarantee theoretically but can be ensured in practice based on various heuristics.

Finally, we applied the developed MPC schemes to two nonlinear examples from the literature in simulation and in a real-world experiment (Section 5.3).

6.2 Discussion

In this thesis, we presented a flexible and general framework for data-driven MPC with closed-loop guarantees on stability and robustness. We considered different problem formulations, including linear and nonlinear systems as well as noise-free and noisy data, and developed suitable MPC schemes to tackle these problems. The proposed data-driven MPC schemes all have in common that they employ (variations of) the Fundamental Lemma for prediction instead of a state-space model. This provides multiple inherent advantages, e.g., they require no explicit model knowledge but only one input-output trajectory and a bound on the lag or the system order, they are simple to apply and do not require an intermediate identification step, they are output-feedback MPC schemes, and the underlying optimization problems can be solved efficiently.

From the perspective of standard (model-based) MPC, we employed three different, well-established approaches to ensure stability:

- *terminal equality constraints w.r.t. the setpoint* for linear systems with noise-free (Section 3.1) and noisy (Sections 4.1 and 4.2) data,
- *general terminal ingredients* for linear systems with noise-free data (Section 3.2), which can be easily extended to noisy data, compare Section 4.5.2, and
- *tracking MPC with artificial setpoint and terminal equality constraints* for linear systems with noise-free (Section 3.3) and noisy (Section 4.3) data as well as nonlinear systems (Chapter 5).

Thus, the contribution of this thesis can also be interpreted as translating existing model-based MPC approaches and their theoretical guarantees into data-driven MPC based on the Fundamental Lemma. When comparing the different data-driven MPC approaches in the respective Chapters 3 and 4, we can draw the same conclusions as in model-based MPC: While being very simple to implement, terminal equality constraints as in Sections 3.1, 4.1, and 4.2 typically possess the weakest closed-loop properties w.r.t. robustness or the size of the region of attraction. Considering general terminal ingredients as in Section 3.2 can significantly improve the robustness and increase the region of attraction, possibly leading to global stability guarantees, at the price of an additional offline design step. A tracking formulation as in Sections 3.3 and 4.3 also leads to a larger region of attraction than simple terminal equality constraints, it allows for online setpoint changes, and the setpoint need not be a feasible equilibrium for the underlying system. The last point is particularly relevant in a data-driven control scenario: In the common case where only an output setpoint is given, finding the corresponding equilibrium input can be challenging if no model is available. The respective advantages of the different proposed MPC formulations were illustrated for the linear noise-free case with a numerical example in Section 3.4. When dealing with nonlinear systems on the other hand (Chapter 5), employing a tracking MPC formulation was crucial. Here, we exploited that, if a certain cost matrix is sufficiently small and the initial state is close to the steady-state manifold, then the artificial setpoint is always close to the current state such that a linearization-based prediction model provides accurate predictions. When using standard terminal ingredients as in Sections 3.1, 3.2, and 4.1 (i.e., without artificial equilibrium), then closed-loop stability of data-driven MPC for nonlinear systems can only be guaranteed in a neighborhood of the setpoint.

We now *close the loop* w.r.t. the introduction (Chapter 1) by discussing the relation of our results to existing data-driven control approaches. Let us start by addressing indirect approaches, i.e., those based on sequential system identification and model-based control. In Section 1.1, we explained that obtaining rigorous theoretical guarantees in indirect data-driven control can be challenging due to the lack of tight and computationally tractable error bounds in system identification when noisy data of finite length are available. On the contrary, the direct data-driven MPC framework proposed in this thesis provides exactly such guarantees: In Chapter 4, we showed that the presented schemes practically exponentially stabilize the closed loop using only noisy input-output data. Moreover, since no intermediate model estimation step is included, the MPC scheme as well as the theoretical analysis directly involve the measured data, thus allowing us to establish direct links between the data quality and the resulting closed-loop performance. Furthermore, as shown in Chapter 5, the presented framework can also handle nonlinear systems, for which obtaining accurate and computationally tractable models via system identification is difficult. In particular, our results in Chapter 5 yield closed-loop stability guarantees when controlling unknown nonlinear systems based only on input-output data. To the best of the author's knowledge, the existing literature does not contain any comparable results under similar assumptions, regardless of whether indirect or direct data-driven control is considered. On the other hand, indirect approaches also possess advantages over our data-driven MPC framework: For example, the computational complexity of the proposed MPC schemes increases with the data length, which may be large in the case of linear systems with noisy data if good performance is desired. While data compression techniques, e.g., from [40, 152], provide one potential route to alleviate the computational burden of data-driven MPC, indirect approaches have an inherent advantage at this point: The online complexity of model-based MPC is always the same, regardless of the length of the data used to identify, e.g., the employed state-space matrices. Finally, as discussed in the introduction (Chapter 1), the recent literature connected to the Fundamental Lemma contains many interesting contributions on data-driven optimal control, however, guarantees are typically only given for *open-loop* settings. Hence, the results in this thesis can also be interpreted as providing *closed-loop* guarantees when applying these optimal control approaches in a receding-horizon implementation.

6.3 Outlook

There are many open research questions that can be addressed based on the results in this thesis.

On a conceptual level, the main contribution of this thesis lies in 1) bridging the theory on model-based MPC with the Fundamental Lemma to develop direct data-driven MPC schemes, and 2) developing tools to analyze the behavior of these schemes when the data do not provide an exact prediction model (e.g., in the noisy or nonlinear case). Loosely speaking, one can apply similar arguments to transform almost any model-based MPC scheme into a data-driven MPC scheme while showing closed-loop guarantees on, e.g., (practical) stability or constraint satisfaction. In fact, the recent literature contains various data-driven MPC approaches following this idea, some of which were inspired by the results in this thesis. For example, [JB17] proves stability and robustness of robust data-driven MPC without terminal ingredients, assuming that the prediction horizon is sufficiently long. Further contributions include data-driven formulations of distributed MPC [5, 6, JB23], encrypted MPC [3, 4], explicit MPC [18, 125], MPC for linear parameter-varying systems [141], resilient MPC under denial-of-service attacks [84], stochastic MPC [66,

111], and moving horizon estimation [145]. Developing further data-driven MPC formulations on the basis of the Fundamental Lemma and analyzing the resulting closed-loop behavior is a highly interesting field for future research. To this end, the continuity result of data-driven MPC shown in Proposition 5.4 may prove useful since it provides a generic, separation-type approach for the theoretical analysis of data-driven MPC.

Additionally, we discussed above that the proposed data-driven MPC framework has potential advantages over the established indirect data-driven control approach based on sequential system identification and model-based control. However, it may also be possible to derive comparable guarantees for indirect data-driven MPC by following similar arguments. As a result, it is unclear whether any of the two approaches will be superior in general. Interesting preliminary results on this issue are provided by [36, 75]. In [36], it is shown that indirect data-driven control can in fact be translated into direct data-driven control based on a suitable regularization. Furthermore, [75] analyzes and compares the performance of indirect and direct data-driven optimal control for linear systems with noisy data, and they conclude that, depending on the specific problem setup, either approach can be superior. These results, however, only address the open-loop behavior. Employing the results in this thesis to investigate the relation between indirect and direct data-driven optimal control in a receding-horizon implementation, i.e., for data-driven MPC, is a largely open research question.

Moreover, almost all data-driven control approaches based on the Fundamental Lemma assume that one or multiple trajectories of the underlying system are collected and used to set up the controller *offline*. In contrast, our nonlinear data-driven MPC in Section 5.2 updates the data online at every time step. In fact, this key ingredient allows us to control nonlinear systems based only on data, and without data updates the performance for nonlinear systems can be very poor, compare Section 5.3. The usefulness of online data updates in data-driven MPC has also been reported in [78] for a building control application. To the best of the author's knowledge, the only approach in the literature related to the Fundamental Lemma which also provides theoretical guarantees when updating the data online is suggested in [121], which handles data-driven control of switched systems. Developing further data-driven control methods based on the Fundamental Lemma which exploit online data updates and deriving theoretical guarantees is another very interesting issue for future research. To this end, recall that our results in Section 5.2 required that the data generated in closed loop are persistently exciting, which can be hard to ensure. Further, we only showed *practical* stability of the closed loop. Tackling these issues via appropriate modifications of the proposed algorithm is an interesting first step in order to explore the role of online data updates in data-driven control.

We conclude by mentioning a number of more obvious but still relevant open problems originating from our theoretical results. The bounds derived in Chapter 4, which were used to guarantee practical stability and constraint satisfaction, are typically not tight and can, in fact, be quite conservative. This is particularly relevant for the constraint tightening in Section 4.2 which can only handle relatively small noise levels. It may be possible to improve these bounds and thus enhance the practical applicability of our framework, possibly relying on existing open-loop robustness results for data-driven optimal control, e.g., [28, 44, 63, 147, 149, 152]. Closely related to this issue, we have shown in Section 4.3.4 that our results imply performance bounds and suboptimality estimates which are similar to bounds in the existing literature [28, 43, 44, 63, 149]. Improving these bounds and investigating the corresponding infinite-horizon performance in more detail is also a relevant problem for future research. Finally, the data-driven MPC approach for nonlinear systems in Section 5.2 required a number of different assumptions, many of which are hard to verify without model knowledge. Developing appropriate tools for the verification of these assumptions and linking them to existing data-driven system analysis methods for nonlinear systems [96, 97, 98] constitutes another interesting next step.

To summarize, the data-driven MPC framework proposed in this thesis paves the way for various interesting research directions, all of which may contribute to the development of a rigorous and practical data-driven control theory.

Appendix A

Technical proofs

A.1 Proof of Theorem 3.4

Proof. **(i) Recursive feasibility**
For the artificial equilibrium, we choose as a candidate at time $t+1$ the previously optimal one, i.e., $u^{s\prime}(t+1) = u^{s*}(t)$, $y^{s\prime}(t+1) = y^{s*}(t)$. Moreover, for the input-output predictions, we consider the standard candidate solution, consisting of the previously optimal solution shifted by one step and appended by $(u^{s\prime}(t+1), y^{s\prime}(t+1))$, i.e.,

$$\bar{u}'_k(t+1) = \bar{u}^*_{k+1}(t), \;\; \bar{y}'_k(t+1) = \bar{y}^*_{k+1}(t), \;\; k \in \mathbb{I}_{[-n,L-1]},$$

and $\bar{u}'_L(t+1) = u^{s\prime}(t+1), \bar{y}'_L(t+1) = y^{s\prime}(t+1)$. Finally, according to Theorem 3.3, there exists an $\alpha'(t+1)$ satisfying (3.26b).

(ii) Constraint satisfaction
This follows directly from recursive feasibility, together with Theorem 3.3 and the constraints of Problem 3.4.

(iii) Exponential stability
We first show that the Lyapunov function candidate V is quadratically lower and upper bounded. Thereafter, we prove that V is non-increasing and decreases exponentially over n time steps, which implies exponential stability for the closed loop.

(iii.a) Lower bound on V
Using that $J^*_{\text{eq}} \leq \|u^s - u^r\|_S^2 + \|y^s - y^r\|_T^2$ for any equilibrium (u^s, y^s) satisfying the constraints of (3.22), V is quadratically lower bounded as

$$\gamma \lambda_{\min}(P_W) \|\xi_t - \xi^{sr}\|_2^2 \leq V(\xi_t).$$

(iii.b) Local upper bound on V

Let x_t satisfy $\|x_t - x^{\mathrm{sr}}\|_2 \overset{(2.4)}{\leq} \|T_{\mathrm{x},\xi}\|_2\|\xi_t - \xi^{\mathrm{sr}}\|_2 \leq \delta$ for a sufficiently small $\delta > 0$. Since $(u^{\mathrm{sr}}, y^{\mathrm{sr}}) \in \mathbb{U}^{\mathrm{s}} \times \mathbb{Y}^{\mathrm{s}} \subseteq \mathrm{int}(\mathbb{U} \times \mathbb{Y})$, $L \geq l + n$, and by controllability, there exists a feasible input-output trajectory steering the state to x^{sr} in $n \leq L - l$ steps, while satisfying

$$\sum_{k=0}^{L} \|\bar{u}_k(t) - u^{\mathrm{sr}}\|_2^2 + \|\bar{y}_k(t) - y^{\mathrm{sr}}\|_2^2 \leq \Gamma_{\mathrm{uy}}\|x_t - x^{\mathrm{sr}}\|_2^2 \tag{A.1}$$

for some $\Gamma_{\mathrm{uy}} > 0$, compare (2.10). For the artificial equilibrium, we consider the candidate solution $u^{\mathrm{s}}(t) = u^{\mathrm{sr}}$, $y^{\mathrm{s}}(t) = y^{\mathrm{sr}}$. Finally, by Theorem 3.3, there exists an $\alpha(t)$ satisfying (3.26b), which implies that the defined trajectory satisfies all constraints of Problem 3.4. Hence, a local quadratic upper bound on V can be obtained as

$$\begin{aligned} V(\xi_t) \leq& (\Gamma_{\mathrm{uy}}\lambda_{\max}(Q, R) + \gamma\lambda_{\max}(P_{\mathrm{W}}))\|x_t - x^{\mathrm{sr}}\|_2^2 \\ \overset{(2.4)}{\leq}& (\Gamma_{\mathrm{uy}}\lambda_{\max}(Q, R) + \gamma\lambda_{\max}(P_{\mathrm{W}}))\|T_{\mathrm{x},\xi}\|_2^2\|\xi_t - \xi^{\mathrm{sr}}\|_2^2. \end{aligned}$$

(iii.c) Exponential decay of V

Define n candidate solutions for $i \in \mathbb{I}_{[1,n]}$, similar to Part (i) of the proof, as

$$\bar{u}'_k(t+i) = \bar{u}^*_{k+1}(t+i-1),\ \bar{y}'_k(t+i) = \bar{y}^*_{k+1}(t+i-1),$$

for $k \in \mathbb{I}_{[-l,L-1]}$, and $\bar{u}'_L(t+i) = u^{\mathrm{s}\prime}(t+i), \bar{y}'_L(t+i) = y^{\mathrm{s}\prime}(t+i)$. The candidate artificial equilibria are defined as

$$u^{\mathrm{s}\prime}(t+i) = u^{\mathrm{s}*}(t+i-1),\ y^{\mathrm{s}\prime}(t+i) = y^{\mathrm{s}*}(t+i-1),$$

and $\alpha'(t+i)$ as a corresponding solution to (3.18). Using this candidate solution, it is readily shown that, for any $i \in \mathbb{I}_{[1,n]}$, the optimal cost is non-increasing with

$$J_L^*(\xi_{t+i}) \leq J_L^*(\xi_{t+i-1}) - \|u_{t+i-1} - u^{\mathrm{s}*}(t+i-1)\|_R^2 - \|y_{t+i-1} - y^{\mathrm{s}*}(t+i-1)\|_Q^2. \tag{A.2}$$

We now derive a decay bound for J_L^* over n steps by studying different cases.
Case 1: Assume

$$\begin{aligned} &\sum_{i=0}^{n-1} \|u_{t+i} - u^{\mathrm{s}*}(t+i)\|_R^2 + \|y_{t+i} - y^{\mathrm{s}*}(t+i)\|_Q^2 \\ \geq& \gamma_1 \sum_{i=0}^{n-1} \left(\|u^{\mathrm{s}*}(t+i) - u^{\mathrm{sr}}\|_T^2 + \|y^{\mathrm{s}*}(t+i) - y^{\mathrm{sr}}\|_S^2\right), \end{aligned} \tag{A.3}$$

for a constant $\gamma_1 > 0$, which will be fixed later in the proof. It follows from (A.2) that the optimal cost over n steps decreases as

$$\begin{aligned} J_L^*(\xi_{t+n}) - J_L^*(\xi_t) &= \sum_{i=0}^{n-1} J_L^*(\xi_{t+i+1}) - J_L^*(\xi_{t+i}) \\ &\leq -\sum_{i=0}^{n-1} \left(\|u_{t+i} - u^{s*}(t+i)\|_R^2 + \|y_{t+i} - y^{s*}(t+i)\|_Q^2 \right), \end{aligned} \tag{A.4}$$

where a telescoping sum argument is used for the first equality. Using $a^2 + b^2 \geq \frac{1}{2}(a+b)^2$, (A.4) implies

$$\begin{aligned} J_L^*(\xi_{t+n}) - J_L^*(\xi_t) &\overset{(A.3)}{\leq} -\frac{\gamma_1}{2} \sum_{i=0}^{n-1} \left(\|u^{s*}(t+i) - u^{sr}\|_T^2 + \|y^{s*}(t+i) - y^{sr}\|_S^2 \right) \\ &\quad -\frac{1}{2} \sum_{i=0}^{n-1} \left(\|u_{t+i} - u^{s*}(t+i)\|_R^2 + \|y_{t+i} - y^{s*}(t+i)\|_Q^2 \right) \\ &\leq -c_3 \frac{\min\{1, \gamma_1\}}{4} \sum_{i=0}^{n-1} \left(\|u_{t+i} - u^{sr}\|_2^2 + \|y_{t+i} - y^{sr}\|_2^2 \right), \end{aligned} \tag{A.5}$$

where $c_3 = \lambda_{\min}(Q, R, S, T)$.

Case 2: Assume

$$\begin{aligned} &\sum_{i=0}^{n-1} \|u_{t+i} - u^{s*}(t+i)\|_R^2 + \|y_{t+i} - y^{s*}(t+i)\|_Q^2 \\ &\leq \gamma_1 \sum_{i=0}^{n-1} \left(\|u^{s*}(t+i) - u^{sr}\|_T^2 + \|y^{s*}(t+i) - y^{sr}\|_S^2 \right). \end{aligned} \tag{A.6}$$

Case 2a: Assume further

$$\sum_{i=0}^{n-1} \|x_{t+i} - x^{s*}(t+i)\|_2^2 \leq \gamma_2 \sum_{i=0}^{n-1} \|x^{s*}(t+i) - x^{sr}\|_2^2, \tag{A.7}$$

for a constant $\gamma_2 > 0$, which will be fixed later in the proof. We consider now a different candidate solution at time $t+1$ with artificial equilibrium $\hat{u}^s(t+1) = \lambda u^{s*}(t) + (1-\lambda)u^{sr}$, $\hat{y}^s(t+1) = \lambda y^{s*}(t) + (1-\lambda)y^{sr}$ for some $\lambda \in (0,1)$, which will be fixed later in the proof. Clearly, this is a feasible equilibrium and it holds for the corresponding equilibrium state that $\hat{x}^s(t+1) = \lambda x^{s*}(t) + (1-\lambda)x^{sr}$. Further,

$$\hat{x}^s(t+1) - x^{s*}(t) = (1-\lambda)(x^{sr} - x^{s*}(t)), \tag{A.8}$$

and similarly for the input and output. Due to compactness of $\mathbb{U}$ and $\mathbb{Y}$, the right-hand side of (A.7) is bounded from above by $\gamma_2 x_{\max}$ for some $x_{\max} > 0$. Thus, if γ_2 is sufficiently small, then $\sum_{i=0}^{n-1} \|x_{t+i} - x^{s*}(t+i)\|_2^2$ is arbitrarily small as well. Hence, if in addition $(1-\lambda)$ is sufficiently small, then, by controllability and since $(\hat{u}^s(t+1), \hat{y}^s(t+1)) \in \mathrm{int}(\mathbb{U} \times \mathbb{Y})$, there exists a feasible input-output trajectory $\hat{u}(t+1), \hat{y}(t+1)$ steering the system to $(\hat{u}^s(t+1), \hat{y}^s(t+1))$ in n steps. Moreover, there exists a corresponding $\hat{\alpha}(t+1)$ satisfying all constraints of Problem 3.4. Further, it holds that

$$\begin{aligned}
&\|\hat{u}^s(t+1) - u^{\mathrm{r}}\|_T^2 - \|u^{s*}(t) - u^{\mathrm{r}}\|_T^2 \\
=&\,(\hat{u}^s(t+1) - u^{s*}(t))^\top T\,(\hat{u}^s(t+1) + u^{s*}(t) - 2u^{\mathrm{r}}) \\
=&(1-\lambda)\,(u^{\mathrm{sr}} - u^{s*}(t))^\top T\,((1+\lambda)u^{s*}(t) + (1-\lambda)u^{\mathrm{sr}} - 2u^{\mathrm{r}}) \\
=&-(1-\lambda^2)\|u^{s*}(t) - u^{\mathrm{sr}}\|_T^2 - 2(1-\lambda)(u^{s*}(t) - u^{\mathrm{sr}})^\top T(u^{\mathrm{sr}} - u^{\mathrm{r}}) \\
\leq&-(1-\lambda^2)\|u^{s*}(t) - u^{\mathrm{sr}}\|_T^2,
\end{aligned} \tag{A.9}$$

where the last inequality follows from noting that $2T(u^{\mathrm{sr}} - u^{\mathrm{r}})$ is the gradient of $\|u - u^{\mathrm{r}}\|_T^2$ evaluated at u^{sr}, and the directional derivative of this function towards any other feasible direction increases, due to convexity of (3.22) by Assumption 3.8 (compare [69] for details). Similarly,

$$\|\hat{y}^s(t+1) - y^{\mathrm{r}}\|_S^2 - \|y^{s*}(t) - y^{\mathrm{r}}\|_S^2 \leq -(1-\lambda^2)\|y^{s*}(t) - y^{\mathrm{sr}}\|_S^2.$$

By controllability, there exists $\Gamma_{\mathrm{uy}} > 0$ as in (A.1) such that

$$\begin{aligned}
&\sum_{k=0}^{L} \|\hat{u}_k(t+1) - \hat{u}^s(t+1)\|_R^2 + \|\hat{y}_k(t+1) - \hat{y}^s(t+1)\|_Q^2 \\
\leq&\,\Gamma_{\mathrm{uy}}\lambda_{\max}(Q,R)\|x_{t+1} - \hat{x}^s(t+1)\|_2^2 \\
\leq&\,2\Gamma_{\mathrm{uy}}\lambda_{\max}(Q,R)(\|x_{t+1} - x^{s*}(t)\|_2^2 + \|x^{s*}(t) - \hat{x}^s(t+1)\|_2^2),
\end{aligned}$$

using again the fact that $(a+b)^2 \leq 2a^2 + 2b^2$. The first term can be bounded as

$$\begin{aligned}
\|x_{t+1} - x^{s*}(t)\|_2^2 \leq&\|A\|_2^2\|x_t - x^{s*}(t)\|_2^2 + \|B\|_2^2\|u_t - u^{s*}(t)\|_2^2 \\
\leq&\underbrace{(\|A\|_2^2 + \|B\|_2^2\Gamma_{\mathrm{uy}})}_{c_4:=}\|x_t - x^{s*}(t)\|_2^2,
\end{aligned}$$

where the last inequality follows again from a controllability argument. Combining the bounds, we infer

$$\begin{aligned}
&J_L^*(\xi_{t+1}) - J_L^*(\xi_t)\\
&\leq 2c_4\Gamma_{\mathrm{uy}}\lambda_{\max}(Q,R)\|x_t - x^{\mathrm{s}*}(t)\|_2^2 + 2\Gamma_{\mathrm{uy}}\lambda_{\max}(Q,R)(1-\lambda)^2\|x^{\mathrm{sr}} - x^{\mathrm{s}*}(t)\|_2^2\\
&\quad - \|u_t - u^{\mathrm{s}*}(t)\|_R^2 - \|y_t - y^{\mathrm{s}*}(t)\|_Q^2 - (1-\lambda^2)(\|u^{\mathrm{s}*}(t) - u^{\mathrm{sr}}\|_T^2 + \|y^{\mathrm{s}*}(t) - y^{\mathrm{sr}}\|_S^2).
\end{aligned}$$

Using a similar candidate solution at time $t+i$ for $i \in \mathbb{I}_{[2,n]}$, it can be shown that

$$\begin{aligned}
&J_L^*(\xi_{t+n}) - J_L^*(\xi_t)\\
\leq &2\Gamma_{\mathrm{uy}}\lambda_{\max}(Q,R)\sum_{i=0}^{n-1}(c_4\|x_{t+i} - x^{\mathrm{s}*}(t+i)\|_2^2 + (1-\lambda)^2\|x^{\mathrm{sr}} - x^{\mathrm{s}*}(t+i)\|_2^2)\\
&- \sum_{i=0}^{n-1}(\|u_{t+i} - u^{\mathrm{s}*}(t+i)\|_R^2 + \|y_{t+i} - y^{\mathrm{s}*}(t+i)\|_Q^2)\\
&- (1-\lambda^2)\sum_{i=0}^{n-1}(\|u^{\mathrm{s}*}(t+i) - u^{\mathrm{sr}}\|_T^2 + \|y^{\mathrm{s}*}(t+i) - y^{\mathrm{sr}}\|_S^2)\\
\overset{(3.25),(A.7)}{\leq} &(2c_4\gamma_2 + 2(1-\lambda)^2)\Gamma_{\mathrm{uy}}\lambda_{\max}(Q,R)c_{\mathrm{x},2}\\
&\cdot\sum_{i=0}^{n-1}(\|u^{\mathrm{s}*}(t+i) - u^{\mathrm{sr}}\|_2^2 + \|y^{\mathrm{s}*}(t+i) - y^{\mathrm{sr}}\|_2^2)\\
&- \sum_{i=0}^{n-1}(\|u_{t+i} - u^{\mathrm{s}*}(t+i)\|_R^2 + \|y_{t+i} - y^{\mathrm{s}*}(t+i)\|_Q^2)\\
&- (1-\lambda^2)\sum_{i=0}^{n-1}(\|u^{\mathrm{s}*}(t+i) - u^{\mathrm{sr}}\|_T^2 + \|y^{\mathrm{s}*}(t+i) - y^{\mathrm{sr}}\|_S^2)\\
\leq &-c_5\sum_{i=0}^{n-1}(\|u_{t+i} - u^{\mathrm{sr}}\|_2^2 + \|y_{t+i} - y^{\mathrm{sr}}\|_2^2)
\end{aligned}$$

for some $c_5 > 0$, where the last inequality holds if γ_2 is sufficiently small and λ is sufficiently close to 1.

Case 2b: Assume

$$\sum_{i=0}^{n-1}\|x_{t+i} - x^{\mathrm{s}*}(t+i)\|_2^2 \geq \gamma_2\sum_{i=0}^{n-1}\|x^{\mathrm{s}*}(t+i) - x^{\mathrm{sr}}\|_2^2. \tag{A.10}$$

This implies the existence of an index $k \in \mathbb{I}_{[0,n-1]}$ such that

$$\|x_{t+k} - x^{\mathrm{s}*}(t+k)\|_2^2 \geq \frac{\gamma_2}{n}\sum_{i=0}^{n-1}\|x^{\mathrm{s}*}(t+i) - x^{\mathrm{sr}}\|_2^2. \tag{A.11}$$

The following auxiliary result will be central for the proof of Case 2b.

Lemma A.1. *There exist $\gamma_3 > 0$ and $j \in \mathbb{I}_{[0,n-1]}$ such that*

$$\begin{aligned}&\|u_{t+j} - u^{s*}(t+k)\|_2^2 + \|y_{t+j} - y^{s*}(t+k)\|_2^2 \\ \geq&\gamma_3 \sum_{i=0}^{n-1} \left(\|u^{s*}(t+i) - u^{sr}\|_2^2 + \|y^{s*}(t+i) - y^{sr}\|_2^2\right),\end{aligned} \tag{A.12}$$

with k as in (A.11).

Proof. Using the system dynamics, it holds that

$$\|x_{t+k} - x^{s*}(t+k)\|_2^2 \leq a_1\|x_t - x^{s*}(t+k)\|_2^2 + a_2\|u_{[t,t+k-1]} - \mathbb{1}_k \otimes u^{s*}(t+k)\|_2^2, \tag{A.13}$$

for suitable $a_1, a_2 > 0$. Further, for the observability matrix Φ_n and a suitable matrix Γ_n, which depends on B, C, and D, we obtain

$$y_{[t,t+n-1]} - \mathbb{1}_n \otimes y^{s*}(t+k) = \Phi_n(x_t - x^{s*}(t+k)) + \Gamma_n(u_{[t,t+n-1]} - \mathbb{1}_n \otimes u^{s*}(t+k)).$$

By observability, we can solve the latter equation for $x_t - x^{s*}(t+k)$, which leads, together with (A.13), to

$$\|x_{t+k} - x^{s*}(t+k)\|_2^2 \leq a_3 \sum_{i=0}^{n-1} (\|u_{t+i} - u^{s*}(t+k)\|_2^2 + \|y_{t+i} - y^{s*}(t+k)\|_2^2), \tag{A.14}$$

for a suitable $a_3 > 0$. Let j be the index, for which

$$\|u_{t+j} - u^{s*}(t+k)\|_2^2 + \|y_{t+j} - y^{s*}(t+k)\|_2^2$$

is maximal, which implies

$$\begin{aligned}&\|u_{t+j} - u^{s*}(t+k)\|_2^2 + \|y_{t+j} - y^{s*}(t+k)\|_2^2 \\ \geq&\frac{1}{n} \sum_{i=0}^{n-1} \left(\|u_{t+i} - u^{s*}(t+k)\|_2^2 + \|y_{t+i} - y^{s*}(t+k)\|_2^2\right),\end{aligned} \tag{A.15}$$

which in turn leads to

$$\|x_{t+k} - x^{s*}(t+k)\|_2^2 \leq a_3 n(\|u_{t+j} - u^{s*}(t+k)\|_2^2 + \|y_{t+j} - y^{s*}(t+k)\|_2^2). \tag{A.16}$$

Combining (A.16) with (A.11) and (3.24) concludes the proof of Lemma A.1. ∎

It follows from (A.6) and (A.12) that $j \neq k$, as long as $\gamma_1 < \gamma_3$, which will be assumed in the following. At time $t+j$, we now define a different candidate solution as a convex combination between the optimal solution and the candidate solution from Case 1, i.e.,

$$\bar{u}''(t+j) = \beta\bar{u}'(t+j) + (1-\beta)\bar{u}^*(t+j),$$

for some $\beta \in [0,1]$, with $\bar{u}'(t+j)$ as in the beginning of Part (iii.c) of the proof. The other variables $\bar{y}''(t+j), u^{s''}(t+j), y^{s''}(t+j), \alpha''(t+j)$ are defined analogously. By convexity, this is a feasible solution to Problem 3.4. Let us write J_L' and J_L'' for the open-loop cost of the candidate solution $\bar{u}'(t+j), \bar{y}'(t+j), u^{s'}(t+j), y^{s'}(t+j), \alpha'(t+j)$ and $\bar{u}''(t+j), \bar{y}''(t+j), u^{s''}(t+j), y^{s''}(t+j), \alpha''(t+j)$, respectively, with initial state ξ_{t+j}. Then,

$$\begin{aligned}
J_L^*(\xi_{t+j}) \leq J_L'' \leq\, & \beta J_L' + (1-\beta)J_L^*(\xi_{t+j}) \\
& - 2\bar{c}\beta(1-\beta)\left(\|u^{s*}(t+j) - u^{s'}(t+j)\|_2^2 + \|y^{s*}(t+j) - y^{s'}(t+j)\|_2^2\right), \\
\stackrel{\text{(A.2)}}{\leq}\, & \beta J_L^*(\xi_{t+j-1}) + (1-\beta)J_L^*(\xi_{t+j}) - 2\bar{c}\beta(1-\beta)\|u^{s*}(t+j) - u^{s'}(t+j)\|_2^2 \\
& - 2\bar{c}\beta(1-\beta)\|y^{s*}(t+j) - y^{s'}(t+j)\|_2^2,
\end{aligned}$$

where the second inequality follows from strong convexity of (3.22) for some $\bar{c} > 0$. Fixing $\beta = \frac{1}{2}$ and dividing by β, we obtain

$$\begin{aligned}
J_L^*(\xi_{t+j}) - J_L^*(\xi_{t+j-1}) \leq & -\bar{c}\|u^{s*}(t+j) - u^{s*}(t+j-1)\|_2^2 \\
& -\bar{c}\|y^{s*}(t+j) - y^{s*}(t+j-1)\|_2^2.
\end{aligned}$$

Suppose now $j > k$. Then, by defining similar candidate solutions at time instants $t+k, \ldots, t+j$, and applying $a^2 + b^2 \geq \frac{1}{2}(a+b)^2$ repeatedly $j-k-1$ times, we obtain

$$\begin{aligned}
& J_L^*(\xi_{t+j}) - J_L^*(\xi_{t+k}) \\
\leq & -\frac{\bar{c}}{2^{j-k-1}}(\|u^{s*}(t+j) - u^{s*}(t+k)\|_2^2 + \|y^{s*}(t+j) - y^{s*}(t+k)\|_2^2).
\end{aligned}$$

Conversely, if $k > j$, we arrive at

$$\begin{aligned}
& J_L^*(\xi_{t+k}) - J_L^*(\xi_{t+j}) \\
\leq & -\frac{\bar{c}}{2^{k-j-1}}(\|u^{s*}(t+j) - u^{s*}(t+k)\|_2^2 + \|y^{s*}(t+j) - y^{s*}(t+k)\|_2^2).
\end{aligned}$$

Combining the two cases and noting that $j \neq k$ (cf. Lemma A.1), it follows from (A.2) that

$$\begin{aligned} & J_L^*(\xi_{t+n}) - J_L^*(\xi_t) && \text{(A.17)} \\ & \leq -\frac{\bar{c}}{2^{|j-k|-1}} \|u^{s*}(t+j) - u^{s*}(t+k)\|_2^2 - \frac{\bar{c}}{2^{|j-k|-1}} \|y^{s*}(t+j) - y^{s*}(t+k)\|_2^2. \end{aligned}$$

Now, we bound the right-hand side of (A.17) by using

$$a + b = \sqrt{(a+b)^2} \geq \sqrt{a^2 + b^2}, \tag{A.18a}$$

$$a + b = \sqrt{(a+b)^2} \leq \sqrt{2a^2 + 2b^2}, \tag{A.18b}$$

which hold for any $a, b \geq 0$. In the following, let γ_1 be sufficiently small such that $\gamma_1 < \min\{1, \frac{1}{8c_6}\}\gamma_3$, where $c_6 = \frac{\lambda_{\max}(S,T)}{\lambda_{\min}(Q,R)} > 0$. Note that this implies $\gamma_1 < \gamma_3$ and hence $j \neq k$, as is required above. Moreover,

$$\begin{aligned} & \|u^{s*}(t+j) - u^{s*}(t+k)\|_2 + \|y^{s*}(t+j) - y^{s*}(t+k)\|_2 \\ \geq & \|u_{t+j} - u^{s*}(t+k)\|_2 - \|u_{t+j} - u^{s*}(t+j)\|_2 \\ & + \|y_{t+j} - y^{s*}(t+k)\|_2 - \|y_{t+j} - y^{s*}(t+j)\|_2 \\ \overset{\text{(A.12),(A.18a)}}{\geq} & \frac{1}{2}\|u_{t+j} - u^{s*}(t+k)\|_2 + \frac{1}{2}\|y_{t+j} - y^{s*}(t+k)\|_2 - \|u_{t+j} - u^{s*}(t+j)\|_2 \\ & - \|y_{t+j} - y^{s*}(t+j)\|_2 + \frac{\sqrt{\gamma_3}}{2}\sqrt{\sum_{i=0}^{n-1} \|u^{s*}(t+i) - u^{sr}\|_2^2 + \|y^{s*}(t+i) - y^{sr}\|_2^2} \\ \overset{\text{(A.6),(A.18b)}}{\geq} & \frac{1}{2}\|u_{t+j} - u^{s*}(t+k)\|_2 + \frac{1}{2}\|y_{t+j} - y^{s*}(t+k)\|_2 \\ & + \left(\frac{\sqrt{\gamma_3}}{2} - \sqrt{2\gamma_1 c_6}\right)\sqrt{\sum_{i=0}^{n-1} \|u^{s*}(t+i) - u^{sr}\|_2^2 + \|y^{s*}(t+i) - y^{sr}\|_2^2} \\ \overset{\text{(A.15)}}{\geq} & \frac{1}{2}\sqrt{\frac{1}{n}\sum_{i=0}^{n-1} \left(\|u_{t+i} - u^{s*}(t+k)\|_2^2 + \|y_{t+i} - y^{s*}(t+k)\|_2^2\right)} \\ & + \left(\frac{\sqrt{\gamma_3}}{2} - \sqrt{2\gamma_1 c_6}\right)\sqrt{\|u^{s*}(t+k) - u^{sr}\|_2^2 + \|y^{s*}(t+k) - y^{sr}\|_2^2} \\ \geq & c_7 \sum_{i=0}^{n-1} \left(\|u_{t+i} - u^{sr}\|_2 + \|y_{t+i} - y^{sr}\|_2\right), \end{aligned}$$

for a suitable $c_7 > 0$, where the last inequality follows from an inequality similar to (A.18b). This, together with (A.17) implies the existence of a constant $c_8 > 0$

such that

$$J_L^*(\xi_{t+n}) - J_L^*(\xi_t) \leq -c_8 \sum_{i=0}^{n-1} \left(\|u_{t+i} - u^{\text{sr}}\|_2^2 + \|y_{t+i} - y^{\text{sr}}\|_2^2 \right).$$

Combination: Combining all cases, there exists some $c_9 > 0$ such that

$$J_L^*(\xi_{t+n}) - J_L^*(\xi_t) \leq -c_9 \sum_{i=0}^{n-1} \left(\|u_{t+i} - u^{\text{sr}}\|_2^2 + \|y_{t+i} - y^{\text{sr}}\|_2^2 \right). \tag{A.19}$$

Furthermore, the IOSS Lyapunov function W satisfies

$$\begin{aligned} &W(\xi_{t+1} - \xi^{\text{sr}}) - W(\xi_t - \xi^{\text{sr}}) \\ =&W(\tilde{A}(\xi_t - \xi^{\text{sr}}) + \tilde{B}(u_t - u^{\text{sr}})) - W(\xi_t - \xi^{\text{sr}}) \\ \overset{(3.28)}{\leq}& - \|\xi_t - \xi^{\text{sr}}\|_2^2 + c_{\text{IOSS},1}\|u_t - u^{\text{sr}}\|_2^2 + c_{\text{IOSS},2}\|y_t - y^{\text{sr}}\|_2^2. \end{aligned} \tag{A.20}$$

Applying this inequality recursively, we arrive at

$$\begin{aligned} &W(\xi_{t+n} - \xi^{\text{sr}}) - W(\xi_t - \xi^{\text{sr}}) \\ \leq& \sum_{i=0}^{n-1} \left(-\|\xi_{t+i} - \xi^{\text{sr}}\|_2^2 + c_{\text{IOSS},1}\|u_{t+i} - u^{\text{sr}}\|_2^2 + c_{\text{IOSS},2}\|y_{t+i} - y^{\text{sr}}\|_2^2 \right). \end{aligned} \tag{A.21}$$

Thus, choosing $\gamma = \frac{c_9}{\max\{c_{\text{IOSS},1}, c_{\text{IOSS},2}\}} > 0$, Inequalities (A.19) and (A.21) can be used to bound the Lyapunov function candidate as

$$V(\xi_{t+n}) - V(\xi_t) \leq -\gamma \sum_{i=0}^{n-1} \|\xi_{t+i} - \xi^{\text{sr}}\|_2^2.$$

Due to Parts (iii.a) and (iii.b) of the proof, V is locally quadratically lower and upper bounded. Thus, the equilibrium ξ^{sr} is exponentially stable by standard Lyapunov arguments [118]. ■

A.2 Proof of Theorem 4.3

Proof. The proof is divided into four parts. We first show the lower and upper bounds (4.79) on the Lyapunov function candidate in Part (i). In Parts (ii) and (iii), we propose two different candidate solutions for Problem 4.4 at time $t+n$ for two complementary scenarios, assuming that Problem 4.4 is feasible at time t. In Part (iv), we combine the bounds to prove (4.80). Throughout the proof, we will make repeated use of the inequalities

$$\|a+b\|_P^2 \leq 2\|a\|_P^2 + 2\|b\|_P^2, \tag{A.22}$$

$$\|a\|_P^2 - \|b\|_P^2 \leq \|a-b\|_P^2 + 2\|a-b\|_P\|b\|_P, \tag{A.23}$$

which hold for any vectors a, b and matrix $P = P^\top \succ 0$.

(i) Lower and upper bound on $V(\xi_t)$

(i.a) Lower bound on $V(\xi_t)$

Using that $(\check{u}^{s*}(t), \check{y}^{s*}(t))$ is feasible for (4.67), we have

$$\begin{aligned}\|\check{y}^{s*}(t) - y^r\|_S^2 - \|y^{sr} - y^r\|_S^2 &\overset{(4.69)}{\geq} \|\check{y}^{s*}(t) - y^{sr}\|_S^2 \\ &\overset{(4.68)}{\geq} \frac{\lambda_{\min}(S)}{2}\left(\|\check{y}^{s*}(t) - y^{sr}\|_2^2 + \frac{1}{c_g}\|\check{u}^{s*}(t) - u^{sr}\|_2^2\right).\end{aligned} \tag{A.24}$$

In combination with (A.22), this implies

$$\begin{aligned}V(\xi_t) &\geq \sum_{k=-n}^{-1} \|u_{t+k} - \check{u}^{s*}(t)\|_R^2 + \|y_{t+k} - \check{y}^{s*}(t)\|_Q^2 + \|\check{y}^{s*}(t) - y^r\|_S^2 - \|y^{sr} - y^r\|_S^2 \\ &\overset{(A.24)}{\geq} c_l \sum_{k=-n}^{-1} (\|u_{t+k} - u^{sr}\|_2^2 + \|y_{t+k} - y^{sr}\|_2^2)\end{aligned} \tag{A.25}$$

and therefore, the lower bound in (4.79) with

$$c_l := \frac{1}{2}\min\left\{\lambda_{\min}(Q,R), \frac{\lambda_{\min}(S)}{2n}\min\left\{1, \frac{1}{c_g}\right\}\right\}.$$

(i.b) Upper bound on $V(\xi_t)$

Suppose $\|\xi_t - \xi^{sr}\|_2 \leq \delta$ for a sufficiently small $\delta > 0$. We define a candidate for the artificial equilibrium by $(u^s(t), y^s(t)) = (u^{sr}, y^{sr})$. Using controllability and $u^{sr} \in \mathbb{U}^s \subseteq \operatorname{int}(\mathbb{U})$, there exists a feasible input-output trajectory $\{\bar{u}(t), \bar{y}(t)\}_{k=-n}^{L}$ for Problem 4.4 with $\tilde{\sigma} = 0$ satisfying the terminal constraint (4.76d) as well as

$$\sum_{k=-n}^{L} \|\bar{u}_k(t) - u^{sr}\|_2^2 + \|\bar{y}_k - y^{sr}\|_2^2 \leq \Gamma_\xi \|\xi_t - \xi^{sr}\|_2^2 \tag{A.26}$$

for some $\Gamma_\xi > 0$. The vector $\alpha(t)$ is chosen as

$$\alpha(t) = H_{\text{ux}}^{\dagger}\begin{bmatrix}\bar{u}(t)\\ x_{t-n}\\ 1\end{bmatrix},$$

where H_{ux} is defined as

$$H_{\mathrm{ux}} := \begin{bmatrix} H_{L+n+1}(u^{\mathrm{d}}) \\ H_1(x^{\mathrm{d}}_{[0,N-L-n-1]}) \\ \mathbb{1}^\top_{N-L-n} \end{bmatrix}. \tag{A.27}$$

This implies that all constraints of Problem 4.4 are satisfied (compare the proof of "only if" in Theorem 3.3) and thus, the Lyapunov function candidate $V(\xi_t)$ is upper bounded as

$$V(\xi_t) \leq \Gamma_\xi \lambda_{\max}(Q,R)\|\xi_t - \xi^{\mathrm{sr}}\|_2^2 + \lambda_\alpha \bar{\varepsilon}^{\beta_\alpha}\|\alpha(t) - \alpha^{\mathrm{sr}}\|_2^2. \tag{A.28}$$

Further, using $\alpha^{\mathrm{sr}} = H_{\mathrm{ux}}^\dagger \begin{bmatrix} \mathbb{1}_{L+n+1} \otimes u^{\mathrm{sr}} \\ x^{\mathrm{sr}} \\ 1 \end{bmatrix}$, we infer

$$\|\alpha(t) - \alpha^{\mathrm{sr}}\|_2^2 \overset{(A.26)}{\leq} \|H_{\mathrm{ux}}^\dagger\|_2^2(\Gamma_\xi\|\xi_t - \xi^{\mathrm{sr}}\|_2^2 + \|x_{t-n} - x^{\mathrm{sr}}\|_2^2). \tag{A.29}$$

Finally, using observability, there exists a matrix M such that

$$x_{t-n} - x^{\mathrm{sr}} = M(\xi_t - \xi^{\mathrm{sr}}). \tag{A.30}$$

Combining (A.28)–(A.30), we deduce that the upper bound in (4.79) holds for all ξ_t satisfying $\|\xi_t - \xi^{\mathrm{sr}}\|_2 \leq \delta$ with

$$c_{\mathrm{u}} := \Gamma_\xi \lambda_{\max}(Q,R) + \lambda_\alpha \bar{\varepsilon}^{\beta_\alpha}\|H_{\mathrm{ux}}^\dagger\|_2^2\left(1 + \Gamma_\xi + \|M\|_2^2\right).$$

Analogous to Part (ii) of the proof of Theorem 4.1 (compare (4.40)), (4.79) thus holds for any ξ_t satisfying $V(\xi_t) \leq V_{\max}$ (with a modified constant c_{u}).

(ii) Candidate solution 1

Assume

$$\sum_{k=-n}^{-1} \|u_{t+k} - \check{u}^{\mathrm{s}*}(t)\|_R^2 + \|y_{t+k} - \check{y}^{\mathrm{s}*}(t)\|_Q^2 \geq \gamma\|\check{y}^{\mathrm{s}*}(t) - y^{\mathrm{sr}}\|_S^2 \tag{A.31}$$

for a constant $\gamma > 0$ which will be fixed later in the proof.

(ii.a) Definition of candidate solution

We choose both the input and output equilibrium candidate as the previously optimal solution $\check{u}^{\mathrm{s}\prime}(t+n) = \check{u}^{\mathrm{s}*}(t)$, $\check{y}^{\mathrm{s}\prime}(t+n) = \check{y}^{\mathrm{s}*}(t)$. The first $L-2n$ elements of the predicted input trajectory are a shifted version of the previously optimal trajectory, i.e., $\check{u}_k'(t+n) = \check{u}_{k+n}^*(t)$ for $k \in \mathbb{I}_{[0,L-2n-1]}$. Over the time steps $k \in \mathbb{I}_{[-n,-1]}$, we let $\check{u}_k'(t+n) = u_{[t+n+k]}$ and $\check{y}_k'(t+n) = y_{t+n+k}$. Denote by $\{y_k'(t+n)\}_{k=0}^{L-n}$ the output resulting from an application of $\check{u}_{[n,L]}^*(t)$ to the system (3.14) initialized at $(u_{[t,t+n-1]}, y_{[t,t+n-1]})$. For $k \in \mathbb{I}_{[0,L-2n-1]}$, we let $\check{y}_k'(t+n) = y_k'(t+n)$. We write $\check{x}_{L-2n}'(t+n)$ for the state at time $L-2n$ corresponding to $\{\check{u}_k'(t+n), \check{y}_k'(t+n)\}_{k=0}^{L-2n-1}$. By controllability, there exists an input-output trajectory $\{\check{u}_k'(t+n), \check{y}_k'(t+n)\}_{k=L-2n}^{L-n-1}$ steering the system to the steady-state $\check{x}^{s*}(t)$ corresponding to $(\check{u}^{s*}(t), \check{y}^{s*}(t))$ while satisfying

$$\sum_{k=L-2n}^{L-n-1} \|\check{u}_k'(t+n) - \check{u}^{s*}(t)\|_2^2 + \|\check{y}_k'(t+n) - \check{y}^{s*}(t)\|_2^2 \leq \Gamma \|\check{x}_{L-2n}'(t+n) - \check{x}^{s*}(t)\|_2^2 \tag{A.32}$$

for some $\Gamma > 0$, compare (2.10). In the following, we show that $\check{u}_k'(t+n) \in \mathbb{U}$, $k \in \mathbb{I}_{[L-2n,L-n-1]}$, if $\bar{\varepsilon}$ is sufficiently small. Recall that $\check{x}^{s*}(t)$ is the steady-state corresponding to $(\check{u}^{s*}(t), \check{y}^{s*}(t))$, whereas the output $y_{[L-2n,L-n-1]}'(t+n)$ results from applying $\check{u}_n^{s*}(t)$ to the system at initial state $\check{x}_{L-2n}'(t+n)$. Hence, using observability, there exists $c_{x,1} > 0$ such that

$$\|\check{x}_{L-2n}'(t+n) - \check{x}^{s*}(t)\|_2^2 \leq c_{x,1} \|y_{[L-2n,L-n-1]}'(t+n) - \mathbb{1}_n \otimes \check{y}^{s*}(t)\|_2^2.$$

The output trajectories $y_{[L-2n,L-n-1]}'(t+n)$ and $\mathbb{1}_n \otimes \check{y}^{s*}(t)$ result from applying the input $\check{u}_{[n,L]}^*(t)$ to (3.14) with initial conditions $(u_{[t,t+n-1]}, y_{[t,t+n-1]})$ and $(\check{u}_{[0,n-1]}^*(t), \check{y}_{[0,n-1]}^*(t))$, respectively. Since the difference between these initial conditions is linear in the disturbance $d_{[t,t+n-1]}$ (compare (4.78)), there exists $c_{x,2} > 0$ such that

$$\|\check{x}_{L-2n}'(t+n) - \check{x}^{s*}(t)\|_2^2 \leq c_{x,2}\bar{\varepsilon}^2. \tag{A.33}$$

Together with (A.32) and $\check{u}^{s*}(t) \in \text{int}(\mathbb{U})$, this shows that $\check{u}_k'(t+n) \in \mathbb{U}$ for $k \in \mathbb{I}_{[L-2n,L-n-1]}$ if $\bar{\varepsilon}$ is sufficiently small. Finally, we let $(\check{u}_k'(t+n), \check{y}_k'(t+n)) = (\check{u}^{s*}(t), \check{y}^{s*}(t))$ for $k \in \mathbb{I}_{[L-n,L]}$. Using Assumption 4.10, we choose

$$\check{\alpha}'(t+n) = H_{ux}^{\dagger} \begin{bmatrix} \check{u}'(t+n) \\ x_t \\ 1 \end{bmatrix} \tag{A.34}$$

with H_{ux} as in (A.27). This implies that (4.76b) and thus, all constraints of Problem 4.4 hold.

(ii.b) Lyapunov function decay
Using the above candidate solution, we have

$$\begin{aligned} &V(\xi_{t+n}) - V(\xi_t) \\ &\leq \sum_{k=-n}^{L} \left(\|\breve{u}'_k(t+n) - \breve{u}^{s*}(t)\|_R^2 + \|\breve{y}'_k(t+n) - \breve{y}^{s*}(t)\|_Q^2 \right) \\ &+ \lambda_\alpha \bar{\varepsilon}^{\beta_\alpha} (\|\breve{\alpha}'(t+n) - \alpha^{sr}\|_2^2 - \|\breve{\alpha}^*(t) - \alpha^{sr}\|_2^2) \\ &- \sum_{k=-n}^{L} (\|\breve{u}^*_k(t) - \breve{u}^{s*}(t)\|_R^2 + \|\breve{y}^*_k(t) - \breve{y}^{s*}(t)\|_Q^2). \end{aligned} \tag{A.35}$$

The terms involving the input are

$$\begin{aligned} &\sum_{k=-n}^{L} \|\breve{u}'_k(t+n) - \breve{u}^{s*}(t)\|_R^2 - \|\breve{u}^*_k(t) - \breve{u}^{s*}(t)\|_R^2 \\ &= - \sum_{k=-n}^{-1} \|u_{t+k} - \breve{u}^{s*}(t)\|_R^2 + \sum_{k=L-2n}^{L-n-1} \|\breve{u}'_k(t+n) - \breve{u}^{s*}(t)\|_R^2 \\ &+ \sum_{k=0}^{n-1} \left(\|u_{t+k} - \breve{u}^{s*}(t)\|_R^2 - \|\breve{u}^*_k(t) - \breve{u}^{s*}(t)\|_R^2 \right). \end{aligned} \tag{A.36}$$

For $k \in \mathbb{I}_{[-n,L]}$, it holds that

$$\|\breve{u}^*_k(t) - \breve{u}^{s*}(t)\|_R^2 \leq V(\xi_t). \tag{A.37}$$

Together with the fact that $\|u_{t+k} - \breve{u}^*_k(t)\|_2 \leq \bar{\varepsilon}$ for $k \in \mathbb{I}_{[0,n-1]}$ (compare (4.78)) and using (A.23), this leads to

$$\sum_{k=0}^{n-1} \|u_{t+k} - \breve{u}^{s*}(t)\|_R^2 - \|\breve{u}^*_k(t) - \breve{u}^{s*}(t)\|_R^2 \leq c_{u,1}\bar{\varepsilon}^2 + c_{u,2}\bar{\varepsilon}\sqrt{V(\xi_t)} \tag{A.38}$$

for some $c_{u,1}, c_{u,2} > 0$. Further, the second term on the right-hand side of (A.36) is bounded as

$$\sum_{k=L-2n}^{L-n-1} \|\breve{u}'_k(t+n) - \breve{u}^{s*}(t)\|_R^2 \overset{(A.32),(A.33)}{\leq} \lambda_{\max}(R)\Gamma c_{x,2}\bar{\varepsilon}^2. \tag{A.39}$$

Next, we analyze the terms in (A.35) depending on the output trajectory. Inequalities (A.32) and (A.33) imply

$$\sum_{k=L-2n}^{L-n-1} \|\breve{y}'_k(t+n) - \breve{y}^{s*}(t)\|_Q^2 \leq \lambda_{\max}(Q)\Gamma c_{x,2}\bar{\varepsilon}^2. \tag{A.40}$$

The trajectories $\{\breve{y}'_k(t+n)\}_{k=0}^{L-2n-1}$ and $\{\breve{y}^*_{k+n}(t)\}_{k=0}^{L-2n-1}$ differ only in terms of their initial conditions which in turn differ linearly in terms of $d_{[t,t+n-1]}$. Hence, following the arguments above leading to (A.38), there exist $c_{y,1}, c_{y,2} > 0$, such that for $k \in \mathbb{I}_{[-n,L-2n-1]}$

$$\|\breve{y}'_k(t+n) - \breve{y}^{s*}(t)\|_Q^2 - \|\breve{y}^*_{k+n}(t) - \breve{y}^{s*}(t)\|_Q^2 \leq c_{y,1}\bar{\varepsilon}^2 + c_{y,2}\bar{\varepsilon}\sqrt{V(\xi_t)}. \tag{A.41}$$

Finally, by definition of $\breve{\alpha}'(t+n)$ in (A.34), we have

$$\begin{aligned}
&\|\breve{\alpha}'(t+n) - \alpha^{sr}\|_2^2 \qquad\qquad (A.42)\\
\leq &\|H_{ux}^\dagger\|_2^2 \left\| \begin{bmatrix} \breve{u}'_{[-n,L]}(t+n) - \mathbb{1}_{L+n+1} \otimes u^{sr} \\ x_t - x^{sr} \\ 0 \end{bmatrix} \right\|_2^2 \\
= &\|H_{ux}^\dagger\|_2^2 (\|\breve{u}^*_{[0,n-1]}(t) + d_{[t,t+n-1]} - \mathbb{1}_n \otimes u^{sr}\|_2^2 + \|\breve{u}^*_{[n,L-n-1]}(t) - \mathbb{1}_{L-2n} \otimes u^{sr}\|_2^2 \\
&+ (n+1)\|\breve{u}^{s*}(t) - u^{sr}\|_2^2 + \|\breve{u}'_{[L-2n,L-n-1]}(t+n) - \mathbb{1}_n \otimes u^{sr}\|_2^2 + \|x_t - x^{sr}\|_2^2).
\end{aligned}$$

Note that

$$\begin{aligned}
\|\breve{u}^{s*}(t) - u^{sr}\|_2^2 &\overset{(4.68)}{\leq} c_g\|\breve{y}^{s*}(t) - y^{sr}\|_2^2 \overset{(A.22)}{\leq} 2c_g(\|\breve{y}^{s*}(t) - y^r\|_2^2 + \|y^{sr} - y^r\|_2^2) \\
&\leq \frac{2c_g}{\lambda_{\min}(S)}(V(\xi_t) + 2J^*_{eq}),
\end{aligned} \tag{A.43}$$

where we exploit $\breve{J}^*_L(\xi_t) = V(\xi_t) + J^*_{eq}$ for the last inequality. Using this inequality as well as (4.79), (A.22), (A.32), (A.33), and (A.37), it is straightforward to verify that the input-dependent terms in (A.42) are bounded by $c_{\alpha,1}J^*_{eq} + c_{\alpha,2}\|\xi_t - \xi^{sr}\|_2^2 + c_{\alpha,3}\bar{\varepsilon}^2$ for some $c_{\alpha,i} > 0$, $i \in \mathbb{I}_{[1,3]}$ i.e.,

$$\|\breve{\alpha}'(t+n) - \alpha^{sr}\|_2^2 \leq c_{\alpha,1}J^*_{eq} + c_{\alpha,2}\|\xi_t - \xi^{sr}\|_2^2 + c_{\alpha,3}\bar{\varepsilon}^2 + \|H_{ux}^\dagger\|_2^2\|x_t - x^{sr}\|_2^2. \tag{A.44}$$

Plugging the bounds (A.38), (A.39) for the input, (A.40), (A.41) for the output, and (A.44) for $\breve{\alpha}'(t+n)$ into (A.35), and using (4.75), we obtain

$$\begin{aligned}
V(\xi_{t+n}) - V(\xi_t) \leq &-\sum_{k=-n}^{-1} (\|u_{t+k} - \breve{u}^{s*}(t)\|_R^2 + \|y_{t+k} - \breve{y}^{s*}(t)\|_Q^2) \qquad (A.45)\\
&+ c_{J,1}\bar{\varepsilon}^{\beta_\alpha}\|\xi_t - \xi^{sr}\|_2^2 + c_{J,2}\bar{\varepsilon}\sqrt{V(\xi_t)} + c_{J,3}\bar{\varepsilon}^{\beta_\alpha}J^*_{eq} + c_{J,4}(\bar{\varepsilon}^2 + \bar{\varepsilon}^{2+\beta_\alpha})
\end{aligned}$$

for some $c_{J,i} > 0$, $i \in \mathbb{I}_{[1,4]}$. Note that

$$\begin{aligned}
&- \sum_{k=-n}^{-1} (\|u_{t+k} - \check{u}^{s*}(t)\|_R^2 + \|y_{t+k} - \check{y}^{s*}(t)\|_Q^2) \\
\overset{(4.68),(A.31)}{\leq} &- \frac{1}{2} \sum_{k=-n}^{-1} (\|u_{t+k} - \check{u}^{s*}(t)\|_R^2 + \|y_{t+k} - \check{y}^{s*}(t)\|_Q^2) \\
&- \frac{\gamma}{2} \left(\frac{\lambda_{\min}(S)}{2c_g} \|\check{u}^{s*}(t) - u^{sr}\|_2^2 + \frac{1}{2} \|\check{y}^{s*}(t) - y^{sr}\|_S^2 \right) \\
\overset{(A.22)}{\leq} &- c_{J,6} \sum_{k=-n}^{-1} (\|u_{t+k} - u^{sr}\|_2^2 + \|y_{t+k} - y^{sr}\|_2^2)
\end{aligned}$$

with some $c_{J,6} > 0$. This implies

$$\begin{aligned}
&V(\xi_{t+n}) - V(\xi_t) \\
\leq &- (c_{J,6} - c_{J,1}\bar{\varepsilon}^{\beta_\alpha}) \|\xi_t - \xi^{sr}\|_2^2 + c_{J,2}\bar{\varepsilon}\sqrt{V(\xi_t)} + c_{J,3}\bar{\varepsilon}^{\beta_\alpha} J_{eq}^* + c_{J,4}(\bar{\varepsilon}^2 + \bar{\varepsilon}^{2+\beta_\alpha}).
\end{aligned} \tag{A.46}$$

(iii) Candidate solution 2

Assume

$$\sum_{k=-n}^{-1} \|u_{t+k} - \check{u}^{s*}(t)\|_R^2 + \|y_{t+k} - \check{y}^{s*}(t)\|_Q^2 \leq \gamma \|\check{y}^{s*}(t) - y^{sr}\|_S^2. \tag{A.47}$$

(iii.a) Definition of candidate solution

We choose the equilibrium candidate as a convex combination of $(\check{u}^{s*}(t), \check{y}^{s*}(t))$ and the optimal reachable equilibrium, i.e.,

$$\begin{aligned}
\hat{u}^s(t+n) &= \lambda \check{u}^{s*}(t) + (1-\lambda) u^{sr}, \\
\hat{y}^s(t+n) &= \lambda \check{y}^{s*}(t) + (1-\lambda) y^{sr}
\end{aligned} \tag{A.48}$$

for some $\lambda \in (0,1)$ which will be fixed later in the proof, and we denote the corresponding state by $\hat{x}^s(t+n)$. By controllability, there exists an input steering the system from x_{t+n} to $\hat{x}^s(t+n)$ in $L-n \geq n$ steps while satisfying

$$\sum_{k=0}^{L} \|\hat{u}_k(t+n) - \hat{u}^s(t+n)\|_2^2 + \|\hat{y}_k(t+n) - \hat{y}^s(t+n)\|_2^2 \leq \Gamma \|x_{t+n} - \hat{x}^s(t+n)\|_2^2 \tag{A.49}$$

for some $\Gamma > 0$, compare (2.10). In the following, we show that $\hat{u}_k(t+n) \in \mathbb{U}$, $k \in \mathbb{I}_{[0,L]}$ if γ, $(1-\lambda)$, and $\bar{\varepsilon}$ are sufficiently small. Denoting the extended state (4.73) corresponding to $(\check{u}^{s*}(t), \check{y}^{s*}(t))$ by $\check{\xi}^{s*}(t)$, we have

$$\begin{aligned}
&\sum_{k=0}^{n-1} \|\check{u}_k^*(t) - \check{u}^{s*}(t)\|_2^2 + \|\check{y}_k^*(t) - \check{y}^{s*}(t)\|_2^2 \qquad \text{(A.50)}\\
\leq &\sum_{k=0}^{L} \|\check{u}_k^*(t) - \check{u}^{s*}(t)\|_2^2 + \|\check{y}_k^*(t) - \check{y}^{s*}(t)\|_2^2\\
\leq &\bar{c}_u(\|\xi_t - \check{\xi}^{s*}(t)\|_2^2 + \|\check{x}^{s*}(t) - x^{sr}\|_2^2 + \|\check{u}^{s*}(t) - u^{sr}\|_2^2)\\
\overset{(4.68),(A.47)}{\leq} &\bar{c}_u \left(\frac{\lambda_{\max}(S)}{\lambda_{\min}(Q,R)}\bar{c}_u\gamma + c_g\right) \|\check{y}^{s*}(t) - y^{sr}\|_2^2
\end{aligned}$$

for a suitable constant $\bar{c}_u > 0$. The second inequality in (A.50) can be shown analogously to the upper bound in Part (i.b) of the proof, using a controllability argument based on (A.47) with a sufficiently small γ and bounding $\|\alpha^{s*}(t) - \alpha^{sr}\|_2^2$ with $\alpha^{s*}(t) := H_{ux}^{\dagger} \begin{bmatrix} \mathbb{1}_{L+n+1} \otimes \check{u}^{s*}(t) \\ \check{x}^{s*}(t) \\ 1 \end{bmatrix}$ in terms of $\|\check{x}^{s*}(t) - x^{sr}\|_2^2$ and $\|\check{u}^{s*}(t) - u^{sr}\|_2^2$. Moreover, for $k \in \mathbb{I}_{[0,n-1]}$, the difference

$$\|u_{t+k} - \check{u}^{s*}(t)\|_2^2 - \|\check{u}_k^*(t) - \check{u}^{s*}(t)\|_2^2$$

is bounded as in (A.38), and similarly for the output, cf. (A.41). Thus, adding and subtracting $\|\check{u}_k^*(t) - \check{u}^{s*}(t)\|_2^2$ and $\|\check{y}_k^*(t) - \check{y}^{s*}(t)\|_2^2$, we obtain

$$\begin{aligned}
\|\xi_{t+n} - \check{\xi}^{s*}(t)\|_2^2 = &\sum_{k=0}^{n-1} \left(\|u_{t+k} - \check{u}^{s*}(t)\|_2^2 - \|\check{u}_k^*(t) - \check{u}^{s*}(t)\|_2^2\right) \qquad \text{(A.51)}\\
&+ \sum_{k=0}^{n-1} \left(\|y_{t+k} - \check{y}^{s*}(t)\|_2^2 - \|\check{y}_k^*(t) - \check{y}^{s*}(t)\|_2^2\right)\\
&+ \sum_{k=0}^{n-1} \left(\|\check{u}_k^*(t) - \check{u}^{s*}(t)\|_2^2 + \|\check{y}_k^*(t) - \check{y}^{s*}(t)\|_2^2\right)\\
\overset{(A.38),(A.41),(A.50)}{\leq} &\bar{c}_{x,1}\gamma\|\check{y}^{s*}(t) - y^{sr}\|_2^2 + \bar{c}_{x,2}\bar{\varepsilon}^2 + \bar{c}_{x,3}\bar{\varepsilon}\sqrt{V(\xi_t)}
\end{aligned}$$

for some $\bar{c}_{x,i} > 0$, $i \in \mathbb{I}_{[1,3]}$. Writing $\hat{\xi}^s(t+n)$ for the extended state (4.73) corresponding to $(\hat{u}^s(t+n), \hat{y}^s(t+n))$, it holds that

$$\begin{aligned}
\|\check{\xi}^{s*}(t) - \hat{\xi}^s(t+n)\|_2^2 \overset{(A.48)}{=} &(1-\lambda)^2\|\check{\xi}^{s*}(t) - \xi^{sr}\|_2^2 \qquad \text{(A.52)}\\
\overset{(4.68),(4.73)}{\leq} &(1-\lambda)^2 n(1+c_g)\|\check{y}^{s*}(t) - y^{sr}\|_2^2.
\end{aligned}$$

Combining these bounds, we obtain

$$\begin{aligned}\|x_{t+n}-\hat{x}^{\mathrm{s}}(t+n)\|_2^2 &\overset{(4.75)}{\leq} \|T_{\mathrm{x}}\|_2^2\|\xi_{t+n}-\hat{\xi}^{\mathrm{s}}(t+n)\|_2^2 \\ &\leq 2\|T_{\mathrm{x}}\|_2^2(\|\xi_{t+n}-\check{\xi}^{\mathrm{s}*}(t)\|_2^2+\|\check{\xi}^{\mathrm{s}*}(t)-\hat{\xi}^{\mathrm{s}}(t+n)\|_2^2) \\ &\overset{(A.51),(A.52)}{\leq} 2\|T_{\mathrm{x}}\|_2^2\Big((\bar{c}_{\mathrm{x},1}\gamma+n(1+c_g)(1-\lambda)^2)\|\check{y}^{\mathrm{s}*}(t)-y^{\mathrm{sr}}\|_2^2 \\ &\quad+\bar{c}_{\mathrm{x},2}\bar{\varepsilon}^2+\bar{c}_{\mathrm{x},3}\bar{\varepsilon}\sqrt{V(\xi_t)}\Big).\end{aligned} \tag{A.53}$$

Moreover, similar to (A.43), we have

$$\|\check{y}^{\mathrm{s}*}(t)-y^{\mathrm{sr}}\|_2^2 \leq \frac{2}{\lambda_{\min}(S)}(V(\xi_t)+2J_{\mathrm{eq}}^*). \tag{A.54}$$

Thus, using $V(\xi_t)\leq V_{\max}$, if γ, $(1-\lambda)$, and $\bar{\varepsilon}$ are sufficiently small, then x_{t+n} is sufficiently close to $\hat{x}^{\mathrm{s}}(t+n)$ such that (A.49) and $\hat{u}^{\mathrm{s}}(t+n)\in\operatorname{int}(\mathbb{U})$ ensure $\hat{u}_k(t+n)\in\mathbb{U}$ for $k\in\mathbb{I}_{[0,L]}$.

Further, choosing the input-output candidate $(\hat{u}(t+n),\hat{y}(t+n))$ such that $(\hat{u}_k(t+n),\hat{y}_k(t+n))=(u_{t+n+k},y_{t+n+k})$ for $k\in\mathbb{I}_{[-n,-1]}$ and $(\hat{u}_k(t+n),\hat{y}_k(t+n))=(\hat{u}^{\mathrm{s}}(t+n),\hat{y}^{\mathrm{s}}(t+n))$ for $k\in\mathbb{I}_{[L-n,L]}$ satisfies (4.76c) and (4.76d), respectively. Finally, with H_{ux} as in (A.27), we choose

$$\hat{\alpha}(t+n)=H_{\mathrm{ux}}^{\dagger}\begin{bmatrix}\hat{u}(t+n)\\ x_t\\ 1\end{bmatrix}, \tag{A.55}$$

which implies that all constraints of Problem 4.4 are fulfilled.

(iii.b) Lyapunov function decay

Using the above candidate solution, we obtain

$$\begin{aligned}V(\xi_{t+n})-V(\xi_t)\leq&\sum_{k=-n}^{L}\Big(\|\hat{u}_k(t+n)-\hat{u}^{\mathrm{s}}(t+n)\|_R^2+\|\hat{y}_k(t+n)-\hat{y}^{\mathrm{s}}(t+n)\|_Q^2\Big)\\ &+\lambda_\alpha\bar{\varepsilon}^{\beta_\alpha}(\|\hat{\alpha}(t+n)-\alpha^{\mathrm{sr}}\|_2^2-\|\check{\alpha}^*(t)-\alpha^{\mathrm{sr}}\|_2^2)\\ &-\sum_{k=0}^{L}(\|\check{u}_k^*(t)-\check{u}^{\mathrm{s}*}(t)\|_R^2+\|\check{y}_k^*(t)-\check{y}^{\mathrm{s}*}(t)\|_Q^2)\\ &+\|\hat{y}^{\mathrm{s}}(t+n)-y^{\mathrm{r}}\|_S^2-\|\check{y}^{\mathrm{s}*}(t)-y^{\mathrm{r}}\|_S^2.\end{aligned} \tag{A.56}$$

Similar to [69, Inequality (19)], strong convexity of the cost in (4.67) implies

$$\|\hat{y}^{\mathrm{s}}(t+n)-y^{\mathrm{r}}\|_S^2-\|\check{y}^{\mathrm{s}*}(t)-y^{\mathrm{r}}\|_S^2\leq-(1-\lambda^2)\|\check{y}^{\mathrm{s}*}(t)-y^{\mathrm{sr}}\|_S^2. \tag{A.57}$$

The definition of $\hat{\alpha}(t+n)$ implies

$$\|\hat{\alpha}(t+n) - \alpha^{\mathrm{sr}}\|_2^2 \leq \|H_{\mathrm{ux}}^{\dagger}\|_2^2(\|\hat{u}(t+n) - \mathbb{1}_{L+n+1} \otimes u^{\mathrm{sr}}\|_2^2 + \|x_t - x^{\mathrm{sr}}\|_2^2). \quad \text{(A.58)}$$

Using $\hat{u}^{\mathrm{s}}(t+n) - u^{\mathrm{sr}} = \lambda(\check{u}^{\mathrm{s}*}(t) - u^{\mathrm{sr}})$, as well as (4.79), (A.22), (A.43), (A.49), (A.53), and (A.54) we obtain

$$\begin{aligned} &\|\hat{\alpha}(t+n) - \alpha^{\mathrm{sr}}\|_2^2 && \text{(A.59)} \\ \leq &\bar{c}_{\alpha,1} J_{\mathrm{eq}}^* + \bar{c}_{\alpha,2}\|\xi_t - \xi^{\mathrm{sr}}\|_2^2 + \bar{c}_{\alpha,3}\bar{\varepsilon}^2 + \bar{c}_{\alpha,4}\bar{\varepsilon}\sqrt{V(\xi_t)} + \|H_{\mathrm{ux}}^{\dagger}\|_2^2\|x_t - x^{\mathrm{sr}}\|_2^2 \end{aligned}$$

for some $\bar{c}_{\alpha,i} > 0$, $i \in \mathbb{I}_{[1,4]}$. Plugging (A.49), (A.57), and (A.59) into (A.56) and using

$$\begin{aligned} &\sum_{k=-n}^{-1} \|\hat{u}_k(t+n) - \hat{u}^{\mathrm{s}}(t+n)\|_R^2 + \|\hat{y}_k(t+n) - \hat{y}^{\mathrm{s}}(t+n)\|_Q^2 \\ \leq &\lambda_{\max}(Q,R)\|\xi_{t+n} - \hat{\xi}^{\mathrm{s}}(t+n)\|_2^2 \end{aligned}$$

as well as (4.75), there exist $\bar{c}_{\mathrm{J},i} > 0$, $i \in \mathbb{I}_{[1,7]}$, such that

$$\begin{aligned} &V(\xi_{t+n}) - V(\xi_t) && \text{(A.60)} \\ \leq &- \sum_{k=-n}^{-1} (\|u_{t+k} - \check{u}^{\mathrm{s}*}(t)\|_R^2 + \|y_{t+k} - \check{y}^{\mathrm{s}*}(t)\|_Q^2) \\ &+ \bar{c}_{\mathrm{J},1}\|\xi_{t+n} - \hat{\xi}^{\mathrm{s}}(t+n)\|_2^2 - (1-\lambda^2)\|\check{y}^{\mathrm{s}*}(t) - y^{\mathrm{sr}}\|_S^2 \\ &+ \lambda_\alpha \bar{\varepsilon}^{\beta_\alpha}\left(\bar{c}_{\alpha,1}J_{\mathrm{eq}}^* + (\bar{c}_{\alpha,2} + \|H_{\mathrm{ux}}^{\dagger}\|_2^2\|T_{\mathrm{x}}\|_2^2)\|\xi_t - \xi^{\mathrm{sr}}\|_2^2 + \bar{c}_{\alpha,3}\bar{\varepsilon}^2 + \bar{c}_{\alpha,4}\bar{\varepsilon}\sqrt{V(\xi_t)}\right) \\ \overset{\text{(A.53)}}{\leq} &- \sum_{k=-n}^{-1} (\|u_{t+k} - \check{u}^{\mathrm{s}*}(t)\|_R^2 + \|y_{t+k} - \check{y}^{\mathrm{s}*}(t)\|_Q^2) \\ &+ (\bar{c}_{\mathrm{J},2}\gamma + \bar{c}_{\mathrm{J},3}(1-\lambda)^2 - \lambda_{\min}(S)(1-\lambda^2))\|\check{y}^{\mathrm{s}*}(t) - y^{\mathrm{sr}}\|_2^2 + \bar{c}_{\mathrm{J},4}\bar{\varepsilon}^{\beta_\alpha}\|\xi_t - \xi^{\mathrm{sr}}\|_2^2 \\ &+ \bar{c}_{\mathrm{J},5}(\bar{\varepsilon} + \bar{\varepsilon}^{1+\beta_\alpha})\sqrt{V(\xi_t)} + \bar{c}_{\mathrm{J},6}\bar{\varepsilon}^{\beta_\alpha}J_{\mathrm{eq}}^* + \bar{c}_{\mathrm{J},7}(\bar{\varepsilon}^2 + \bar{\varepsilon}^{2+\beta_\alpha}). \end{aligned}$$

If γ and $(1-\lambda)$ are sufficiently small such that

$$\bar{c}_{\mathrm{J},2}\gamma + \bar{c}_{\mathrm{J},3}(1-\lambda)^2 - \bar{c}_{\mathrm{J},4}(1-\lambda^2) < 0,$$

then (4.68) and (A.22) lead to

$$\begin{aligned} &V(\xi_{t+n}) - V(\xi_t) && \text{(A.61)} \\ \leq &-(\bar{c}_{\mathrm{J},9} - \bar{c}_{\mathrm{J},4}\bar{\varepsilon}^{\beta_\alpha})\|\xi_t - \xi^{\mathrm{sr}}\|_2^2 + \bar{c}_{\mathrm{J},6}\bar{\varepsilon}^{\beta_\alpha}J_{\mathrm{eq}}^* + \bar{c}_{\mathrm{J},5}(\bar{\varepsilon} + \bar{\varepsilon}^{1+\beta_\alpha})\sqrt{V(\xi_t)} + \bar{c}_{\mathrm{J},7}(\bar{\varepsilon}^2 + \bar{\varepsilon}^{2+\beta_\alpha}) \end{aligned}$$

for some $\tilde{c}_{J,9} > 0$.

(iv) Practical stability

Using (A.46) and (A.61) and letting $\bar{\varepsilon} < 1$, there exist $\tilde{c}_{J,i} > 0$, $i \in \mathbb{I}_{[1,5]}$, such that

$$V(\xi_{t+n}) - V(\xi_t) \leq -(\tilde{c}_{J,4} - \tilde{c}_{J,5}\bar{\varepsilon}^{\beta_\alpha})\|\xi_t - \xi^{\mathrm{sr}}\|_2^2 + \tilde{c}_{J,1}\bar{\varepsilon}\sqrt{V(\xi_t)} + \tilde{c}_{J,2}\bar{\varepsilon}^{\beta_\alpha} J_{\mathrm{eq}}^* + \tilde{c}_{J,3}\bar{\varepsilon}^2. \tag{A.62}$$

If $\bar{\varepsilon}_{\max} < \left(\frac{\tilde{c}_{J,4}}{\tilde{c}_{J,5}}\right)^{\frac{1}{\beta_\alpha}}$, then this together with (4.79) and $V(\xi_t) \leq V_{\max}$ implies (4.80) for $\check{c}_{\mathrm{V}} := 1 - \frac{\tilde{c}_{J,4} - \tilde{c}_{J,5}\bar{\varepsilon}^{\beta_\alpha}}{c_{\mathrm{u}}} < 1$ and some $\beta_{\mathrm{d}} \in \mathcal{K}_\infty$. If $\bar{\varepsilon}_{\max}$ is sufficiently small, then $V(\xi_{t+n}) \leq V_{\max}$ such that the MPC scheme is recursively feasible and Inequalities (4.79) and (4.80) hold for all $t = ni$, $i \in \mathbb{I}_{\geq 0}$. ■

A.3 Proof of Proposition 5.2

In the following, we provide a proof of Proposition 5.2 by considering two complementary cases with two different candidate solutions at time $t + n$. First, in Proposition A.1, we prove the statement under the assumption that the tracking cost w.r.t. the artificial steady-state is large, quantified via a suitable inequality. Thereafter, we consider the complementary case, which, together with Proposition A.1, proves the full statement of Proposition 5.2.

Proof of Proposition 5.2 - candidate 1

Proposition A.1. *Suppose Assumptions 5.1–5.7 hold. Then, there exist* $V_{\max}, J_{\mathrm{eq}}^{\max} > 0$ *such that, if* $V(x_t) \leq V_{\max}$, $J_{\mathrm{eq,Lin}}^*(x_t) \leq J_{\mathrm{eq}}^{\max}$, $J_{\mathrm{eq,Lin}}^*(x_{t+n}) \leq J_{\mathrm{eq}}^{\max}$, *and if there exists* $\gamma_1 > 0$ *such that*

$$\sum_{k=0}^{n-1} \|\bar{x}_k^*(t) - x^{\mathrm{s}*}(t)\|_2^2 + \|\bar{u}_k^*(t) - u^{\mathrm{s}*}(t)\|_2^2 \geq \gamma_1 \|x^{\mathrm{s}*}(t) - x_{\mathrm{Lin}}^{\mathrm{sr}}(x_t)\|_2^2, \tag{A.63}$$

then Problem 5.1 is feasible at time $t + n$ *and there exists a constant* $0 < c_{\mathrm{V}1} < 1$ *such that* $V(x_{t+n}) \leq c_{\mathrm{V}1} V(x_t)$.

Proof. First, we define the candidate equilibrium $x^{\mathrm{s}\prime}(t+n), u^{\mathrm{s}\prime}(t+n)$ and the first $L - n$ components of the candidate input $\bar{u}'(t+n)$ (Part (i)). Thereafter, in Part (ii), we derive useful bounds involving this candidate trajectory. Next, we show that the state of this candidate solution at time $L-n$ is sufficiently close to the candidate artificial steady-state and thus, we can construct a local deadbeat controller to steer the system to this steady-state (Part (iii)). Finally, we show $V(x_{t+n}) \leq c_{V1} V(x_t)$ in Part (iv).

Note that x_t lies in the set $\{x \in \mathbb{R}^n \mid V(x) \leq V_{\max}\}$, which is compact due to the lower bound (5.19) and Assumption 5.6. We define the set X as the union of the L-step reachable sets of the linearized and the nonlinear dynamics (compare [127, Definition 2]), starting at x_t. Using that the dynamics (5.1) are Lipschitz continuous and the input constraints are compact, we conclude that X is compact. Throughout this proof, whenever we apply Inequality (5.8), we use the fact that all involved states lie in X and we use the corresponding constant c_X.

(i) Definition of candidate solution for $k \in \mathbb{I}_{[0,L-n]}$

We choose the candidate equilibrium input as the old solution, i.e., $u^{s\prime}(t+n) = u^{s*}(t)$. According to Assumption 5.5, there exists a unique equilibrium state $x^{s\prime}(t+n)$ for the system linearized at x_{t+n} such that

$$x^{s\prime}(t+n) = A_{x_{t+n}} x^{s\prime}(t+n) + B u^{s\prime}(t+n) + e_{x_{t+n}}.$$

The corresponding output $y^{s\prime}(t+n)$ is computed via (5.18f). Further, for $k \in \mathbb{I}_{[0,L-n-1]}$, we choose the candidate input as the previously optimal one, i.e., $\bar{u}'_k(t+n) = \bar{u}^*_{k+n}(t)$. This leads to the state trajectory candidate $\bar{x}'_k(t+n)$, $k \in \mathbb{I}_{[0,L-n]}$, resulting from an open-loop application of $\bar{u}'(t+n)$ with initial condition x_{t+n} to the dynamics linearized at time $t+n$, i.e., $\bar{x}'_0(t+n) = x_{t+n}$ and

$$\bar{x}'_{k+1}(t+n) = f_{x_{t+n}}(\bar{x}'_k(t+n), \bar{u}'_k(t+n)) = A_{x_{t+n}} \bar{x}'_k(t+n) + B\bar{u}'_k(t+n) + e_{x_{t+n}},$$

for $k \in \mathbb{I}_{[0,L-n-1]}$.

(ii) Bounds on candidate solution

Throughout Part (ii) of the proof, let $k \in \mathbb{I}_{[0,L-n]}$. Further, abbreviate $\underline{q} := \lambda_{\min}(Q)$, $\bar{q} := \lambda_{\max}(Q)$, and similarly for $\underline{s}$, $\bar{s}$, $\underline{r}$, $\bar{r}$. It clearly holds that

$$\begin{aligned}
&\sum_{k=0}^{L-1} \|\bar{x}^*_k(t) - x^{s*}(t)\|_Q^2 \qquad\qquad\qquad\qquad \text{(A.64)}\\
\leq &\sum_{k=0}^{L-1} \|\bar{x}^*_k(t) - x^{s*}(t)\|_Q^2 + \|y^{s*}(t) - y^{\mathrm{r}}\|_S^2 - J^*_{\mathrm{eq,Lin}}(x_t)\\
\leq &J^*_L(x_t) - J^*_{\mathrm{eq,Lin}}(x_t) = V(x_t),
\end{aligned}$$

and hence,

$$\sum_{k=0}^{L-1} \|\bar{x}_k^*(t) - x^{s*}(t)\|_2^2 \leq \frac{1}{\underline{q}} V(x_t). \tag{A.65}$$

Next, we bound several expressions involving the optimal solution at time t and the candidate solution at time $t+n$.

(ii.a) Bound on $\|\bar{x}_n^*(t) - x_{t+n}\|_2$

Using (5.8), which holds by Assumption 5.2, and (5.4), we obtain

$$\begin{aligned}
&\|\bar{x}_n^*(t) - x_{t+n}\|_2 \qquad \text{(A.66)}\\
=&\|f_{x_t}(\bar{x}_{n-1}^*(t), \bar{u}_{n-1}^*(t)) - f_{x_{t+n-1}}(x_{t+n-1}, \bar{u}_{n-1}^*(t))\|_2\\
\overset{(5.8)}{\leq}& c_X \|\bar{x}_{n-1}^*(t) - x_t\|_2^2 + L_f \|\bar{x}_{n-1}^*(t) - x_{t+n-1}\|_2\\
\overset{(5.21)}{\leq}& 2c_X \|\bar{x}_{n-1}^*(t) - x^{s*}(t)\|_2^2 + 2c_X \|x^{s*}(t) - x_t\|_2^2 + L_f \|\bar{x}_{n-1}^*(t) - x_{t+n-1}\|_2\\
\overset{(A.65)}{\leq}& 2\frac{c_X}{\underline{q}} V(x_t) + L_f \|\bar{x}_{n-1}^*(t) - x_{t+n-1}\|_2 \leq \cdots \leq 2\frac{c_X}{\underline{q}} V(x_t) \sum_{k=0}^{n-2} L_f^k,
\end{aligned}$$

where the summand for $k = n-1$ vanishes since $\|\bar{x}_1^*(t) - x_{t+1}\|_2 = 0$.

(ii.b) Bound on $\|\bar{x}_k'(t+n) - \bar{x}_{k+n}^*(t)\|_2$

Define $\{a_k\}_{k=0}^{L-n}$ recursively in dependence of $V(x_t)$ as

$$\begin{aligned}
a_0 &:= 2\frac{c_X}{\underline{q}} V(x_t) \sum_{k=0}^{n-2} L_f^k,\\
a_k &:= 2c_X a_{k-1}^2 + L_f a_{k-1} + 16\frac{c_X}{\underline{q}} V(x_t) + 8c_X \left(2\frac{c_X}{\underline{q}} \sum_{k=0}^{n-2} L_f^k\right)^2 V(x_t)^2
\end{aligned}$$

for $k = 1, \ldots, L-n$. In the following, we prove that for any $k \in \mathbb{I}_{[0,L-n]}$

$$\|\bar{x}_k'(t+n) - \bar{x}_{k+n}^*(t)\|_2 \leq a_k. \tag{A.67}$$

According to (A.66) and using $\bar{x}_0'(t+n) = x_{t+n}$, Inequality (A.67) holds for $k = 0$.

Using an induction argument over k, we have

$$\begin{aligned}
&\|\bar{x}'_k(t+n) - \bar{x}^*_{k+n}(t)\|_2 \\
=&\|f_{x_{t+n}}(\bar{x}'_{k-1}(t+n), \bar{u}^*_{k+n-1}(t)) - f_{x_t}(\bar{x}^*_{k+n-1}(t), \bar{u}^*_{k+n-1}(t))\|_2 \\
\overset{(5.8)}{\leq}& c_X\|\bar{x}'_{k-1}(t+n) - x_{t+n}\|_2^2 + c_X\|\bar{x}^*_{k+n-1}(t) - x_t\|_2^2 \\
&+ L_f\|\bar{x}'_{k-1}(t+n) - \bar{x}^*_{k+n-1}(t)\|_2 \\
\overset{(5.21)}{\leq}& 2c_X\|\bar{x}'_{k-1}(t+n) - \bar{x}^*_{k+n-1}(t)\|_2^2 + L_f\|\bar{x}'_{k-1}(t+n) - \bar{x}^*_{k+n-1}(t)\|_2 \\
&+ 2c_X\|\bar{x}^*_{k+n-1}(t) - x_{t+n}\|_2^2 + c_X\|\bar{x}^*_{k+n-1}(t) - x_t\|_2^2 \\
\overset{(5.21),(A.67)}{\leq}& 2c_X a_{k-1}^2 + L_f a_{k-1} + 6c_X\|\bar{x}^*_{k+n-1}(t) - x^{s*}(t)\|_2^2 \\
&+ 4c_X\|x^{s*}(t) - x_{t+n}\|_2^2 + 2c_X\|x^{s*}(t) - x_t\|_2^2 \\
\overset{(5.21),(A.65)}{\leq}& 2c_X a_{k-1}^2 + L_f a_{k-1} + 8\frac{c_X}{\underline{q}}V(x_t) + 8c_X\|x^{s*}(t) - \bar{x}^*_n(t)\|_2^2 \\
&+ 8c_X\|\bar{x}^*_n(t) - x_{t+n}\|_2^2 \\
\overset{(A.65),(A.66)}{\leq}& 2c_X a_{k-1}^2 + L_f a_{k-1} + 16\frac{c_X}{\underline{q}}V(x_t) + 8c_X\left(2\frac{c_X}{\underline{q}}\sum_{k=0}^{n-2} L_f^k\right)^2 V(x_t)^2 = a_k,
\end{aligned}$$

which proves (A.67). Note that a_k is a polynomial in $V(x_t)$ which becomes arbitrarily small if $V(x_t)$ is sufficiently small.

(ii.c) Bound on $\|x^{s*}(t) - x^{s\prime}(t+n)\|_2$

Note that

$$\begin{aligned}
(I - A_{x_t})x^{s*}(t) &= Bu^{s*}(t) + e_{x_t}, \\
(I - A_{x_{t+n}})x^{s\prime}(t+n) &= Bu^{s\prime}(t+n) + e_{x_{t+n}}.
\end{aligned}$$

Using additionally $u^{s\prime}(t+n) = u^{s*}(t)$, this implies

$$\begin{aligned}
(I - A_{x_{t+n}})x^{s\prime}(t+n) =& x^{s*}(t) - x^{s*}(t) + Bu^{s\prime}(t+n) + e_{x_{t+n}} \\
=&(I - A_{x_t})x^{s*}(t) + e_{x_{t+n}} - e_{x_t}.
\end{aligned}$$

Using that $(I - A_{x_{t+n}})$ is invertible by Assumption 5.5, we obtain

$$\begin{aligned}
x^{s\prime}(t+n) &= (I - A_{x_{t+n}})^{-1}\left((I - A_{x_t})x^{s*}(t) + e_{x_{t+n}} - e_{x_t}\right) \\
&= x^{s*}(t) + (I - A_{x_{t+n}})^{-1}\left((A_{x_{t+n}} - A_{x_t})x^{s*}(t) + e_{x_{t+n}} - e_{x_t}\right).
\end{aligned}$$

Moreover, Assumption 5.5 implies $\|(I - A_{x_{t+n}})^{-1}\|_2 \leq \frac{1}{\underline{\sigma}}$ and hence, we arrive at

$$\begin{aligned} &\|x^{s\prime}(t+n) - x^{s*}(t)\|_2 \qquad\qquad (A.68)\\ \leq &\frac{1}{\underline{\sigma}}\|f_{x_{t+n}}(x^{s*}(t), u^{s*}(t)) - f_{x_t}(x^{s*}(t), u^{s*}(t))\|_2 \\ \overset{(5.8)}{\leq} &\frac{c_X}{\underline{\sigma}}\left(\|x^{s*}(t) - x_{t+n}\|_2^2 + \|x^{s*}(t) - x_t\|_2^2\right). \end{aligned}$$

Together with (A.65) and (A.66), this implies

$$\begin{aligned} &\|x^{s\prime}(t+n) - x^{s*}(t)\|_2 \qquad\qquad (A.69)\\ \overset{(5.21),(A.65)}{\leq} &\frac{c_X}{\underline{\sigma}}\left(2\|x^{s*}(t) - \bar{x}_n^*(t)\|_2^2 + 2\|\bar{x}_n^*(t) - x_{t+n}\|_2^2 + \frac{V(x_t)}{\underline{q}}\right) \\ \overset{(A.65)}{\leq} &\frac{c_X}{\underline{\sigma}}\left(3\frac{V(x_t)}{\underline{q}} + 2\|\bar{x}_n^*(t) - x_{t+n}\|_2^2\right) \\ \overset{(A.66)}{\leq} &\frac{c_X}{\underline{\sigma}}\left(3\frac{V(x_t)}{\underline{q}} + 2\left(2\frac{c_X}{\underline{q}}\sum_{k=0}^{n-2} L_f^k\right)^2 V(x_t)^2\right) \\ =: &c_1 V(x_t)^2 + c_2 V(x_t). \end{aligned}$$

(ii.d) Bound on $\|y^{s\prime}(t+n) - y^{s*}(t)\|_2$

By using an inequality of the form (5.8) for the vector field h (note that h is sufficiently smooth by Assumption 5.2), there exist constants $c_{Xh}, L_h \geq 0$ such that

$$\begin{aligned} &\|y^{s\prime}(t+n) - y^{s*}(t)\|_2 \\ = &\|h_{x_{t+n}}(x^{s\prime}(t+n), u^{s*}(t)) - h_{x_t}(x^{s*}(t), u^{s*}(t))\|_2 \\ \leq &c_{Xh}\|x^{s\prime}(t+n) - x_{t+n}\|_2^2 + c_{Xh}\|x^{s*}(t) - x_t\|_2^2 + L_h\|x^{s\prime}(t+n) - x^{s*}(t)\|_2 \\ \overset{(A.65),(A.69)}{\leq} &c_{Xh}\|x^{s\prime}(t+n) - x_{t+n}\|_2^2 + \frac{c_{Xh}}{\underline{q}}V(x_t) + L_h(c_1 V(x_t)^2 + c_2 V(x_t)). \end{aligned}$$

Moreover, using $\|a+b+c\|_2^2 \leq 2\|a\|_2^2 + 4\|b\|_2^2 + 4\|c\|_2^2$, which holds for arbitrary a, b, c due to (5.21), we obtain

$$\begin{aligned} &\|x^{s\prime}(t+n) - x_{t+n}\|_2^2 \\ \leq &2\|x^{s\prime}(t+n) - x^{s*}(t)\|_2^2 + 4\|x^{s*}(t) - \bar{x}_n^*(t)\|_2^2 + 4\|\bar{x}_n^*(t) - x_{t+n}\|_2^2 \\ \overset{(A.65),(A.66),(A.69)}{\leq} &2(c_1 V(x_t)^2 + c_2 V(x_t))^2 + \frac{4}{\underline{q}}V(x_t) + 4\left(2\frac{c_X}{\underline{q}}V(x_t)\sum_{k=0}^{n-2} L_f^k\right)^2. \end{aligned}$$

Hence, using $V(x_t) \leq V_{\max}$, there exists $c_3 > 0$ such that

$$\|y^{s\prime}(t+n) - y^{s*}(t)\|_2 \leq c_3 V(x_t). \tag{A.70}$$

(iii) Appending deadbeat controller

In the following, we show that for $V_{\max}$ sufficiently small $x^{s\prime}(t+n)$ is sufficiently close to $\bar{x}'_{L-n}(t+n)$ such that we can append a deadbeat controller steering the state to $x^{s\prime}(t+n)$ in n steps. To be precise, combining (A.67) with $k = L-n$ and (A.69), we obtain

$$\begin{aligned}
&\|\bar{x}'_{L-n}(t+n) - x^{s\prime}(t+n)\|_2 \qquad\qquad\qquad\qquad\qquad\qquad\qquad\qquad\qquad (A.71)\\
\leq &\|\bar{x}'_{L-n}(t+n) - \bar{x}^*_L(t)\|_2 + \|\bar{x}^*_L(t) - x^{s*}(t)\|_2 + \|x^{s*}(t) - x^{s\prime}(t+n)\|_2\\
\overset{(A.67),(A.69)}{\leq} & a_{L-n} + c_1 V(x_t)^2 + c_2 V(x_t),
\end{aligned}$$

where for the second inequality we used that $\bar{x}^*_L(t) = x^{s*}(t)$ due to the terminal equality constraint (5.18c). Assumption 5.4 implies the existence of an input $\bar{u}'_k(t+n), k \in \mathbb{I}_{[L-n,L-1]}$, steering the state to $\bar{x}'_L(t+n) = x^{s\prime}(t+n)$ while satisfying

$$\begin{aligned}
&\sum_{k=L-n}^{L-1} \|\bar{x}'_k(t+n) - x^{s\prime}(t+n)\|_2 + \|\bar{u}'_k(t+n) - u^{s\prime}(t+n)\|_2 \qquad\qquad (A.72)\\
\overset{(5.17)}{\leq} & \Gamma\|\bar{x}'_{L-n}(t+n) - x^{s\prime}(t+n)\|_2 \overset{(A.71)}{\leq} \Gamma(a_{L-n} + c_1 V(x_t)^2 + c_2 V(x_t)).
\end{aligned}$$

If $V_{\max}$ and hence $V(x_t)$ and a_{L-n} are sufficiently small, then $\bar{u}'_k(t+n) \in \mathbb{U}$ for $k \in \mathbb{I}_{[L-n,L-1]}$ (note that $u^{s\prime}(t+n) \in \text{int}(\mathbb{U})$), i.e., the input candidate satisfies the input constraints.

(iv) Invariance of $V(x_t) \leq V_{\max}$

So far, we have only shown that the MPC scheme is feasible at time $t+n$. It remains to be shown that there exists a constant $0 < c_{V1} < 1$ such that $V(x_{t+n}) \leq c_{V1} V(x_t)$. Note that

$$\begin{aligned}
J^*_L(x_{t+n}) - J^*_L(x_t) \leq &\sum_{k=0}^{L-1} \left(\|\bar{x}'_k(t+n) - x^{s\prime}(t+n)\|_Q^2 + \|\bar{u}'_k(t+n) - u^{s\prime}(t+n)\|_R^2\right)\\
&- \sum_{k=0}^{L-1} (\|\bar{x}^*_k(t) - x^{s*}(t)\|_Q^2 + \|\bar{u}^*_k(t) - u^{s*}(t)\|_R^2) \qquad (A.73)\\
&+ \|y^{s\prime}(t+n) - y^{r}\|_S^2 - \|y^{s*}(t) - y^{r}\|_S^2.
\end{aligned}$$

We now bound several terms on the right-hand side of (A.73) separately. The definition of the input candidate implies

$$\begin{aligned}
&\sum_{k=0}^{L-1} \|\bar{u}_k'(t+n) - u^{s\prime}(t+n)\|_R^2 - \|\bar{u}_k^*(t) - u^{s*}(t)\|_R^2 \\
&\overset{(A.72)}{\leq} -\sum_{k=0}^{n-1} \|\bar{u}_k^*(t) - u^{s*}(t)\|_R^2 + \bar{r} \cdot \Gamma^2 (a_{L-n} + c_1 V(x_t)^2 + c_2 V(x_t))^2.
\end{aligned}$$

Using $\|y^{s*}(t) - y^r\|_S \leq \sqrt{J_L^*(x_t)}$, we obtain

$$\begin{aligned}
&\|y^{s\prime}(t+n) - y^r\|_S^2 - \|y^{s*}(t) - y^r\|_S^2 \\
&\overset{(5.22)}{\leq} \|y^{s\prime}(t+n) - y^{s*}(t)\|_S^2 + 2\|y^{s\prime}(t+n) - y^{s*}(t)\|_S \|y^{s*}(t) - y^r\|_S \\
&\leq \|y^{s\prime}(t+n) - y^{s*}(t)\|_S^2 + 2\|y^{s\prime}(t+n) - y^{s*}(t)\|_S \sqrt{V(x_t) + J_{\text{eq,Lin}}^*(x_t)} \\
&\overset{(A.70)}{\leq} \bar{s} c_3^2 V(x_t)^2 + 2\sqrt{\bar{s}} c_3 V(x_t) \sqrt{V(x_t) + J_{\text{eq,Lin}}^*(x_t)}.
\end{aligned} \tag{A.74}$$

Finally, note that

$$\begin{aligned}
&\sum_{k=0}^{L-1} \|\bar{x}_k'(t+n) - x^{s\prime}(t+n)\|_Q^2 - \|\bar{x}_k^*(t) - x^{s*}(t)\|_Q^2 \\
&= -\sum_{k=0}^{n-1} \|\bar{x}_k^*(t) - x^{s*}(t)\|_Q^2 + \underbrace{\sum_{k=L-n}^{L-1} \|\bar{x}_k'(t+n) - x^{s\prime}(t+n)\|_Q^2}_{\overset{(A.72)}{\leq} \bar{q}\Gamma^2(a_{L-n}+c_1 V(x_t)^2 + c_2 V(x_t))^2} \\
&\quad + \sum_{k=0}^{L-n-1} \left(\|\bar{x}_k'(t+n) - x^{s\prime}(t+n)\|_Q^2 - \|\bar{x}_{k+n}^*(t) - x^{s*}(t)\|_Q^2 \right).
\end{aligned}$$

We bound the last sum on the right-hand side further as

$$
\begin{aligned}
&\sum_{k=0}^{L-n-1} \|\bar{x}_k'(t+n) - x^{s\prime}(t+n)\|_Q^2 - \|\bar{x}_{k+n}^*(t) - x^{s*}(t)\|_Q^2 \\
\overset{(5.22)}{\leq} &\sum_{k=0}^{L-n-1} \|\bar{x}_k'(t+n) - x^{s\prime}(t+n) - \bar{x}_{k+n}^*(t) + x^{s*}(t)\|_Q^2 \\
&+ 2\|\bar{x}_k'(t+n) - x^{s\prime}(t+n) - \bar{x}_{k+n}^*(t) + x^{s*}(t)\|_Q \|\bar{x}_{k+n}^*(t) - x^{s*}(t)\|_Q \\
\overset{(5.21),(A.64)}{\leq} &\sum_{k=0}^{L-n-1} 2\|\bar{x}_k'(t+n) - \bar{x}_{k+n}^*(t)\|_Q^2 + 2\|x^{s*}(t) - x^{s\prime}(t+n)\|_Q^2 \\
&+ 2\sqrt{V(x_t)}\left(\|\bar{x}_k'(t+n) - \bar{x}_{k+n}^*(t)\|_Q + \|x^{s*}(t) - x^{s\prime}(t+n)\|_Q\right) \\
\overset{(A.67),(A.69)}{\leq} &\sum_{k=0}^{L-n-1} 2\bar{q}(a_k^2 + (c_1 V(x_t)^2 + c_2 V(x_t))^2) \\
&+ 2\sqrt{\bar{q} V(x_t)}(a_k + c_1 V(x_t)^2 + c_2 V(x_t)).
\end{aligned}
$$

Inserting all of the derived bounds into (A.73), we arrive at

$$
\begin{aligned}
&J_L^*(x_{t+n}) - J_L^*(x_t) \qquad\qquad (A.75)\\
\leq &-\sum_{k=0}^{n-1}(\|\bar{x}_k^*(t) - x^{s*}(t)\|_Q^2 + \|\bar{u}_k^*(t) - u^{s*}(t)\|_R^2) \\
&+ \sum_{k=0}^{L-n-1}\left(2\bar{q}(a_k^2 + (c_1 V(x_t)^2 + c_2 V(x_t))^2) + 2\sqrt{\bar{q} V(x_t)}(a_k + c_1 V(x_t)^2 + c_2 V(x_t)\right) \\
&+ (\bar{q} + \bar{r})\Gamma^2(a_{L-n} + c_1 V(x_t)^2 + c_2 V(x_t))^2 + \bar{s} c_3^2 V(x_t)^2 \\
&+ 2\sqrt{\bar{s}} c_3 V(x_t)\sqrt{V(x_t) + J_{\text{eq,Lin}}^*(x_t)}.
\end{aligned}
$$

Note that (A.63) implies

$$
\begin{aligned}
&-\sum_{k=0}^{n-1}(\|\bar{x}_k^*(t) - x^{s*}(t)\|_Q^2 + \|\bar{u}_k^*(t) - u^{s*}(t)\|_R^2) \qquad (A.76)\\
\overset{(A.63)}{\leq} &-\frac{1}{2}\sum_{k=0}^{n-1}(\|\bar{x}_k^*(t) - x^{s*}(t)\|_Q^2 + \|\bar{u}_k^*(t) - u^{s*}(t)\|_R^2) \\
&- \frac{\gamma_1 \min\{\underline{q}, \underline{r}\}}{2}\|x^{s*}(t) - x_{\text{Lin}}^{\text{sr}}(x_t)\|_2^2 \\
\overset{(5.21)}{\leq} &-\frac{\min\{\underline{q}, \underline{r}\} \cdot \min\{1, \gamma_1\}}{4}\|x_t - x_{\text{Lin}}^{\text{sr}}(x_t)\|_2^2.
\end{aligned}
$$

The local upper bound (5.20), which holds for $\|x_t - x^{\text{sr}}_{\text{Lin}}(x_t)\|_2 \leq \delta$, implies that for all $x_t \in X$

$$V(x_t) \leq c_{\text{u,V}} \|x_t - x^{\text{sr}}_{\text{Lin}}(x_t)\|_2^2, \tag{A.77}$$

where $c_{\text{u,V}} := \max\{\frac{V_{\max}}{\delta^2}, c_{\text{u}}\}$. Thus, we obtain

$$\begin{aligned} &-\sum_{k=0}^{n-1}(\|\bar{x}^*_k(t) - x^{\text{s}*}(t)\|_Q^2 + \|\bar{u}^*_k(t) - u^{\text{s}*}(t)\|_R^2) \\ \overset{(A.76),(A.77)}{\leq} &-\frac{\min\{\underline{q}, \underline{r}\} \cdot \min\{1, \gamma_1\}}{4c_{\text{u,V}}} V(x_t). \end{aligned} \tag{A.78}$$

Note that all positive terms on the right-hand side of (A.75) are either at least of order $V(x_t)^2$, or they are of order $V(x_t)$ but are multiplied by $\sqrt{V(x_t) + J^*_{\text{eq,Lin}}(x_t)}$. Hence, if we plug (A.78) into (A.75) and choose $V_{\max}$ and $J^{\max}_{\text{eq}}$ sufficiently small, then we obtain

$$J^*_L(x_{t+n}) - J^*_L(x_t) \leq (\tilde{c}_{\text{V1}} - 1)V(x_t),$$

for some $0 < \tilde{c}_{\text{V1}} < 1$. This implies

$$\begin{aligned} V(x_{t+n}) = &J^*_L(x_{t+n}) - J^*_{\text{eq,Lin}}(x_{t+n}) \\ \leq &J^*_L(x_t) + (\tilde{c}_{\text{V1}} - 1)V(x_t) - J^*_{\text{eq,Lin}}(x_{t+n}) \\ = &\tilde{c}_{\text{V1}} V(x_t) + J^*_{\text{eq,Lin}}(x_t) - J^*_{\text{eq,Lin}}(x_{t+n}). \end{aligned} \tag{A.79}$$

As the last step in the proof, we now derive a bound on $J^*_{\text{eq,Lin}}(x_t) - J^*_{\text{eq,Lin}}(x_{t+n})$. First, we define a candidate solution to the optimization problem (5.11) with optimal cost $J^*_{\text{eq,Lin}}(x_t)$. The input candidate is defined as $\tilde{u}^{\text{s}} = u^{\text{sr}}_{\text{Lin}}(x_{t+n})$, i.e., the optimal reachable equilibrium input for the linearized system at x_{t+n}. The state and output candidates are chosen as the corresponding equilibria for the dynamics linearized at x_t, i.e.,

$$\begin{aligned} \tilde{x}^{\text{s}} &= A_{x_t}\tilde{x}^{\text{s}} + B\tilde{u}^{\text{s}} + e_{x_t}, \\ \tilde{y}^{\text{s}} &= C_{x_t}\tilde{x}^{\text{s}} + D\tilde{u}^{\text{s}} + r_{x_t}. \end{aligned}$$

Note that such $\tilde{x}^{\text{s}}$ (and hence also $\tilde{y}^{\text{s}}$) exists due to Assumption 5.5. Following the same steps leading to (A.70), it can be shown that there exists $\tilde{c} > 0$ such that

$$\|\tilde{y}^{\text{s}} - y^{\text{sr}}_{\text{Lin}}(x_{t+n})\|_S \leq \tilde{c}V(x_t). \tag{A.80}$$

Hence, by optimality, we obtain

$$\begin{aligned} & J^*_{\mathrm{eq,Lin}}(x_t) - J^*_{\mathrm{eq,Lin}}(x_{t+n}) \\ \leq & \|\tilde{y}^{\mathrm{s}} - y^{\mathrm{r}}\|_S^2 - \|y^{\mathrm{sr}}_{\mathrm{Lin}}(x_{t+n}) - y^{\mathrm{r}}\|_S^2 \\ \overset{(5.22)}{\leq} & \|\tilde{y}^{\mathrm{s}} - y^{\mathrm{sr}}_{\mathrm{Lin}}(x_{t+n})\|_S^2 + 2\|\tilde{y}^{\mathrm{s}} - y^{\mathrm{sr}}_{\mathrm{Lin}}(x_{t+n})\|_S \|y^{\mathrm{sr}}_{\mathrm{Lin}}(x_{t+n}) - y^{\mathrm{r}}\|_S \\ \overset{(A.80)}{\leq} & \tilde{c}^2 V(x_t)^2 + 2\tilde{c}V(x_t)\sqrt{J^*_{\mathrm{eq,Lin}}(x_{t+n})} \\ \leq & \tilde{c}^2 V(x_t)^2 + 2\tilde{c}V(x_t)\sqrt{J^{\max}_{\mathrm{eq}}}. \end{aligned} \tag{A.81}$$

Combining (A.79) and (A.81), we conclude that, for $V_{\max}, J^{\max}_{\mathrm{eq}}$ sufficiently small, there exists $0 < c_{\mathrm{V}1} < 1$ such that $V(x_{t+n}) \leq c_{\mathrm{V}1} V(x_t)$. ∎

Proof of Proposition 5.2 - candidate 2

Proof. In the following, we prove Proposition 5.2, i.e., we prove the result in Proposition A.1 without assuming Inequality (A.63). This is done by showing that the statement remains true if (A.63) does not hold. More precisely, we show that there exists $\gamma_1 > 0$ such that the statement of Proposition 5.2 holds if

$$\sum_{k=0}^{n-1} \|\bar{x}^*_k(t) - x^{\mathrm{s}*}(t)\|_2^2 + \|\bar{u}^*_k(t) - u^{\mathrm{s}*}(t)\|_2^2 \leq \gamma_1 \|x^{\mathrm{s}*}(t) - x^{\mathrm{sr}}_{\mathrm{Lin}}(x_t)\|_2^2. \tag{A.82}$$

Note that (A.82) implies the existence of some $\tilde{\gamma} > 0$ such that

$$\sum_{k=0}^{n-1} \|\bar{x}^*_k(t) - x^{\mathrm{s}*}(t)\|_2 + \|\bar{u}^*_k(t) - u^{\mathrm{s}*}(t)\|_2 \leq \tilde{\gamma}\sqrt{\gamma_1} \|x^{\mathrm{s}*}(t) - x^{\mathrm{sr}}_{\mathrm{Lin}}(x_t)\|_2. \tag{A.83}$$

(i) Definition of candidate solution

We consider now a different candidate solution at time $t+n$, where the artificial equilibrium input is defined as a convex combination of the optimal artificial equilibrium at time t and the optimal reachable equilibrium input given the system dynamics linearized at x_t, i.e.,

$$\hat{u}^{\mathrm{s}}(t+n) = \lambda u^{\mathrm{s}*}(t) + (1-\lambda) u^{\mathrm{sr}}_{\mathrm{Lin}}(x_t)$$

for some $\lambda \in (0,1)$ which will be fixed later in the proof. We choose the artificial equilibrium $\hat{x}^{\mathrm{s}}(t+n)$ as the corresponding equilibrium state satisfying (5.18e) (note that $\hat{x}^{\mathrm{s}}(t+n)$ exists due to Assumption 5.5) and the output $\hat{y}^{\mathrm{s}}(t+n)$ such that (5.18f) holds. In the following, we show that x_{t+n} is sufficiently close to $\hat{x}^{\mathrm{s}}(t+n)$ such that we can steer the system to $\hat{x}^{\mathrm{s}}(t+n)$ in L steps. It follows from the proof of Proposition A.1 that

$$\begin{aligned}\|x_{t+n}-x^{\mathrm{s}*}(t)\|_2^2 &\overset{(5.21)}{\leq} 2\|x_{t+n}-\bar{x}_n^*(t)\|_2^2+2\|\bar{x}_n^*(t)-x^{\mathrm{s}*}(t)\|_2^2 \\ &\overset{(A.65),(A.66)}{\leq} 2\left(2\frac{c_X}{\underline{q}}V(x_t)\sum_{k=0}^{n-2}L_f^k\right)^2+\frac{2}{\underline{q}}V(x_t).\end{aligned} \tag{A.84}$$

We define $\tilde{x}^{\mathrm{s}} := \lambda x^{\mathrm{s}*}(t)+(1-\lambda)x_{\mathrm{Lin}}^{\mathrm{sr}}(x_t)$ as the steady-state corresponding to the input $\hat{u}^{\mathrm{s}}(t+n)$ for the dynamics linearized at x_t. Note that

$$\|x^{\mathrm{s}*}(t)-\tilde{x}^{\mathrm{s}}\|_2=(1-\lambda)\|x^{\mathrm{s}*}(t)-x_{\mathrm{Lin}}^{\mathrm{sr}}(x_t)\|_2\leq\bar{c}_1(1-\lambda)\|y^{\mathrm{s}*}(t)-y_{\mathrm{Lin}}^{\mathrm{sr}}(x_t)\|_2 \tag{A.85}$$

for some $\bar{c}_1>0$ due to the existence of a linear (and hence Lipschitz continuous) map $\hat{g}_{x_t}$ as in (5.14). Note that $\hat{x}^{\mathrm{s}}(t+n)$ and $\tilde{x}^{\mathrm{s}}$ both correspond to the same equilibrium input $\tilde{u}^{\mathrm{s}}=\hat{u}^{\mathrm{s}}(t+n)$, but to different dynamics linearized at x_{t+n} and x_t, respectively. Therefore, following the same steps as in (A.68) and (A.69), we can derive

$$\begin{aligned}&\|\tilde{x}^{\mathrm{s}}-\hat{x}^{\mathrm{s}}(t+n)\|_2 \\ &\overset{(A.68)}{\leq}\frac{c_X}{\underline{\sigma}}\left(\|\tilde{x}^{\mathrm{s}}-x_{t+n}\|_2^2+\|\tilde{x}^{\mathrm{s}}-x_t\|_2^2\right) \\ &\overset{(5.21),(A.85)}{\leq}2\frac{c_X}{\underline{\sigma}}(2\bar{c}_1^2(1-\lambda)^2\|y^{\mathrm{s}*}(t)-y_{\mathrm{Lin}}^{\mathrm{sr}}(x_t)\|_2^2+\|x^{\mathrm{s}*}(t)-x_{t+n}\|_2^2+\|x^{\mathrm{s}*}(t)-x_t\|_2^2) \\ &\overset{(A.69)}{\leq}4\frac{c_X\bar{c}_1^2}{\underline{\sigma}}(1-\lambda)^2\|y^{\mathrm{s}*}(t)-y_{\mathrm{Lin}}^{\mathrm{sr}}(x_t)\|_2^2+2(c_1V(x_t)^2+c_2V(x_t)).\end{aligned} \tag{A.86}$$

Combining (A.84)–(A.86), we see that, if $(1-\lambda)$ and $V_{\max}$ are sufficiently small, then x_{t+n} is arbitrarily close to $\hat{x}^{\mathrm{s}}(t+n)$. Hence, by Assumption 5.4, there exists an input-state trajectory $\hat{u}(t+n),\hat{x}(t+n)$ steering the system from $\hat{x}_0(t+n)=x_{t+n}$ to $\hat{x}_L(t+n)=\hat{x}^{\mathrm{s}}(t+n)$ while satisfying $\hat{u}_k(t+n)\in\mathbb{U},k\in\mathbb{I}_{[0,L-1]}$ (note that $\hat{u}^{\mathrm{s}}(t+n)\in\mathrm{int}(\mathbb{U})$) and

$$\sum_{k=0}^{L-1}\|\hat{x}_k(t+n)-\hat{x}^{\mathrm{s}}(t+n)\|_2+\|\hat{u}_k(t+n)-\hat{u}^{\mathrm{s}}(t+n)\|_2\leq\Gamma\|\hat{x}^{\mathrm{s}}(t+n)-x_{t+n}\|_2. \tag{A.87}$$

(ii) Bounds on candidate solution

In the following, we derive multiple bounds on the candidate solution that will be useful in the remainder of the proof.

(ii.a) Bound on $\|\hat{x}^{\mathrm{s}}(t+n) - x_{t+n}\|_2$

Note that

$$\|\hat{x}^{\mathrm{s}}(t+n) - x_{t+n}\|_2 \leq \|\hat{x}^{\mathrm{s}}(t+n) - x^{\mathrm{s}*}(t)\|_2 + \|x^{\mathrm{s}*}(t) - x_{t+n}\|_2.$$

Inequality (A.84) provides a bound on $\|x^{\mathrm{s}*}(t) - x_{t+n}\|_2$. In the following, we derive a more sophisticated bound which will be required to find a useful bound on $\|\hat{x}^{\mathrm{s}}(t+n) - x_{t+n}\|_2$. Define $x^{\mathrm{s}\prime}(t+n-1)$ as the steady-state for the linearized dynamics at x_{t+n-1} and with input $u^{\mathrm{s}*}(t)$. Then, we have

$$\|x^{\mathrm{s}*}(t) - x_{t+n}\|_2 \leq \|x^{\mathrm{s}*}(t) - x^{\mathrm{s}\prime}(t+n-1)\|_2 + \|x^{\mathrm{s}\prime}(t+n-1) - x_{t+n}\|_2. \quad \text{(A.88)}$$

In the following, we bound the terms on the right-hand side of (A.88). First, we obtain

$$\begin{aligned}
&\|x^{\mathrm{s}\prime}(t+n-1) - x_{t+n}\|_2 \qquad \text{(A.89)}\\
=\,& \|A_{x_{t+n-1}}(x^{\mathrm{s}\prime}(t+n-1) - x_{t+n-1}) + B(u^{\mathrm{s}*}(t) - u_{t+n-1})\|_2\\
\leq\,& \|A_{x_{t+n-1}}\|_2 \|x^{\mathrm{s}\prime}(t+n-1) - x_{t+n-1}\|_2 + \|B\|_2 \|u^{\mathrm{s}*}(t) - u_{t+n-1}\|_2\\
\leq\,& \|A_{x_{t+n-1}}\|_2 \Big(\|x^{\mathrm{s}\prime}(t+n-1) - x^{\mathrm{s}*}(t)\|_2 + \|x^{\mathrm{s}*}(t) - x_{t+n-1}\|_2 \Big)\\
&+ \|B\|_2 \|u^{\mathrm{s}*}(t) - \bar{u}^*_{n-1}(t)\|_2.
\end{aligned}$$

Using the triangle inequality, it holds that

$$\|x^{\mathrm{s}*}(t) - x_{t+n-1}\|_2 \leq \|x^{\mathrm{s}*}(t) - \bar{x}^*_{n-1}(t)\|_2 + \|\bar{x}^*_{n-1}(t) - x_{t+n-1}\|_2. \quad \text{(A.90)}$$

Defining $c_{\mathrm{A}} := \|A_{x_{t+n-1}}\|_2$, $c_{\mathrm{B}} := \|B\|_2$, $c_{\mathrm{AB}} := \max\{c_{\mathrm{A}}, c_{\mathrm{B}}\}$, and using (A.83), (A.88), (A.89), and (A.90), we have

$$\begin{aligned}
\|x^{\mathrm{s}*}(t) - x_{t+n}\|_2 \leq& (1 + c_{\mathrm{A}}) \|x^{\mathrm{s}*}(t) - x^{\mathrm{s}\prime}(t+n-1)\|_2 \qquad \text{(A.91)}\\
&+ c_{\mathrm{AB}} \tilde{\gamma} \sqrt{\gamma_1} \|x^{\mathrm{s}*}(t) - x^{\mathrm{sr}}_{\mathrm{Lin}}(x_t)\|_2 + c_{\mathrm{A}} \|\bar{x}^*_{n-1}(t) - x_{t+n-1}\|_2.
\end{aligned}$$

Next, following the same steps as in Part (ii.c) of the proof of Proposition A.1 (note that $x^{\mathrm{s}\prime}(t+n-1)$ is defined as an equilibrium of the linearization at x_{t+n-1} with the same input $u^{\mathrm{s}*}(t)$ as $x^{\mathrm{s}*}(t)$), it can be shown that

$$\|x^{\mathrm{s}*}(t) - x^{\mathrm{s}\prime}(t+n-1)\|_2 \leq c_1 V(x_t)^2 + c_2 V(x_t). \quad \text{(A.92)}$$

Moreover, similar to (A.66), it is readily derived that

$$\|\bar{x}_{n-1}^*(t) - x_{t+n-1}\|_2 \leq 2\frac{c_X}{\underline{q}}V(x_t)\sum_{k=0}^{n-3} L_f^k. \tag{A.93}$$

Finally, it follows from (A.85) and (A.86) that

$$\begin{aligned}\|\hat{x}^s(t+n) - x^{s*}(t)\|_2 \leq& \bar{c}_1(1-\lambda)\|y^{s*}(t) - y_{\text{Lin}}^{\text{sr}}(x_t)\|_2 \\ &+ 2(c_1V(x_t)^2 + c_2V(x_t)) + 4\frac{c_X\bar{c}_1^2}{\underline{\sigma}}(1-\lambda)^2\|y^{s*}(t) - y_{\text{Lin}}^{\text{sr}}(x_t)\|_2^2.\end{aligned} \tag{A.94}$$

Inserting the bounds (A.92) and (A.93) into (A.91), and combining this with (A.94), we obtain

$$\begin{aligned}\|\hat{x}^s(t+n) - x_{t+n}\|_2 \leq& \|\hat{x}^s(t+n) - x^{s*}(t)\|_2 + \|x^{s*}(t) - x_{t+n}\|_2 \\ \leq& \bar{c}_1(1-\lambda)\|y^{s*}(t) - y_{\text{Lin}}^{\text{sr}}(x_t)\|_2 + 2(c_1V(x_t)^2 + c_2V(x_t)) \\ &+ 4\frac{c_X\bar{c}_1^2}{\underline{\sigma}}(1-\lambda)^2\|y^{s*}(t) - y_{\text{Lin}}^{\text{sr}}(x_t)\|_2^2 \\ &+ (1+c_A)(c_1V(x_t)^2 + c_2V(x_t)) + 2c_A\frac{c_X}{\underline{q}}V(x_t)\sum_{k=0}^{n-3} L_f^k \\ &+ c_{AB}\tilde{\gamma}\sqrt{\gamma_1}\|x^{s*}(t) - x_{\text{Lin}}^{\text{sr}}(x_t)\|_2 \\ \stackrel{(A.85)}{\leq}& \bar{c}_1((1-\lambda) + c_{AB}\tilde{\gamma}\sqrt{\gamma_1})\|y^{s*}(t) - y_{\text{Lin}}^{\text{sr}}(x_t)\|_2 + c_4V(x_t) \\ &+ 4\frac{c_X\bar{c}_1^2}{\underline{\sigma}}(1-\lambda)^2\|y^{s*}(t) - y_{\text{Lin}}^{\text{sr}}(x_t)\|_2^2\end{aligned} \tag{A.95}$$

for some $c_4 > 0$, using that $V(x_t) \leq V_{\text{max}}$ in the last inequality.

(ii.b) Bound on $\|\hat{y}^s(t+n) - y^r\|_S^2 - \|y^{s*}(t) - y^r\|_S^2$

Similar to the proof of Theorem 3.4 in Section 3.3, compare (A.9), it is straightforward to exploit the convexity condition (5.12) in order to derive

$$\|\tilde{y}^s - y^r\|_S^2 - \|y^{s*}(t) - y^r\|_S^2 \leq -(1-\lambda^2)\|y^{s*}(t) - y_{\text{Lin}}^{\text{sr}}(x_t)\|_S^2, \tag{A.96}$$

where $\tilde{y}^s = \lambda y^{s*}(t) + (1-\lambda)y_{\text{Lin}}^{\text{sr}}(x_t)$. Moreover, (5.22) implies

$$\|\hat{y}^s(t+n) - y^r\|_S^2 - \|\tilde{y}^s - y^r\|_S^2 \leq \|\hat{y}^s(t+n) - \tilde{y}^s\|_S^2 + 2\|\hat{y}^s(t+n) - \tilde{y}^s\|_S\|\tilde{y}^s - y^r\|_S. \tag{A.97}$$

Recall that $\tilde{u}^s = \lambda u^{s*}(t) + (1-\lambda)u^{sr}_{Lin}(x_t) = \hat{u}^s(t+n)$. Hence, using an inequality of the form (5.8) for the vector field h, there exist constants $c_{Xh}, L_h \geq 0$ such that

$$\begin{aligned}
&\|\hat{y}^s(t+n) - \tilde{y}^s\|_2 \qquad\qquad \text{(A.98)}\\
=&\|h_{x_{t+n}}(\hat{x}^s(t+n), \tilde{u}^s) - h_{x_t}(\tilde{x}^s, \tilde{u}^s)\|_2\\
\leq& c_{Xh}\|\hat{x}^s(t+n) - x_{t+n}\|_2^2 + c_{Xh}\|\tilde{x}^s - x_t\|_2^2 + L_h\|\hat{x}^s(t+n) - \tilde{x}^s\|_2.
\end{aligned}$$

The second term on the right-hand side is bounded as

$$\begin{aligned}
\|\tilde{x}^s - x_t\|_2^2 &\overset{(5.21)}{\leq} 2\|\tilde{x}^s - x^{s*}(t)\|_2^2 + 2\|x^{s*}(t) - x_t\|_2^2\\
&\overset{(A.65),(A.85)}{\leq} 2\bar{c}_1^2(1-\lambda)^2\|y^{s*}(t) - y^{sr}_{Lin}(x_t)\|_2^2 + \frac{2}{\underline{q}}V(x_t).
\end{aligned}$$

Using in addition the bounds (A.86) and (A.95), this implies

$$\begin{aligned}
&\|\hat{y}^s(t+n) - \tilde{y}^s\|_2 \qquad\qquad \text{(A.99)}\\
\leq& c_{Xh}\Big(\bar{c}_1((1-\lambda) + c_{AB}\tilde{\gamma}\sqrt{\gamma_1})\|y^{s*}(t) - y^{sr}_{Lin}(x_t)\|_2\\
&+ c_4 V(x_t) + 4\frac{c_X\bar{c}_1^2}{\underline{\sigma}}(1-\lambda)^2\|y^{s*}(t) - y^{sr}_{Lin}(x_t)\|_2^2\Big)^2\\
&+ c_{Xh}\left(2\bar{c}_1^2(1-\lambda)^2\|y^{s*}(t) - y^{sr}_{Lin}(x_t)\|_2^2 + \frac{2}{\underline{q}}V(x_t)\right)\\
&+ 4L_h\frac{c_X\bar{c}_1^2}{\underline{\sigma}}(1-\lambda)^2\|y^{s*}(t) - y^{sr}_{Lin}(x_t)\|_2^2 + 2L_h(c_1V(x_t)^2 + c_2V(x_t)).
\end{aligned}$$

Further, by convexity (recall that $\tilde{y}^s$ is defined as a convex combination of $y^{s*}(t)$ and $y^{sr}_{Lin}(x_t)$) we have

$$\begin{aligned}
\|\tilde{y}^s - y^r\|_S^2 \leq& \lambda\|y^{s*}(t) - y^r\|_S^2 + (1-\lambda)\|y^{sr}_{Lin}(x_t) - y^r\|_S^2 \qquad\qquad \text{(A.100)}\\
\leq& \lambda(V(x_t) + J^{max}_{eq}) + (1-\lambda)J^{max}_{eq}.
\end{aligned}$$

The bound (A.97) together with the subsequently derived bounds will play an important role in the remainder of the proof. To this end, using (A.96), we conclude

$$\begin{aligned}
&\|\hat{y}^s(t+n) - y^r\|_S^2 - \|y^{s*}(t) - y^r\|_S^2 \qquad\qquad \text{(A.101)}\\
=&\|\hat{y}^s(t+n) - y^r\|_S^2 - \|\tilde{y}^s - y^r\|_S^2 + \|\tilde{y}^s - y^r\|_S^2 - \|y^{s*}(t) - y^r\|_S^2\\
\overset{(A.96)}{\leq}& -(1-\lambda^2)\|y^{s*}(t) - y^{sr}_{Lin}(x_t)\|_S^2 + \|\hat{y}^s(t+n) - y^r\|_S^2 - \|\tilde{y}^s - y^r\|_S^2.
\end{aligned}$$

(iii) Invariance of $V(x_t) \leq V_{\max}$
It follows directly from (A.87), (A.95), and (A.101), and using the inequality $a^2 + b^2 \leq (a+b)^2$ for $a, b \geq 0$, that

$$\begin{aligned} J_L^*(x_{t+n}) - J_L^*(x_t) \leq & \lambda_{\max}(Q,R)\Gamma^2\Big(4\frac{c_X \bar{c}_1^2}{\underline{\sigma}}(1-\lambda)^2\|y^{s*}(t) - y^{sr}_{\text{Lin}}(x_t)\|_2^2 \\ & + \bar{c}_1((1-\lambda) + c_{AB}\tilde{\gamma}\sqrt{\gamma_1})\|y^{s*}(t) - y^{sr}_{\text{Lin}}(x_t)\|_2 + c_4 V(x_t)\Big)^2 \\ & - \|x_t - x^{s*}(t)\|_Q^2 - (1-\lambda^2)\|y^{s*}(t) - y^{sr}_{\text{Lin}}(x_t)\|_S^2 \\ & + \|\hat{y}^s(t+n) - y^r\|_S^2 - \|\tilde{y}^s - y^r\|_S^2. \end{aligned}$$

If $V_{\max}$, $J_{\text{eq}}^{\max}$, γ_1 and $(1-\lambda)$ all are sufficiently small, then using the bounds derived in Part (ii.b) of the proof, it follows directly that

$$J_L^*(x_{t+n}) - J_L^*(x_t) \leq -\|x_t - x^{s*}(t)\|_Q^2 - c_5\|y^{s*}(t) - y^{sr}_{\text{Lin}}(x_t)\|_S^2 + c_6 V(x_t)$$

with some $c_5, c_6 > 0$, where c_6 becomes arbitrarily small if $V_{\max}$ and $J_{\text{eq}}^{\max}$ are sufficiently small. It follows from the existence of a linear map $\hat{g}_x$ as in (5.14) that $\|y^{s*}(t) - y^{sr}_{\text{Lin}}(x_t)\|_S^2 \geq \frac{1}{\bar{c}_1}\|x^{s*}(t) - x^{sr}_{\text{Lin}}(x_t)\|_S^2$ with $\bar{c}_1 > 0$. Hence, using the upper bound (5.20), we obtain

$$J_L^*(x_{t+n}) - J_L^*(x_t) \leq -\frac{\min\{\underline{q}, \frac{c_5}{\bar{c}_1}\underline{s}\}}{2}\|x_t - x^{sr}_{\text{Lin}}(x_t)\|_2^2 + c_6 V(x_t) \overset{(5.20)}{\leq} (\tilde{c}_{V2} - 1)V(x_t)$$

for some $0 < \tilde{c}_{V2} < 1$, assuming that $V_{\max}$ and $J_{\text{eq}}^{\max}$ and hence c_6 are sufficiently small. This leads to

$$V(x_{t+n}) \leq \tilde{c}_{V2}V(x_t) + J^*_{\text{eq,Lin}}(x_t) - J^*_{\text{eq,Lin}}(x_{t+n}).$$

Finally, following the same steps as in the proof of Proposition A.1, we can show that this implies the existence of a constant $0 < c_{V2} < 1$ such that $V(x_{t+n}) \leq c_{V2}V(x_t)$. Combining this with the statement of Proposition A.1, we obtain $V(x_{t+n}) \leq c_V V(x_t)$ for $c_V := \max\{c_{V1}, c_{V2}\} < 1$. ■

A.4 Proof of Proposition 5.4

Proof. Throughout the proof, we will make repeated use of the inequalities

$$\|a+b\|_P^2 \leq 2\|a\|_P^2 + 2\|b\|_P^2, \tag{A.102}$$
$$\|a\|_P^2 - \|b\|_P^2 \leq \|a-b\|_P^2 + 2\|a-b\|_P\|b\|_P, \tag{A.103}$$

which hold for any vectors a, b and matrix $P = P^\top \succ 0$.

(i) Proof of (5.55)

We construct a candidate solution for Problem 5.2 based on the optimal solution of Problem 5.3 for $\tilde{\sigma} = 0$. To this end, we let

$$\bar{u}(t) = \begin{bmatrix} u_{[t-n,t-1]} \\ \check{u}^*_{[0,L]}(t) \end{bmatrix}, \quad \bar{y}(t) = \begin{bmatrix} y_{[t-n,t-1]} \\ \check{y}^*_{[0,L]}(t) \end{bmatrix}, \tag{A.104}$$

and $u^{\mathrm{s}}(t) = \check{u}^{\mathrm{s}*}(t)$, $y^{\mathrm{s}}(t) = \check{y}^{\mathrm{s}*}(t)$. Further, we define

$$\alpha(t) = H_{ux,t}^\dagger \begin{bmatrix} \bar{u}(t) \\ x_{t-n} \\ 1 \end{bmatrix}$$

with $H_{ux,t}$ as in (5.46). Moreover, we consider the candidate

$$\sigma(t) = H_{L+n+1}(y_{[t-N,t-1]})\alpha(t) - \bar{y}(t), \tag{A.105}$$

which fulfills all constraints of Problem 5.2. Note that $\check{\alpha}^*(t) - \alpha^{\mathrm{sr}}_{\mathrm{Lin}}(\mathcal{D}_t)$ satisfies

$$\begin{bmatrix} H_{L+n+1}(u_{[t-N,t-1]}) \\ H_{L+n+1}(y'(t)) \\ \mathbb{1}_{N-L-n}^\top \end{bmatrix} (\check{\alpha}^*(t) - \alpha^{\mathrm{sr}}_{\mathrm{Lin}}(\mathcal{D}_t)) = \begin{bmatrix} \check{u}^*(t) - \mathbb{1}_{L+n+1} \otimes u^{\mathrm{sr}}_{\mathrm{Lin}}(x_t) \\ \check{y}^*(t) - \mathbb{1}_{L+n+1} \otimes y^{\mathrm{sr}}_{\mathrm{Lin}}(x_t) \\ 0 \end{bmatrix}. \tag{A.106}$$

Due to the minimization in (5.52a), $\check{\alpha}^*(t) - \alpha^{\mathrm{sr}}_{\mathrm{Lin}}(\mathcal{D}_t)$ is the vector satisfying (A.106) with minimum norm, which implies

$$\check{\alpha}^*(t) = H_{ux,t}^\dagger \begin{bmatrix} \check{u}^*(t) \\ x_{t-n} \\ 1 \end{bmatrix}.$$

Thus, it holds that

$$\begin{aligned} \|\alpha(t) - \check{\alpha}^*(t)\|_2^2 &\overset{(A.104)}{\leq} \|H_{ux,t}^\dagger\|_2^2 \, \|u_{[t-n,t-1]} - u'_{[t-n,t-1]}\|_2^2 \\ &\overset{(5.53)}{\leq} \|H_{ux,t}^\dagger\|_2^2 \, \bar{\varepsilon}^2. \end{aligned} \tag{A.107}$$

Using $\|a+b\|_2^2 \leq (1+\bar{\varepsilon}^{\beta_\sigma})\|a\|_2^2 + (1+\frac{1}{\bar{\varepsilon}^{\beta_\sigma}})\|b\|_2^2$ for arbitrary a, b, this implies

$$\begin{aligned} &\|\alpha(t) - \alpha^{\mathrm{sr}}_{\mathrm{Lin}}(\mathcal{D}_t)\|_2^2 - \|\check{\alpha}^*(t) - \alpha^{\mathrm{sr}}_{\mathrm{Lin}}(\mathcal{D}_t)\|_2^2 \\ \leq &\bar{\varepsilon}^{\beta_\sigma} \|\check{\alpha}^*(t) - \alpha^{\mathrm{sr}}_{\mathrm{Lin}}(\mathcal{D}_t)\|_2^2 + \|H_{ux,t}^\dagger\|_2^2 (\bar{\varepsilon}^2 + \bar{\varepsilon}^{2-\beta_\sigma}). \end{aligned} \tag{A.108}$$

Note that the output trajectories $H_{L+n+1}(y'(t))\alpha(t)$ and $\check{y}^*(t)$ differ only due to the difference in the first n components of the corresponding input trajectory, which is in turn bounded by $\bar{\varepsilon}$, cf. (5.53) and (A.104). Thus, there exists $c_1^{\mathrm{u}} > 0$ such that

$$\|H_{L+n+1}(y'(t))\alpha(t) - \check{y}^*(t)\|_2^2 \overset{(5.53)}{\leq} c_1^{\mathrm{u}}\bar{\varepsilon}^2. \tag{A.109}$$

Hence, there exist $c_2^{\mathrm{u}}, c_3^{\mathrm{u}} > 0$ such that

$$\begin{aligned}\|\sigma(t)\|_2^2 &\overset{\text{(A.102),(A.105)}}{\leq} 2\left\|\left(H_{L+n+1}(y_{[t-N,t-1]}) - H_{L+n+1}(y'(t))\right)\alpha(t)\right\|_2^2 \\ &\quad + 4\|H_{L+n+1}(y'(t))\alpha(t) - \check{y}^*(t)\|_2^2 + 4\|\check{y}^*(t) - \bar{y}(t)\|_2^2 \\ &\overset{\text{(5.53),(A.104),(A.109)}}{\leq} c_2^{\mathrm{u}}\bar{\varepsilon}^2 + c_3^{\mathrm{u}}\bar{\varepsilon}^2\|\alpha(t)\|_2^2.\end{aligned} \tag{A.110}$$

Applying $\|a+b\|_R^2 \leq (1+\bar{\varepsilon}^{\beta_\sigma})\|a\|_R^2 + (1+\frac{1}{\bar{\varepsilon}^{\beta_\sigma}})\|b\|_R^2$, which holds for arbitrary vectors a, b, we have

$$\begin{aligned}&\sum_{k=-n}^{-1}\|u_{t+k} - \check{u}^{\mathrm{s}*}(t)\|_R^2 \\ \leq &\sum_{k=-n}^{-1}\left(1+\frac{1}{\bar{\varepsilon}^{\beta_\sigma}}\right)\|u_{t+k} - u'_{t+k}\|_R^2 + (1+\bar{\varepsilon}^{\beta_\sigma})\|u'_{t+k} - \check{u}^{\mathrm{s}*}(t)\|_R^2,\end{aligned} \tag{A.111}$$

and analogously for the output. Using additionally

$$\begin{aligned}&\sum_{k=-n}^{-1}\|u_{t+k} - u'_{t+k}\|_R^2 + \|y_{t+k} - y'_{t+k}\|_Q^2 \overset{(5.53)}{\leq} \lambda_{\max}(Q,R)\bar{\varepsilon}^2, \\ &\sum_{k=-n}^{-1}\|u'_{t+k} - \check{u}^{\mathrm{s}*}(t)\|_R^2 + \|y'_{t+k} - \check{y}^{\mathrm{s}*}(t)\|_Q^2 \leq \check{J}_L^*(\xi'_t, \mathcal{D}_t),\end{aligned}$$

we obtain

$$\begin{aligned}&\sum_{k=-n}^{-1}\left(\|u_{t+k} - \check{u}^{\mathrm{s}*}(t)\|_R^2 - \|u'_{t+k} - \check{u}^{\mathrm{s}*}(t)\|_R^2\right) \\ &+ \sum_{k=-n}^{-1}\left(\|y_{t+k} - \check{y}^{\mathrm{s}*}(t)\|_Q^2 - \|y'_{t+k} - \check{y}^{\mathrm{s}*}(t)\|_Q^2\right) \\ \leq &\bar{\varepsilon}^{\beta_\sigma}\check{J}_L^*(\xi'_t, \mathcal{D}_t) + c_4^{\mathrm{u}}(\bar{\varepsilon}^2 + \bar{\varepsilon}^{2-\beta_\sigma})\end{aligned} \tag{A.112}$$

for a suitably defined $c_4^{\mathrm{u}} > 0$. Using the constructed candidate solution, we have

$$\begin{aligned} J_L^*(\xi_t) - \check{J}_L^*(\xi_t', \mathcal{D}_t) &\overset{(\mathrm{A.112})}{\leq} c_4^{\mathrm{u}}(\bar{\varepsilon}^2 + \bar{\varepsilon}^{2-\beta_\sigma}) + \bar{\varepsilon}^{\beta_\sigma} \check{J}_L^*(\xi_t', \mathcal{D}_t) + \frac{\bar{\lambda}_\sigma}{\bar{\varepsilon}^{\beta_\sigma}} \|\sigma(t)\|_2^2 \\ &\quad + \bar{\lambda}_\alpha \bar{\varepsilon}^{\beta_\alpha} (\|\alpha(t) - \alpha_{\mathrm{Lin}}^{\mathrm{sr}}(x_t)\|_2^2 - \|\check{\alpha}^*(t) - \alpha_{\mathrm{Lin}}^{\mathrm{sr}}(\mathcal{D}_t)\|_2^2) \\ &\overset{(\mathrm{A.108}),(\mathrm{A.110})}{\leq} c_4^{\mathrm{u}}(\bar{\varepsilon}^2 + \bar{\varepsilon}^{2-\beta_\sigma}) + 2\bar{\varepsilon}^{\beta_\sigma} \check{J}_L^*(\xi_t', \mathcal{D}_t) + \bar{\lambda}_\sigma c_2^{\mathrm{u}} \bar{\varepsilon}^{2-\beta_\sigma} \\ &\quad + \bar{\lambda}_\alpha \bar{\varepsilon}^{\beta_\alpha} \|H_{ux,t}^\dagger\|_2^2 (\bar{\varepsilon}^2 + \bar{\varepsilon}^{2-\beta_\sigma}) + \bar{\lambda}_\sigma c_3^{\mathrm{u}} \bar{\varepsilon}^{2-\beta_\sigma} \|\alpha(t)\|_2^2, \end{aligned} \tag{A.113}$$

where we also use $\bar{\lambda}_\alpha \bar{\varepsilon}^{\beta_\alpha} \|\check{\alpha}^*(t) - \alpha_{\mathrm{Lin}}^{\mathrm{sr}}(\mathcal{D}_t)\|_2^2 \leq \check{J}_L^*(\xi_t', \mathcal{D}_t)$ for the second inequality. The term $\|\alpha(t)\|_2^2$ is bounded as

$$\begin{aligned} \|\alpha(t)\|_2^2 &\overset{(\mathrm{A.102})}{\leq} 2\|\alpha(t) - \check{\alpha}^*(t)\|_2^2 + 4\|\check{\alpha}^*(t) - \alpha_{\mathrm{Lin}}^{\mathrm{sr}}(\mathcal{D}_t)\|_2^2 + 4\|\alpha_{\mathrm{Lin}}^{\mathrm{sr}}(\mathcal{D}_t)\|_2^2 \\ &\overset{(\mathrm{A.107})}{\leq} 2\|H_{ux,t}^\dagger\|_2^2 \bar{\varepsilon}^2 + 4\frac{1}{\bar{\lambda}_\alpha \bar{\varepsilon}^{\beta_\alpha}} \check{J}_L^*(\xi_t', \mathcal{D}_t) + 4\|\alpha_{\mathrm{Lin}}^{\mathrm{sr}}(\mathcal{D}_t)\|_2^2, \end{aligned}$$

where $\|\alpha_{\mathrm{Lin}}^{\mathrm{sr}}(\mathcal{D}_t)\|_2^2$ is uniformly bounded by assumption. Plugging this into (A.113) and using $\beta_\alpha + 2\beta_\sigma < 2$, the term with the smallest exponent of $\bar{\varepsilon}$ is proportional to $\bar{\varepsilon}^{2-\beta_\sigma}(1 + \|H_{ux,t}^\dagger\|_2^2 \bar{\varepsilon}^{\beta_\alpha})$. Thus, letting $\bar{\varepsilon}_{\max} < 1$, we obtain (5.55) for appropriately defined constants $c_{\mathrm{J,a}}, c_{\mathrm{J,b}} > 0$.

(ii) Proof of (5.56)

We note that

$$\|\bar{u}_{[-n,-1]}^*(t) - \check{u}_{[-n,-1]}^*(t)\|_2 = \|u_{[t-n,t-1]} - u_{[t-n,t-1]}'\|_2 \overset{(5.53)}{\leq} \bar{\varepsilon}, \tag{A.114}$$

which implies a bound of the form (5.56) for the first n elements. Therefore, we focus on $k \in \mathbb{I}_{[0,L]}$ in the following.

(ii.a) Strong convexity of Problem 5.3

Exploiting the initial condition (5.52c) and the terminal constraint (5.52d), the terms in (5.52a) involving $\bar{u}(t)$ and $u^{\mathrm{s}}(t)$ are

$$\sum_{k=-n}^{-1} \|u_{t+k}' + F_k \tilde{\sigma}_{\mathrm{init}} - u^{\mathrm{s}}(t)\|_R^2 + \sum_{k=0}^{L-n-1} \|\bar{u}_k(t) - u^{\mathrm{s}}(t)\|_R^2,$$

where the matrix F_k picks the k-th input component of $\tilde{\sigma}_{\mathrm{init}}$. The Hessian w.r.t. $\begin{bmatrix} \bar{u}_{[0,L-n-1]}(t) \\ u^{\mathrm{s}}(t) \end{bmatrix}$ is equal to

$$2\begin{bmatrix} I_{L-n} \otimes R & -\mathbb{1}_{L-n} \otimes R \\ -\mathbb{1}_{L-n}^\top \otimes R & LR \end{bmatrix}.$$

This matrix is positive definite since its Schur complement w.r.t. the left upper block is

$$2LR - 2(\mathbb{1}_{L-n}^\top \otimes R)(I_{L-n} \otimes R)^{-1}(\mathbb{1}_{L-n} \otimes R) = 2nR \succ 0.$$

Thus, Problem 5.3 is strongly convex in $\begin{bmatrix} \bar{u}_{[0,L-n-1]}(t) \\ u^s(t) \end{bmatrix}$ and hence, using $\bar{u}_k(t) = u^s(t)$ for $k \in \mathbb{I}_{[L-n,L]}$ due to (5.52d), it is strongly convex in $\bar{u}_{[0,L]}(t)$.

(ii.b) Bound on $\|\bar{u}^*(t) - \tilde{u}(t)\|_2^2$

We denote the optimal solution of Problem 5.3 with

$$\tilde{\sigma} = \tilde{\sigma}_\varepsilon := \begin{bmatrix} u_{[t-n,t-1]} - u'_{[t-n,t-1]} \\ y_{[t-n,t-1]} - y'_{[t-n,t-1]} \\ \sigma^*(t) + H_{L+n+1}\left(y'(t) - y_{[t-N,t-1]}\right)\alpha^*(t) \end{bmatrix} \tag{A.115}$$

by $\tilde{\alpha}(t), \tilde{u}^s(t), \tilde{y}^s(t), \tilde{u}(t), \tilde{y}(t)$, and the corresponding cost by $\tilde{J}_L$. Since $\tilde{J}_L^*(\xi_t', \mathcal{D}_t) \leq \bar{J}$ by assumption and using (5.55) by Part (i), we infer $J_L^*(\xi_t) < \infty$, i.e., Problem 5.2 is feasible. The optimal solution of Problem 5.2 is a feasible (but not necessarily optimal) solution of Problem 5.3 with $\tilde{\sigma} = \tilde{\sigma}_\varepsilon$, i.e.,

$$\tilde{J}_L \leq J_L^*(\xi_t) - \frac{\bar{\lambda}_\sigma}{\bar{\varepsilon}^{\beta_\sigma}}\|\sigma^*(t)\|_2^2. \tag{A.116}$$

Here, the term $\frac{\bar{\lambda}_\sigma}{\bar{\varepsilon}^{\beta_\sigma}}\|\sigma^*(t)\|_2^2$ occurs due to the fact that, in contrast to Problem 5.2, Problem 5.3 does not include the slack variable $\sigma(t)$ in the cost. We now construct a candidate solution for Problem 5.2 based on the optimal solution of Problem 5.3 with $\tilde{\sigma} = \tilde{\sigma}_\varepsilon$. To this end, we choose $\bar{u}(t) = \tilde{u}(t)$, $\bar{y}(t) = \tilde{y}(t)$, $\alpha(t) = \tilde{\alpha}(t)$, $u^s(t) = \tilde{u}^s(t)$, $y^s(t) = \tilde{y}^s(t)$, and

$$\sigma(t) = \sigma^*(t) + H_{L+n+1}(\varepsilon^d)(\tilde{\alpha}(t) - \alpha^*(t)). \tag{A.117}$$

Note that $(\alpha(t), u^s(t), y^s(t), \bar{u}(t), \bar{y}(t))$ is a feasible solution of Problem 5.2, whose cost we denote by $\bar{J}_L'$, which satisfies $\bar{J}_L' \geq J_L^*(\xi_t)$ by optimality. By definition,

$$\bar{J}_L' - \tilde{J}_L = \frac{\bar{\lambda}_\sigma}{\bar{\varepsilon}^{\beta_\sigma}}\|\sigma(t)\|_2^2. \tag{A.118}$$

Moreover, it follows from the definition of $J_L^*(\xi_t)$ that

$$\|\alpha^*(t) - \alpha_{\text{Lin}}^{\text{sr}}(\mathcal{D}_t)\|_2^2 \leq \frac{J_L^*(\xi_t)}{\bar{\lambda}_\alpha \bar{\varepsilon}^{\beta_\alpha}}, \tag{A.119}$$

$$\|\sigma^*(t)\|_2^2 \leq \bar{\varepsilon}^{\beta_\sigma}\frac{J_L^*(\xi_t)}{\bar{\lambda}_\sigma}. \tag{A.120}$$

Similarly, we have

$$\|\tilde{\alpha}(t) - \alpha^{\text{sr}}_{\text{Lin}}(\mathcal{D}_t)\|_2^2 \leq \frac{\tilde{J}_L}{\bar{\lambda}_\alpha \bar{\varepsilon}^{\beta_\alpha}} \overset{\text{(A.116)}}{\leq} \frac{J_L^*(\xi_t)}{\bar{\lambda}_\alpha \bar{\varepsilon}^{\beta_\alpha}}. \tag{A.121}$$

With the definition of $\sigma(t)$ in (A.117), we obtain

$$\begin{aligned} &\|\sigma(t)\|_2^2 - \|\sigma^*(t)\|_2^2 \qquad\qquad \text{(A.122)}\\ \overset{\text{(A.103),(A.117)}}{\leq} \; & c_{\sigma,1}^2 \bar{\varepsilon}^2 \|\tilde{\alpha}(t) - \alpha^*(t)\|_2^2 + 2c_{\sigma,1}\bar{\varepsilon}\|\tilde{\alpha}(t) - \alpha^*(t)\|_2 \|\sigma^*(t)\|_2 \\ \overset{\text{(A.102),(A.119)–(A.121)}}{\leq} \; & c_{\sigma,2}\bar{\varepsilon}^{2-\beta_\alpha} J_L^*(\xi_t) + c_{\sigma,3}\bar{\varepsilon}^{1+\frac{\beta_\sigma - \beta_\alpha}{2}} J_L^*(\xi_t) \end{aligned}$$

for suitably defined $c_{\sigma,i} > 0$, $i \in \mathbb{I}_{[1,3]}$. Recall that

$$\tilde{J}_L + \frac{\bar{\lambda}_\sigma}{\bar{\varepsilon}^{\beta_\sigma}} \|\sigma^*(t)\|_2^2 \overset{\text{(A.116)}}{\leq} J_L^*(\xi_t) \leq \bar{J}_L'. \tag{A.123}$$

Using that the solution with cost $\bar{J}_L'$ is feasible for Problem 5.2, strong convexity as in Part (ii.a) implies

$$\|\bar{u}(t) - \bar{u}^*(t)\|_2^2 \leq c_{\text{u,c}}(\bar{J}_L' - J_L^*(\xi_t)) \tag{A.124}$$

for some $c_{\text{u,c}} > 0$, compare [69, Inequality (11)]. Together with $\bar{u}(t) = \tilde{u}(t)$, this implies

$$\begin{aligned} \|\tilde{u}(t) - \bar{u}^*(t)\|_2^2 &\overset{\text{(A.124)}}{\leq} c_{\text{u,c}}(\bar{J}_L' - J_L^*(\xi_t)) \overset{\text{(A.123)}}{\leq} c_{\text{u,c}} \left(\bar{J}_L' - \tilde{J}_L - \frac{\bar{\lambda}_\sigma}{\bar{\varepsilon}^{\beta_\sigma}} \|\sigma^*(t)\|_2^2 \right) \quad \text{(A.125)} \\ &\overset{\text{(A.118),(A.122)}}{\leq} c_{\text{u,c}} \bar{\lambda}_\sigma \left(c_{\sigma,2} \bar{\varepsilon}^{2-\beta_\alpha-\beta_\sigma} + c_{\sigma,3}\bar{\varepsilon}^{\frac{2-\beta_\alpha-\beta_\sigma}{2}} \right) J_L^*(\xi_t). \end{aligned}$$

(ii.c) Bound on $\|\tilde{u}(t) - \check{u}^*(t)\|_2^2$

Using strong convexity (compare Part (ii.a) of the proof) and the LICQ (Assumption 5.15), the optimal solution of Problem 5.3 is continuous and piecewise affine in $(\xi_t', \tilde{\sigma})$, compare [13]. In particular, using that the partition of the optimal solution admits finitely many regions, it is uniformly Lipschitz continuous in $\tilde{\sigma}$. Moreover, Problem 5.3 is feasible for $\tilde{\sigma} = 0$ by assumption, i.e., by $\check{J}_L^*(\xi_t', \mathcal{D}_t) \leq \bar{J}$, and for $\tilde{\sigma} = \tilde{\sigma}_\varepsilon$ since Problem 5.2 is feasible, cf. Part (i). Thus, there exists $c_{\sigma,4} > 0$ such that

$$\|\tilde{u}(t) - \check{u}^*(t)\|_2 \leq c_{\sigma,4}\|\tilde{\sigma}_\varepsilon\|_2. \tag{A.126}$$

The definition of $\tilde{\sigma}_\varepsilon$ in (A.115) together with (5.53), (A.119), (A.120), and the triangle inequality implies

$$\|\tilde{\sigma}_\varepsilon\|_2 \leq c_{\sigma,4}\bar{\varepsilon} + \bar{\varepsilon}^{\frac{\beta_\sigma}{2}}\sqrt{\frac{J_L^*(\xi_t)}{\bar{\lambda}_\sigma}} + c_{\sigma,5}\bar{\varepsilon}^{1-\frac{\beta_\alpha}{2}}\sqrt{J_L^*(\xi_t)} \tag{A.127}$$

for some $c_{\sigma,4}, c_{\sigma,5} > 0$. Combining (5.55), (A.126), (A.127), and $\check{J}_L^*(\xi_t', \mathcal{D}_t) \leq \bar{J}$, the term $\|\tilde{u}(t) - \check{u}^*(t)\|_2$ is bounded by $\beta_{\mathrm{u}}'(\bar{\varepsilon})$ with some function β_{u}'. It is simple to verify that $\beta_{\mathrm{u}}' \in \mathcal{K}_\infty$ if $\beta_\alpha + 2\beta_\sigma < 2$. Together with (5.55), (A.114), (A.125), and $\check{J}_L^*(\xi_t', \mathcal{D}_t) \leq \bar{J}$, this implies the existence of $\beta_{\mathrm{u}} \in \mathcal{K}_\infty$ such that (5.56) holds. ■

A.5 Proof of Theorem 5.2

Proof. After stating some preliminaries in Part (i), we show in Part (ii) that, if $\|\xi_{k+n} - \xi_{k+j}\|_2$, $j \in \mathbb{I}_{[0,n-1]}$, is bounded by some constant for $k = t-N, t-N+n, \ldots, t-n$, then it is bounded by the same constant for $k = t$. In combination with Lemma 5.3, this serves as a bound on the "perturbation" of the data in Problem 5.2, which we combine with the continuity of data-driven MPC in Section 5.2.3 and the results in Section 5.1 to prove practical stability in Part (iii).

For simplicity, we assume w.l.o.g. that $\frac{N}{n} \in \mathbb{I}_{\geq 0}$, i.e., N is divisible by n. According to Theorem 4.3, the Lyapunov function candidate $V(\xi_t', \mathcal{D}_t)$ admits quadratic lower and upper bounds of the form

$$c_{\mathrm{l}}\|\xi_t' - \xi_{\mathrm{Lin}}^{\mathrm{sr}}(x_t)\|_2^2 \leq V(\xi_t', \mathcal{D}_t) \leq c_{\mathrm{u}}\|\xi_t' - \xi_{\mathrm{Lin}}^{\mathrm{sr}}(x_t)\|_2^2 \tag{A.128}$$

for some $c_{\mathrm{l}}, c_{\mathrm{u}} > 0$. Note that ξ_N' lies in the set $\{\xi' \mid V(\xi', \mathcal{D}_N) \leq V_{\max}\}$, which is compact due to the lower bound (A.128) and Assumption 5.11, i.e., compactness of the (linearized) steady-state manifold (Assumption 5.6). Similar to the proof of Proposition 5.2, Lipschitz continuity of the dynamics (5.36) and compactness of $\mathbb{U}$ imply that the union of the N-step reachable sets of the linearized and the nonlinear dynamics (compare [127]) starting in $\{\xi' \mid V(\xi', \mathcal{D}_N) \leq V_{\max}\}$, which we denote by X, is compact. In the proof, we show that $\{\xi' \mid V(\xi', \mathcal{D}_N) \leq V_{\max}\}$ is positively invariant and thus, the bound in Lemma 5.3 as well as Lipschitz continuity of the map (5.42) hold uniformly throughout the proof. Similarly, whenever we apply Proposition 5.4, we use that the bounds hold uniformly for all linearized dynamics considered in the proof (after potential modification of some constants).

(i) Preliminaries

Let Problem 5.2 be feasible at time $t = ni \geq N$ and suppose

$$\|\xi_{k+n} - \xi_{k+j}\|_2 \leq c_{\xi,0}\theta, \quad j \in \mathbb{I}_{[0,n-1]}, \tag{A.129}$$

for $k = t-N, t-N+n, \ldots, t-n$ with some $c_{\xi,0} > 0$. Note that this holds at initial time $t = N$ with $c_{\xi,0} = 2$ due to $\|\xi_N - \xi_k\|_2 \leq \theta$ for $k \in \mathbb{I}_{[0,N-1]}$. Clearly, (A.129) implies

$$\|\xi_t - \xi_k\|_2 \leq c_{\theta,1}\theta, \quad k \in \mathbb{I}_{[t-N,t-1]}, \tag{A.130}$$

for some $c_{\theta,1} > 0$. Using Lemma 5.3, the difference between y_{t+k} and $y'_k(t)$ is bounded for $k \in \mathbb{I}_{[-N,-1]}$ by

$$\begin{aligned} \|y_{t+k} - y'_k(t)\|_2 = \|\Delta_{t,k}\|_2 &\overset{(5.47)}{\leq} c_\Delta \sum_{i=t-N}^{t-1} \|x_t - x_i\|_2^2 \\ &\overset{(5.42)}{\leq} c_\mathrm{L} c_\Delta \sum_{i=t-N}^{t-1} \|\xi_t - \xi_i\|_2^2 \overset{(A.130)}{\leq} c_\mathrm{L} c_\Delta N c_{\theta,1}^2 \theta^2 =: c_{\theta,2}\theta^2 \end{aligned} \tag{A.131}$$

for some $c_\mathrm{L} > 0$, using that T_L in (5.42) is Lipschitz continuous.

Next, we provide an extended state $\xi'_t = \begin{bmatrix} u'_{[t-n,t-1]} \\ y'_{[t-n,t-1]} \end{bmatrix}$ satisfying (5.43) with a suitable bound on $\|\xi_t - \xi'_t\|_2$. We write $\hat{x}_{[t-n,t]}$ for the state trajectory resulting from an application of the closed-loop input $u_{[t-n,t-1]}$ to the initial condition x_{t-n} for the dynamics linearized at x_t. We define the input component $u'_{[t-n,t-1]}$ of ξ'_t as

$$u'_{[t-n,t-1]} = u_{[t-n,t-1]} + \Gamma_\mathrm{c}(x_t)^\dagger(x_t - \hat{x}_t), \tag{A.132}$$

where $\Gamma_\mathrm{c}(x_t)$ is the controllability matrix of the dynamics linearized at x_t. According to Assumption 5.11 (i.e., Assumption 5.4), the Moore-Penrose inverse $\Gamma_\mathrm{c}(x_t)^\dagger$ is uniformly bounded. Further, we define $y'_{[t-n,t-1]}$ as the output trajectory for the dynamics linearized at x_t with input $u'_{[t-n,t-1]}$ and initial condition x_{t-n}. Similar to Lemma 5.3, we can show that

$$\|x_t - \hat{x}_t\|_2 \leq c_{\theta,3} \sum_{j=t-n}^{t-1} \|x_t - x_j\|_2^2 \overset{(5.42),(A.129)}{\leq} c_{\theta,4}\theta^2 \tag{A.133}$$

for some $c_{\theta,3}, c_{\theta,4} > 0$. Combining this with (A.132), we obtain

$$\|u_{[t-n,t-1]} - u'_{[t-n,t-1]}\|_2 \leq c_{\theta,5}\theta^2 \tag{A.134}$$

with some $c_{\theta,5} > 0$. We write $\underline{x}_{[t-n,t-1]}$ and $\underline{y}_{[t-n,t-1]}$ for the state and output resulting from an application of $u'_{[t-n,t-1]}$ to the nonlinear system (5.36) with initial state x_{t-n}. Using Lipschitz continuity of (5.36) by Assumption 5.11 (i.e., Assumption 5.2), there exists $c_{\theta,6} > 0$ such that

$$\|y_{[t-n,t-1]} - \underline{y}_{[t-n,t-1]}\|_2 \leq c_{\theta,6}\|u_{[t-n,t-1]} - u'_{[t-n,t-1]}\|_2. \tag{A.135}$$

Using again similar arguments as in Lemma 5.3, we can show

$$\begin{aligned}
&\|\underline{y}_{[t-n,t-1]} - y'_{[t-n,t-1]}\|_2 \\
\leq & c_{\theta,7} \sum_{j=t-n}^{t-1} \|x_t - \underline{x}_j\|_2^2 \overset{(A.102)}{\leq} 2c_{\theta,7} \sum_{j=t-n}^{t-1} (\|x_t - x_j\|_2^2 + \|x_j - \underline{x}_j\|_2^2) \\
\overset{(5.42),(A.129)}{\leq} & c_{\theta,8}\theta^2 + c_{\theta,9}\|u_{[t-n,t-1]} - u'_{[t-n,t-1]}\|_2^2
\end{aligned} \tag{A.136}$$

for suitably defined $c_{\theta,i} > 0$, $i \in \mathbb{I}_{[7,9]}$, where the last inequality also uses Lipschitz continuity of the nonlinear dynamics (5.36). Combining (A.134)–(A.136) and letting $\bar{\theta} < 1$ such that $\theta^4 < \theta^2$, there exists $c_{\theta,10} > 0$ such that

$$\|\xi_t - \xi'_t\|_2 \leq c_{\theta,10}\theta^2. \tag{A.137}$$

(ii) Bound on $\|\xi_{t+n} - \xi_{t+j}\|_2$, $i \in \mathbb{I}_{[0,n-1]}$

In this part of the proof, we show that, under suitable conditions, (A.129) holds for $k = t$ if it holds for $k = t-N, t-N+n, \ldots, t-n$. We denote the extended state (5.37) corresponding to $(u^{s*}(t), y^{s*}(t))$ by $\xi^{s*}(t)$. For $j \in \mathbb{I}_{[0,n-1]}$, it holds that

$$\begin{aligned}
\|\xi_{t+n} - \xi_{t+j}\|_2 \leq & \|\xi_{t+n} - \xi^{s*}(t)\|_2 + \|\xi_{t+j} - \xi^{s*}(t)\|_2 \\
\leq & c_{\xi,1}(\|\xi_{t+n} - \xi^{s*}(t)\|_2 + \|\xi_t - \xi^{s*}(t)\|_2)
\end{aligned} \tag{A.138}$$

for some $c_{\xi,1} > 0$, where we use that ξ_{t+n} and ξ_t contain all elements of ξ_{t+j}. Using optimality in Problem 5.2, we have

$$\|\xi_t - \xi^{s*}(t)\|_2^2 \leq \frac{1}{\lambda_{\min}(Q,R)} J_L^*(\xi_t). \tag{A.139}$$

Moreover, using $u_{t+k} = \bar{u}_k^*(t)$ for $k \in \mathbb{I}_{[0,n-1]}$ due to the n-step scheme, cf. Algorithm 5.2, we can derive

$$\begin{aligned}\|\xi_{t+n} - \xi^{s*}(t)\|_2 &= \sum_{k=0}^{n-1} \|u_{t+k} - u^{s*}(t)\|_2 + \|y_{t+k} - y^{s*}(t)\|_2 \\ &\leq \sum_{k=0}^{n-1} \|\bar{u}_k^*(t) - u^{s*}(t)\|_2 + \|y_{t+k} - \bar{y}_k^*(t)\|_2 + \|\bar{y}_k^*(t) - y^{s*}(t)\|_2 \\ &\leq \sum_{k=0}^{n-1} \left(\|y_{t+k} - y_k'(t)\|_2 + \|y_k'(t) - \bar{y}_k^*(t)\|_2\right) + \frac{\sqrt{J_L^*(\xi_t)}}{\sqrt{\lambda_{\min}(Q,R)}}.\end{aligned} \tag{A.140}$$

Similar to (A.131), there exists $c_{\xi,2} > 0$ such that

$$\begin{aligned}\sum_{k=0}^{n-1} \|y_{t+k} - y_k'(t)\|_2 &\leq c_{\xi,2} \sum_{k=-N}^{n-1} \|\xi_{t+k} - \xi_t\|_2^2 \\ &\overset{(A.130)}{\leq} c_{\xi,2} \sum_{k=0}^{n-1} \|\xi_{t+k} - \xi_t\|_2^2 + c_{\xi,2} N c_{\theta,1}^2 \theta^2.\end{aligned} \tag{A.141}$$

We now use a bound on the deviation between the predicted output $\bar{y}^*(t)$ in Problem 5.2 and the output of the linearized dynamics $y'(t)$. This bound is shown for data-driven MPC of LTI systems in Lemma 4.2 but the result remains true for affine systems since only differences between trajectories are considered. With the "noise bound" $\bar{\varepsilon} = c_{\theta,2}\theta^2$ (cf. (A.131)), Lemma 4.2 implies

$$\|y_k'(t) - \bar{y}_k^*(t)\|_2 \leq c_{\xi,3}\theta^2 \|\alpha^*(t)\|_2 + c_{\xi,4}\theta^2 + c_{\xi,5}\|\sigma^*(t)\|_2 \tag{A.142}$$

for $k \in \mathbb{I}_{[0,n-1]}$ with some $c_{\xi,i} > 0$, $i \in \mathbb{I}_{[4,6]}$. Finally, using the definition of $J_L^*(\xi_t)$, we have

$$\|\alpha^*(t)\|_2 \leq \sqrt{\frac{1}{\lambda_\alpha} J_L^*(\xi_t)} + \|\alpha_{\text{Lin}}^{\text{sr}}(\mathcal{D}_t)\|_2, \tag{A.143}$$

$$\|\sigma^*(t)\|_2 \leq \sqrt{\frac{1}{\lambda_\sigma} J_L^*(\xi_t)}, \tag{A.144}$$

where $\|\alpha_{\text{Lin}}^{\text{sr}}(\mathcal{D}_t)\|_2$ is uniformly bounded by assumption. Let now $\lambda_\alpha = \bar{\lambda}_\alpha \theta^{2\beta_\alpha}$ and $\lambda_\sigma = \frac{\bar{\lambda}_\sigma}{\theta^{2\beta_\sigma}}$ for some $\bar{\lambda}_\alpha, \bar{\lambda}_\sigma, \beta_\alpha, \beta_\sigma > 0$ with $\beta_\alpha > 1$, $\beta_\alpha + 2\beta_\sigma < 2$, compare Proposition 5.4. There exist $c_{\xi,6}, c_{\xi,7} > 0$ such that, for $k \in \mathbb{I}_{[0,n-1]}$,

$$\|y_k'(t) - \bar{y}_k^*(t)\|_2 \leq c_{\xi,6}\theta^2 + c_{\xi,7}(\theta^{2-\beta_\alpha} + \theta^{\beta_\sigma})\sqrt{J_L^*(\xi_t)}. \tag{A.145}$$

Combining (A.138)–(A.141) and (A.145), we infer for $j \in \mathbb{I}_{[0,n-1]}$

$$\|\xi_{t+n} - \xi_{t+j}\|_2 \leq c_{\xi,1}\Bigg(\frac{2}{\sqrt{\lambda_{\min}(Q,R)}}\sqrt{J_L^*(\xi_t)} + c_{\xi,2}\sum_{k=0}^{n-1}\|\xi_{t+k} - \xi_t\|_2^2 \\ + c_{\xi,2}Nc_{\theta,1}^2\theta^2 + n\Big(c_{\xi,6}\theta^2 + c_{\xi,7}(\theta^{2-\beta_\alpha} + \theta^{\beta_\sigma})\sqrt{J_L^*(\xi_t)}\Big)\Bigg). \tag{A.146}$$

Let now

$$\bar{S} = \theta^4, \; V_{\max} = \theta^{2+\eta} \tag{A.147}$$

for some $0 < \eta < 2(\beta_\alpha - 1)$ and suppose $\bar{\theta} < 1$. This implies $\theta^{\eta_1} < \theta^{\eta_2}$ if $\eta_1 > \eta_2$, which will be used throughout the proof. Using Assumption 5.11, i.e., Assumptions 5.5 and 5.6, $J_{\text{eq,Lin}}^*(x_t)$ is uniformly bounded by $\bar{S}$, i.e.,

$$J_{\text{eq,Lin}}^*(x_t) \leq c_{\text{J},S}\bar{S} = c_{\text{J},S}\theta^4 \tag{A.148}$$

with $c_{\text{J},S} > 0$. Hence, $\check{J}_L^*(\bar{\xi}_t, \mathcal{D}_t) \leq V_{\max} + J_{\text{eq,Lin}}^*(x_t)$ implies

$$\check{J}_L^*(\xi_t', \mathcal{D}_t) \leq \check{c}_{\text{J},\theta}\theta^{2+\eta} \tag{A.149}$$

for some $\check{c}_{\text{J},\theta} > 0$. We now apply Proposition 5.4, where the noise / disturbance is bounded as in (5.53) by $\bar{\varepsilon} = c_\theta\theta^2$ for some $c_\theta > 0$, compare (A.131) for the "noise" entering the output Hankel matrix and (A.137) for the "disturbance" in the initial conditions. Inequalities (A.148) and (A.149) together with Proposition 5.4 imply

$$J_L^*(\xi_t) \leq \Big(1 + c_{\text{J,a}}(c_\theta\theta^2)^{\beta_\sigma}\Big)\check{c}_{\text{J},\theta}\theta^{2+\eta} + c_{\text{J,b}}(c_\theta\theta^2)^{2-\beta_\sigma}\Big(1 + \|H_{ux,t}^\dagger\|_2^2(c_\theta\theta^2)^{\beta_\alpha}\Big) \\ \leq c_{\text{J},\theta}\theta^{2+\eta} \tag{A.150}$$

for some $c_{\text{J},\theta} > 0$, where we use $\|H_{ux,t+n}^\dagger\|_2^2 \leq \frac{\bar{c}_{\text{H}}^2}{\theta^2}$ by assumption, $\bar{\theta} < 1$ as well as

$$0 < \eta < 2(\beta_\alpha - 1) < 2(1 - 2\beta_\sigma) < 2(1 - \beta_\sigma).$$

Plugging this into (A.146), we obtain

$$\|\xi_{t+n} - \xi_{t+j}\|_2 \leq \beta_1(\theta) + c_{\xi,1}c_{\xi,2}\sum_{k=0}^{n-1}\|\xi_{t+k} - \xi_t\|_2^2 \tag{A.151}$$

for a suitably defined $\beta_1 \in \mathcal{K}_\infty$ containing only terms with order strictly larger than 1. Each of the terms $\|\xi_{t+k} - \xi_t\|_2$, $k \in \mathbb{I}_{[0,n-1]}$, can be bounded analogous to

$\|\xi_{t+n} - \xi_{t+j}\|_2$ in (A.151), using the same bounds as above leading to (A.151). To be precise, following the same steps as above, there exist $\hat{\beta} \in \mathcal{K}_\infty$, $\hat{c}_\xi > 0$ such that, for all $k \in \mathbb{I}_{[0,n-1]}$,

$$\|\xi_{t+k} - \xi_t\|_2^2 \leq \hat{\beta}(\theta) + \hat{c}_\xi \sum_{s=0}^{k-1} \|\xi_{t+s} - \xi_t\|_2^2,$$

where $\hat{\beta}$ contains only terms with order strictly larger than 1. Applying this bound recursively $n-1$ times and plugging the result into (A.151), we obtain

$$\|\xi_{t+n} - \xi_{t+j}\|_2 \leq \beta_2(\theta) \tag{A.152}$$

for $j \in \mathbb{I}_{[0,n-1]}$ with $\beta_2 \in \mathcal{K}_\infty$ containing only terms with order strictly larger than 1. Hence, if $\bar{\theta}$ is sufficiently small such that $\beta_2(\theta) \leq c_{\xi,0}\theta$, then (A.129) holds for $k = t$. To conclude, we have shown that (A.129) holds recursively, assuming that $V(\xi_t', \mathcal{D}_t) \leq V_{\max}$. Thus, using (A.129) for $k = t-N+n, t-N+2n, \ldots, t$, we infer (cf. (A.130), (A.137))

$$\|\xi_{t+n} - \xi_k\|_2 \leq c_{\theta,1}\theta, \quad k \in \mathbb{I}_{[t+n-N,t+n-1]}, \tag{A.153}$$

$$\|\xi_{t+n}' - \xi_{t+n}\|_2 \leq c_{\theta,10}\theta^2. \tag{A.154}$$

With $c_{\theta,2}$ as in (A.131), we have for $k \in \mathbb{I}_{[-N,-1]}$

$$\|y_{t+n+k} - y_k'(t+n)\|_2 \overset{(5.47)}{\leq} c_\Delta \sum_{i=t+n-N}^{t+n-1} \|x_{t+n} - x_i\|_2^2 \leq c_{\theta,2}\theta^2, \tag{A.155}$$

i.e., (A.131) also holds recursively with t replaced by $t+n$.

(iii) Invariance and Lyapunov function decay

(iii.a) Application of model-based results from Section 5.1

Recall that the prediction error due to the inexact model in Problem 5.2 at time t (compared to the nominal MPC problem with cost $\check{J}_L^*(\xi_t', \mathcal{D}_t)$) is induced by output measurement noise and perturbed initial conditions with bound $c_\theta\theta^2$ for some $c_\theta > 0$ (compare (A.131) and (A.137)). According to Proposition 5.4, this noise translates into an input disturbance for the resulting closed loop with bound $\beta_{\mathrm{u}}(c_\theta\theta^2)$, i.e.,

$$\|\check{u}_{[0,n-1]}^*(t) - u_{[t,t+n-1]}\|_2 \leq \beta_{\mathrm{u}}(c_\theta\theta^2), \tag{A.156}$$

where $\check{u}^*(t)$ is the nominal optimal input corresponding to the optimal cost $\check{J}_L^*(\xi_t', \mathcal{D}_t)$. Since the nominal MPC problem (Problem 5.3) with $\bar{\sigma} = 0$ contains an exact model of the linearization, we can apply the main technical result[1] in Section 5.1 (Proposition 5.2) to arrive at

$$\begin{aligned} &V(\xi_{t+n}', \mathcal{D}_{t+n}) \\ \leq &c_{V,1} V(\xi_t', \mathcal{D}_t) + \bar{\lambda}_\alpha (c_\theta \theta^2)^{\beta_\alpha} \|\tilde{\alpha}(t+n) - \alpha_{\text{Lin}}^{\text{sr}}(\mathcal{D}_{t+n})\|_2^2 + c_{d,1}\beta_u(c_\theta\theta^2)^2 \end{aligned} \tag{A.157}$$

for some $0 < c_{V,1} < 1$, $c_{d,1} > 0$. Here, $\tilde{\alpha}(t+n)$ is a later specified candidate solution corresponding to the optimal control problem with optimal cost $\check{J}_L^*(\xi_{t+n}', \mathcal{D}_{t+n})$. The additional term depending on $\tilde{\alpha}(t+n)$ is due to the regularization of α in the cost, which is not present in the model-based MPC scheme in Section 5.1. Further, the term $c_{d,1}\beta_u(c_\theta\theta^2)^2$ is due to the fact that x_{t+n} results from applying the optimal input $u_{[t,t+n-1]} = \bar{u}_{[0,n-1]}^*(t)$ of Problem 5.2 to the state x_t, whereas Proposition 5.2 considers the closed loop under the nominal MPC input $\check{u}_{[0,n-1]}^*(t)$. To be more precise, with minor adaptations of the proof of Proposition 5.2, it can be shown that, if the closed-loop state at time $t+n$ results from applying an input which differs from $\check{u}^*(t)$ by no more than $\beta_u(c_\theta\theta^2)$ (compare (A.156)), then the Lyapunov function decay shown in Proposition 5.2 remains true if the squared disturbance bound $c_{d,1}\beta_u(c_\theta\theta^2)^2$ is added on the right-hand side (i.e., (A.157) holds). This is possible since the main proof idea of Proposition 5.2 does not rely on the exact state value at time $t+n$, but rather on the fact that it remains close to x_t.

(iii.b) Bound on $\|\tilde{\alpha}(t+n) - \alpha_{\text{Lin}}^{\text{sr}}(\mathcal{D}_{t+n})\|_2^2$

We bound this term using the candidate solution of the model-based MPC in Section 5.1. We write $\tilde{u}(t+n)$ for a candidate input used in the proof of Proposition 5.2, which corresponds to a feasible candidate solution to Problem 5.3 with $\bar{\sigma} = 0$ for $\tilde{\alpha}(t+n) = H_{ux,t+n}^\dagger \begin{bmatrix} \tilde{u}(t+n)^\top & x_t^\top & 1 \end{bmatrix}^\top$. Thus, we infer

$$\begin{aligned} &\|\tilde{\alpha}(t+n) - \alpha_{\text{Lin}}^{\text{sr}}(\mathcal{D}_{t+n})\|_2^2 \\ \leq &\|H_{ux,t+n}^\dagger\|_2^2 \Big(\|\tilde{u}(t+n) - \mathbb{1}_{L+n+1} \otimes u_{\text{Lin}}^{\text{sr}}(x_{t+n})\|_2^2 + \|x_t - x_{\text{Lin}}^{\text{sr}}(x_{t+n})\|_2^2 \Big). \end{aligned} \tag{A.158}$$

The analysis in Proposition 5.2 implies that both candidate solutions used in

[1]The result can be applied despite the fact that the cost of Problem 5.3 depends on the output, i.e., is in general only positive semidefinite in the state, since we perform an n-step analysis and the cost is positive definite over n steps.

Section 5.1 satisfy

$$\|\tilde{u}(t+n) - \mathbb{1}_{L+n+1} \otimes u^{\mathrm{sr}}_{\mathrm{Lin}}(x_{t+n})\|_2^2 \leq c_{\mathrm{V},2} V(\xi'_t, \mathcal{D}_t) \tag{A.159}$$

for some $c_{\mathrm{V},2} > 0$. Further, there exist $c_{\mathrm{V},3}, c_{\mathrm{V},4} > 0$ such that

$$\begin{aligned}\|x_t - x^{\mathrm{sr}}_{\mathrm{Lin}}(x_{t+n})\|_2^2 &\overset{(\mathrm{A}.102)}{\leq} 2\|x_t - x_{t+n}\|_2^2 + 2\|x_{t+n} - x^{\mathrm{sr}}_{\mathrm{Lin}}(x_{t+n})\|_2^2 \\ &\overset{(5.42),(\mathrm{A}.153),(\mathrm{A}.128)}{\leq} c_{\mathrm{V},3}\theta^2 + c_{\mathrm{V},4} V(\xi'_{t+n}, \mathcal{D}_{t+n}).\end{aligned} \tag{A.160}$$

Assumption 5.16 and $c_{\mathrm{H}} = \frac{\bar{c}_{\mathrm{H}}}{\theta}$ imply $\|H^{\dagger}_{ux,t+n}\|_2^2 \leq \frac{\bar{c}^2_{\mathrm{H}}}{\theta^2}$. Plugging (A.158)–(A.160) into (A.157), we thus infer

$$\begin{aligned}V(\xi'_{t+n}, \mathcal{D}_{t+n}) \leq &(c_{\mathrm{V},1} + c_{\mathrm{V},5}\theta^{2\beta_\alpha - 2}) V(\xi'_t, \mathcal{D}_t) + c_{\mathrm{d},1}\beta_{\mathrm{u}}(c_\theta\theta^2)^2 \\ &+ c_{\mathrm{V},6}\theta^{2\beta_\alpha} + c_{\mathrm{V},7}\theta^{2\beta_\alpha - 2} V(\xi'_{t+n}, \mathcal{D}_{t+n})\end{aligned} \tag{A.161}$$

for suitably defined $c_{V,i} > 0$, $i \in \mathbb{I}_{[5,7]}$. Given that $\beta_\alpha > 1$, $c_{\mathrm{V},1} < 1$, we can choose $\bar{\theta}$ sufficiently small such that

$$c_{\mathrm{V},7}\theta^{2\beta_\alpha - 2} < 1, \quad \frac{c_{\mathrm{V},1} + c_{\mathrm{V},5}\theta^{2\beta_\alpha - 2}}{1 - c_{\mathrm{V},7}\theta^{2\beta_\alpha - 2}} < 1.$$

Multiplication of both sides of (A.161) by $\frac{1}{1 - c_{\mathrm{V},7}\theta^{2\beta_\alpha - 2}}$ yields

$$V(\xi'_{t+n}, \mathcal{D}_{t+n}) \leq c_{\mathrm{V},8} V(\xi'_t, \mathcal{D}_t) + c_{\mathrm{V},9}(\beta_{\mathrm{u}}(c_\theta\theta^2)^2 + \theta^{2\beta_\alpha}) \tag{A.162}$$

for suitably defined $0 < c_{\mathrm{V},8} < 1$, $c_{\mathrm{V},9} > 0$.

(iii.c) Practical stability

Plugging (A.150) and $\bar{\varepsilon} = c_\theta\theta^2$ into (A.125)–(A.127), straightforward algebraic calculations reveal that (5.56) also holds with

$$\beta_{\mathrm{u}}(c_\theta\theta^2) \leq c_{\mathrm{V},10}\left(\theta^{2 - \frac{\beta_\alpha + \beta_\sigma}{2} + \frac{\eta}{2}} + \theta^2 + \theta^{1 + \beta_\sigma + \frac{\eta}{2}}\right) =: \beta_{\mathrm{V},1}(\theta)$$

for some $c_{\mathrm{V},10} > 0$. Using $\beta_\alpha + 2\beta_\sigma < 2$ and $\eta < 2(\beta_\alpha - 1)$, it is straightforward to verify that the smallest exponent appearing in $(\beta_{\mathrm{V},1}(\theta))^2 + \theta^{2\beta_\alpha}$ is strictly larger than $2 + \eta$. This implies that, if $\bar{\theta}$ is sufficiently small, then (A.162) yields $V(\xi'_{t+n}, \mathcal{D}_{t+n}) \leq \theta^{2+\eta} = V_{\max}$. Hence, all bounds in this proof hold recursively for $t = N + ni$, $i \in \mathbb{I}_{\geq 0}$. In particular, the nominal MPC problem with cost $\check{J}^*_L(\xi_t, \mathcal{D}_t)$ and, using

Proposition 5.4, Problem 5.2 are recursively feasible. Finally, a recursive application of (A.162) yields

$$V(\xi_t', \mathcal{D}_t) \leq c_{\mathrm{V},8}^i V(\xi_N', \mathcal{D}_N) + \beta_{\mathrm{V},3}(\theta) \tag{A.163}$$

for some $\beta_{\mathrm{V},3} \in \mathcal{K}_\infty$ and any $t = N + ni$, $i \in \mathbb{I}_{\geq 0}$. Exploiting the lower and upper bounds in (A.128), we obtain

$$\|\xi_t' - \xi_{\mathrm{Lin}}^{\mathrm{sr}}(x_t)\|_2^2 \leq \frac{c_\mathrm{u}}{c_\mathrm{l}} c_{\mathrm{V},8}^i \|\xi_N' - \xi_{\mathrm{Lin}}^{\mathrm{sr}}(x_N)\|_2^2 + \frac{1}{c_\mathrm{l}} \beta_{\mathrm{V},3}(\theta).$$

Using Assumption 5.13, i.e., (5.44), as well as (A.137), we obtain (5.59) with $c_\mathrm{V} = c_{\mathrm{V},8}$ and some $C > 0$, $\beta_\theta \in \mathcal{K}_\infty$. ■

Bibliography

[1] H. S. Abbas, R. Tóth, N. Meskin, J. Mohammadpour, and J. Hanema. "A robust MPC for input-output LPV models." In: *IEEE Trans. Automat. Control* 61.12 (2016), pp. 4183–4188.

[2] V. Adetola and M. Guay. "Robust adaptive MPC for constrained uncertain nonlinear systems." In: *Int. J. Adaptive Control and Signal Processing* 25.2 (2011), pp. 155–167.

[3] A. B. Alexandru, A. Tsiamis, and G. J. Pappas. "Data-driven control on encrypted data." In: *arXiv:2008.12671* (2020).

[4] A. B. Alexandru, A. Tsiamis, and G. J. Pappas. "Towards private data-driven control." In: *Proc. 59th IEEE Conf. Decision and Control (CDC)*. 2020, pp. 5449–5456.

[5] A. Allibhoy and J. Cortés. "Data-based receding horizon control of linear network systems." In: *IEEE Control Systems Lett.* 5.4 (2021), pp. 1207–1212.

[6] C. A. Alonso, F. Yang, and N. Matni. "Data-driven distributed and localized model predictive control." In: *arXiv:2112.12229* (2021).

[7] D. Alpago, F. Dörfler, and J. Lygeros. "An extended Kalman filter for data-enabled predictive control." In: *IEEE Control Systems Lett.* 4.4 (2020), pp. 994–999.

[8] J. Anderson, J. C. Doyle, S. H. Low, and N. Matni. "System level synthesis." In: *Annual Reviews in Control* 47 (2019), pp. 364–393.

[9] J. A. E. Andersson, J. Gillis, G. Horn, J. B. Rawlings, and M. Diehl. "CasADi: a software framework for nonlinear optimization and control." In: *Mathematical Programming Computation* 11.1 (2019), pp. 1–36.

[10] M. ApS. *The MOSEK optimization toolbox for MATLAB manual. Version 9.0.* 2019.

[11] A. Aswani, H. Gonzalez, S. S. Sastry, and C. Tomlin. "Provably safe and robust learning-based model predictive control." In: *Automatica* 49.5 (2013), pp. 1216–1226.

[12] G. Belforte, B. Bona, and V. Cerone. "Parameter estimation algorithms for a set-membership description of uncertainty." In: *Automatica* 26.5 (1990), pp. 887–898.

[13] A. Bemporad, M. Morari, V. Dua, and E. N. Pistikopoulos. "The explicit linear quadratic regulator for constrained systems." In: *Automatica* 38.1 (2002), pp. 3–20.

[14] F. Berkenkamp, M. Turchetta, A. Schoellig, and A. Krause. "Safe model-based reinforcement learning with stability guarantees." In: *Advances in Neural Information Processing Systems*. 2017, pp. 908–918.

[15] G. Bianchin, M. Vaquero, J. Cortés, and E. Dall'Anese. "Data-driven synthesis of optimization-based controllers for regulation of unknown linear systems." In: *Proc. 60th IEEE Conf. Decision and Control (CDC)*. 2021, pp. 5783–5788.

[16] A. Bisoffi, C. De Persis, and P. Tesi. "Data-based stabilization of unknown bilinear systems with guaranteed basin of attraction." In: *Syst. Contr. Lett.* 145 (2020), p. 104788.

[17] A. Bisoffi, C. De Persis, and P. Tesi. "Trade-offs in learning controllers from noisy data." In: *Syst. Contr. Lett.* 154 (2021), p. 104985.

[18] V. Breschi, A. Sassella, and S. Formentin. "On the design of regularized explicit predictive controllers from input-output data." In: *arXiv:2110.11808* (2021).

[19] S. Brüggemann and R. R. Bitmead. "Forward-looking persistent excitation in model predictive control." In: *Automatica* 136 (2022), p. 110033.

[20] C. Cai and A. R. Teel. "Input–output-to-state stability for discrete-time systems." In: *Automatica* 44.2 (2008), pp. 326–336.

[21] J.-P. Calliess. "Conservative decision-making and inference in uncertain dynamical systems." PhD thesis. University of Oxford, 2014.

[22] M. C. Campi, A. Lecchini, and S. M. Savaresi. "Virtual reference feedback tuning: a direct method for the design of feedback controllers." In: *Automatica* 38.8 (2002), pp. 742–753.

[23] M. Cannon, J. Buerger, B. Kouvaritakis, and S. Rakovic. "Robust tubes in nonlinear model predictive control." In: *IEEE Trans. Automat. Control* 56.8 (2011), pp. 1942–1947.

[24] P. G. Carlet, A. Favato, S. Bolognani, and F. Dörfler. "Data-driven continuous-set predictive current control for synchronous motor drives." In: *IEEE Trans. Power Electronics* (2022). doi: 10.1109/TPEL.2022.3142244.

[25] P. G. Carlet, A. Favato, S. Bolognani, and F. Dörfler. "Data-driven predictive current control for synchronous motor drives." In: *Proc. IEEE Energy Conversion Congress and Exposition*. 2020, pp. 5148–5154.

[26] H. Chen and F. Allgöwer. "A quasi-infinite horizon nonlinear model predictive control scheme with guaranteed stability." In: *Automatica* 34.10 (1998), pp. 1205–1217.

[27] J. Coulson, J. Lygeros, and F. Dörfler. "Data-enabled predictive control: in the shallows of the DeePC." In: *Proc. European Control Conf. (ECC)*. 2019, pp. 307–312.

[28] J. Coulson, J. Lygeros, and F. Dörfler. "Distributionally robust chance constrained data-enabled predictive control." In: *IEEE Trans. Automat. Control* (2021). doi: 10.1109/TAC.2021.3097706.

[29] J. Coulson, J. Lygeros, and F. Dörfler. "Regularized and distributionally robust data-enabled predictive control." In: *Proc. 58th IEEE Conf. Decision and Control (CDC)*. 2019, pp. 2696–2701.

[30] C. De Persis and P. Tesi. "Formulas for data-driven control: stabilization, optimality and robustness." In: *IEEE Trans. Automat. Control* 65.3 (2020), pp. 909–924.

[31] C. De Persis and P. Tesi. "Low-complexity learning of linear quadratic regulators from noisy data." In: *Automatica* 128 (2021), p. 109548.

[32] S. Dean, H. Mania, N. Matni, B. Recht, and S. Tu. "On the sample complexity of the linear quadratic regulator." In: *Foundations of Computational Mathematics* (2019). https://doi.org/10.1007/s10208-019-09426-y.

[33] S. Dean, S. Tu, N. Matni, and B. Recht. "Safely learning to control the constrained linear quadratic regulator." In: *Proc. American Control Conf. (ACC)*. 2019, pp. 5582–5588.

[34] C. A. Desoer and M. Vidyasagar. "Feedback systems: input-output properties." In: *SIAM* (1975).

[35] M. Diehl, R. Findeisen, F. Allgower, H. G. Bock, and J. P. Schloder. "Nominal stability of real-time iteration scheme for nonlinear model predictive control." In: *IEE Proceedings - Control Theory and Applications* 152.3 (2005), pp. 296–308.

[36] F. Dörfler, J. Coulson, and I. Markovsky. "Bridging direct & indirect data-driven control formulations via regularizations and relaxations." In: *IEEE Trans. Automat. Control* (2022). doi: 10.1109/TAC.2022.3148374.

[37] F. Dörfler, P. Tesi, and C. De Persis. "On the certainty-equivalence approach to direct data-driven LQR design." In: *arXiv:2109.06643* (2021).

[38] A. Dutta, E. Hartley, J. Maciejowski, and R. De Keyser. "Certification of a class of industrial predictive controllers without terminal conditions." In: *Proc. 53rd IEEE Conf. Decision and Control (CDC)*. 2014, pp. 6695–6700.

[39] E. Elokda, J. Coulson, J. Lygeros, and F. Dörfler. "Data-enabled predictive control for quadcopters." In: *Int. J. Robust and Nonlinear Control* 31.18 (2021), pp. 8916–8936.

[40] F. Fabiani and P. J. Goulart. "The optimal transport paradigm enables data compression in data-driven robust control." In: *Proc. American Control Conf. (ACC)*. 2021, pp. 2408–2413.

[41] W. Favoreel, B. D. Moor, and M. Gevers. "SPC: Subspace predictive control." In: *IFAC Proceedings Volumes* 32.2 (1999), pp. 4004–4009.

[42] A. Ferramosca, D. Limón, I. Alvarado, T. Alamo, F. Castaño, and E. F. Camacho. "Optimal MPC for tracking of constrained linear systems." In: *Int. J. Systems Science* 42.8 (2011), pp. 1265–1276.

[43] L. Furieri, B. Guo, A. Martin, and G. Ferrari Trecate. "A behavioral input-output parametrization of control policies with suboptimality guarantees." In: *Proc. 60th IEEE Conf. Decision and Control (CDC)*. 2021, pp. 2539–2544.

[44] L. Furieri, B. Guo, A. Martin, and G. Ferrari Trecate. "Near-optimal design of safe output feedback controllers from noisy data." In: *arXiv:2105.10280* (2021).

[45] L. Furieri, Y. Zheng, A. Papachristodoulou, and M. Kamgarpour. "An input-output parametrization of stabilizing controllers: amidst Youla and system level synthesis." In: *IEEE Control Systems Lett.* 3.4 (2019), pp. 1014–1019.

[46] G. C. Goodwin and K. S. Sin. *Adaptive filtering prediction and control.* Courier Corporation, 2014.

[47] G. Grimm, M. J. Messina, S. E. Tuna, and A. R. Teel. "Model predictive control: for want of a local control Lyapunov function, all is not lost." In: *IEEE Trans. Automat. Control* 50.5 (2005), pp. 546–558.

[48] L. Grüne. "NMPC without terminal constraints." In: *Proc. IFAC Conf. Nonlinear Model Predictive Control.* 2012, pp. 1–13.

[49] L. Grüne and V. G. Palma. "Robustness of performance and stability for multistep and updated multistep MPC schemes." In: *Discrete and Continuous Dynamical Systems* 35.9 (2015), pp. 4385–4414.

[50] L. Grüne and J. Pannek. *Nonlinear Model Predictive Control.* Springer, 2017.

[51] L. Grüne and A. Rantzer. "On the infinite horizon performance of receding horizon controllers." In: *IEEE Trans. Automat. Control* 53.9 (2011), pp. 2100–2111.

[52] L. Grüne and M. Stieler. "Asymptotic stability and transient optimality of economic MPC without terminal conditions." In: *J. Proc. Contr.* 24 (2014), pp. 1187–1196.

[53] M. Guay, V. Adetola, and D. DeHaan. *Robust and adaptive model predictive control of non-linear systems.* Control, Robotics, Sensors. Institution of Engineering and Technology, 2015.

[54] M. Guo, C. De Persis, and P. Tesi. "Data-driven stabilization of nonlinear polynomial systems with noisy data." In: *IEEE Trans. Automat. Control* (2021). doi: 10.1109/TAC.2021.3115436.

[55] M. Herceg, M. Kvasnica, C. Jones, and M. Morari. "Multi-Parametric Toolbox 3.0." In: *Proc. European Control Conf. (ECC).* 2013, pp. 502–510.

[56] L. Hewing, J. Kabzan, and M. N. Zeilinger. "Cautious model predictive control using Gaussian process regression." In: *IEEE Trans. Control Systems Technology* 28.6 (2020), pp. 2736–2743.

[57] L. Hewing, K. P. Wabersich, M. Menner, and M. N. Zeilinger. "Learning-based model predictive control: Toward safe learning in control." In: *Ann. Rev. Control, Robotics, and Autonomous Systems* 3 (2020), pp. 269–296.

[58] H. Hjalmarsson, M. Gevers, S. Gunnarsson, and O. Lequin. "Iterative feedback tuning: theory and applications." In: *IEEE Control Systems Magazine* 18.4 (1998), pp. 26–41.

[59] Z.-S. Hou and Z. Wang. "From model-based control to data-driven control: Survey, classification and perspective." In: *Information Sciences* 235 (2013), pp. 3–35.

[60] L. Huang, J. Coulson, J. Lygeros, and F. Dörfler. "Data-enabled predictive control for grid-connected power converters." In: *Proc. 58th IEEE Conf. Decision and Control (CDC)*. 2019, pp. 8130–8135.

[61] L. Huang, J. Coulson, J. Lygeros, and F. Dörfler. "Decentralized data-enabled predictive control for power system oscillation damping." In: *IEEE Trans. Control Systems Technology* (2021). doi: 10.1109/TCST.2021.3088638.

[62] L. Huang, J. Zhen, J. Lygeros, and F. Dörfler. "Quadratic regularization of data-enabled predictive control: theory and application to power converter experiments." In: *IFAC-PapersOnLine* 54.7 (2021), pp. 192–197.

[63] L. Huang, J. Zhen, J. Lygeros, and F. Dörfler. "Robust data-enabled predictive control: tractable formulations and performance guarantees." In: *arXiv:2105.07199* (2021).

[64] A. Iannelli, M. Yin, and R. S. Smith. "Design of input for data-driven simulation with Hankel and Page matrices." In: *Proc. 60th IEEE Conf. Decision and Control (CDC)*. 2021, pp. 139–145.

[65] A. Iannelli, M. Yin, and R. S. Smith. "Experiment design for impulse response identification with signal matrix models." In: *IFAC-PapersOnLine* 54.7 (2021), pp. 625–630.

[66] S. Kerz, J. Teutsch, T. Brüdigam, D. Wollherr, and M. Leibold. "Data-driven stochastic model predictive control." In: *arXiv:2112.04439* (2021).

[67] H. K. Khalil. *Nonlinear systems*. Macmillan, New York, 2002.

[68] J. Köhler. "Analysis and design of MPC frameworks for dynamic operation of nonlinear constrained systems." PhD thesis. University of Stuttgart, 2021.

[69] J. Köhler, M. A. Müller, and F. Allgöwer. "A nonlinear tracking model predictive control scheme for dynamic target signals." In: *Automatica* 118 (2020), p. 109030.

[70] J. Köhler, M. A. Müller, and F. Allgöwer. "Constrained nonlinear output regulation using model predictive control." In: *IEEE Trans. Automat. Control* (2021). doi: 10.1109/TAC.2021.3081080.

[71] J. Köhler, R. Soloperto, M. A. Müller, and F. Allgöwer. "A computationally efficient robust model predictive control framework for uncertain nonlinear system." In: *IEEE Trans. Automat. Control* 66.2 (2021), pp. 794–801.

[72] M. Korda and I. Mezić. "Linear predictors for nonlinear dynamical systems: Koopman operator meets model predictive control." In: *Automatica* 93 (2018), pp. 149–160.

[73] R. L. Kosut. "Uncertainty model unfalsification for robust adaptive control." In: *Annual Reviews in Control* 25 (2001), pp. 65–76.

[74] B. Kouvaritakis and M. Cannon. *Model Predictive Control: Classical, Robust and Stochastic*. Springer, 2016.

[75] V. Krishnan and F. Pasqualetti. "On direct vs indirect data-driven predictive control." In: *Proc. 60th IEEE Conf. Decision and Control (CDC)*. 2021, pp. 736–741.

[76] M. Ławryńczuk and P. Tatjewski. "A computationally efficient nonlinear predictive control algorithm with RBF neural models and its application." In: *Rough Sets and Intelligent Systems Paradigms*. Springer Berlin Heidelberg, 2007, pp. 603–612.

[77] Y. Lian and C. N. Jones. "Nonlinear data-enabled prediction and control." In: *Proc. 3rd Conference on Learning for Dynamics and Control*. Vol. 144. PMLR, 2021, pp. 523–534.

[78] Y. Lian, J. Shi, M. P. Koch, and C. N. Jones. "Adaptive robust data-driven building control via bi-level reformulation: an experimental result." In: *arXiv:2106.05740* (2021).

[79] Y. Lian, R. Wang, and C. N. Jones. "Koopman based data-driven predictive control." In: *arXiv:2102.05122* (2021).

[80] D. Liao-McPherson, M. M. Nicotra, and I. Kolmanovsky. "Time-distributed optimization for real-time model predictive control: stability, robustness, and constraint satisfaction." In: *Automatica* 117 (2020), p. 108973.

[81] D. Limón, T. Alamo, F. Salas, and E. F. Camacho. "On the stability of constrained MPC without terminal constraint." In: *IEEE Trans. Automat. Control* 51.5 (2006), pp. 832–836.

[82] D. Limón, I. Alvarado, T. Alamo, and E. F. Camacho. "MPC for tracking piecewise constant references for constrained linear systems." In: *Automatica* 44.9 (2008), pp. 2382–2387.

[83] D. Limón, A. Ferramosca, I. Alvarado, and T. Alamo. "Nonlinear MPC for Tracking Piece-Wise Constant Reference Signals." In: *IEEE Trans. Automat. Control* 63.11 (2018), pp. 3735–3750.

[84] W. Liu, J. Sun, G. Wang, F. Bullo, and J. Chen. "Data-driven resilient predictive control under denial-of-service." In: *arXiv:2110.12766* (2021).

[85] L. Ljung. *System Identification: Theory for the User*. Prentice-Hall, Englewood Cliffs, NJ, 1987.

[86] J. Löfberg. "YALMIP: a toolbox for modeling and optimization in MATLAB." In: *Proc. IEEE Int. Conf. Robotics and Automation*. 2004, pp. 284–289.

[87] M. Lorenzen, M. Cannon, and F. Allgöwer. "Robust MPC with recursive model update." In: *Automatica* 103 (2019), pp. 461–471.

[88] X. Lu, M. Cannon, and D. Koksal-Rivet. "Robust adaptive model predictive control: performance and parameter estimation." In: *Int. J. Robust and Nonlinear Control* 31.18 (2021), pp. 8703–8724.

[89] A. Luppi, C. De Persis, and P. Tesi. "On data-driven stabilization of systems with quadratic nonlinearities." In: *arXiv:2103.15631* (2021).

[90] L. Magni and R. Scattolini. "On the solution of the tracking problem for non-linear systems with MPC." In: *Int. J. Systems Science* 36 (2005), pp. 477–484.

[91] G. Marafioti, R. R. Bitmead, and M. Hovd. "Persistently exciting model predictive control." In: *Int. J. Adaptive Control and Signal Processing* 28.6 (2014), pp. 536–552.

[92] I. Markovsky and F. Dörfler. "Behavioral systems theory in data-driven analysis, signal processing, and control." In: *Annual Reviews in Control* 52 (2021), pp. 42–64.

[93] I. Markovsky and F. Dörfler. "Identifiability in the behavioral setting." In: *Available online: http://homepages.vub.ac.be/~imarkovs/publications/identifiability.pdf* (2020).

[94] I. Markovsky and P. Rapisarda. "Data-driven simulation and control." In: *Int. J. Control* 81.12 (2008), pp. 1946–1959.

[95] I. Markovsky and P. Rapisarda. "On the linear quadratic data-driven control." In: *Proc. European Control Conf. (ECC)*. 2007, pp. 5313–5318.

[96] T. Martin and F. Allgöwer. "Data-driven inference on optimal input-output properties of polynomial systems with focus on nonlinearity measures." In: *arXiv:2103.10306* (2021).

[97] T. Martin and F. Allgöwer. "Data-driven system analysis of nonlinear systems using polynomial approximation." In: *arXiv:2108.11298* (2021).

[98] T. Martin and F. Allgöwer. "Dissipativity verification with guarantees for polynomial systems from noisy input-state data." In: *IEEE Control Systems Lett.* 5.4 (2021), pp. 1399–1404.

[99] A. Martinelli, M. Gargiani, M. Draskovic, and J. Lygeros. "Data-driven optimal control of affine systems: a linear programming perspective." In: *arXiv:2203.12044* (2022).

[100] N. Matni, A. Proutiere, A. Rantzer, and S. Tu. "From self-tuning regulators to reinforcement learning and back again." In: *Proc. 58th IEEE Conf. Decision and Control (CDC)*. 2019, pp. 3724–3740.

[101] N. Matni and S. Tu. "A tutorial on concentration bounds for system identification." In: *Proc. 58th IEEE Conf. Decision and Control (CDC)*. 2019, pp. 3741–3749.

[102] T. M. Maupong, J. C. Mayo-Maldonado, and P. Rapisarda. "On Lyapunov functions and data-driven dissipativity." In: *IFAC-PapersOnLine* 50.1 (2017), pp. 7783–7788.

[103] D. Q. Mayne, E. C. Kerrigan, E. van Wyk, and P. Falugi. "Tube-based robust nonlinear model predictive control." In: *Int. J. Robust and Nonlinear Control* 21 (2011), pp. 1341–1353.

[104] D. Q. Mayne, J. B. Rawlings, C. V. Rao, and P. O. M. Scokaert. "Constrained model predictive control: Stability and optimality." In: *Automatica* 36.6 (2000), pp. 789–814.

[105] M. Milanese and C. Novara. "Set membership identification of nonlinear systems." In: *Automatica* 40.6 (2004), pp. 957–975.

[106] M. Milanese and A. Vicino. "Optimal estimation theory for dynamic systems with set membership uncertainty: an overview." In: *Automatica* 27.6 (1991), pp. 997–1009.

[107] V. K. Mishra, I. Markovsky, A. Fazzi, and P. Dreesen. "Data-driven simulation for NARX systems." In: *Proc. European Signal Processing Conference*. 2021, pp. 1055–1059.

[108] V. K. Mishra, I. Markovsky, and B. Grossmann. "Data-driven tests for controllability." In: *IEEE Control Systems Lett.* 5.2 (2021), pp. 517–522.

[109] M. Morari and U. Maeder. "Nonlinear offset-free model predictive control." In: *Automatica* 48.9 (2012), pp. 2059–2067.

[110] B. Nortmann and T. Mylvaganam. "Direct data-driven control of linear time-varying systems." In: *arXiv:2111.02342* (2021).

[111] G. Pan, R. Ou, and T. Faulwasser. "On a stochastic fundamental lemma and its use for data-driven MPC." In: *arXiv:2111.13636* (2021).

[112] D. Papadimitriou, U. Rosolia, and F. Borrelli. "Control of unknown nonlinear systems with linear time-varying MPC." In: *Proc. 59th IEEE Conf. Decision and Control (CDC)*. 2020, pp. 2258–2263.

[113] J. W. Polderman and J. C. Willems. *Introduction to Mathematical Systems Theory*. Springer, New York, 1998.

[114] *Quanser coupled tanks system data sheet*. [Online] `https://www.quanser.com/products/coupled-tanks/`. Accessed: 2022-02-13.

[115] M. Radulescu and S. Radulescu. "Global inversion theorems and applications to differential equations." In: *Nonlinear Analysis, Theory, Methods & Applications* 4.4 (1980), pp. 951–965.

[116] T. Raff, S. Huber, Z. K. Nagy, and F. Allgöwer. "Nonlinear model predictive control of a four tank system: An experimental stability study." In: *Proc. IEEE Int. Conf. Control Applications (CCA)*. 2006, pp. 237–242.

[117] D. M. Raimondo, D. Limón, M. Lazar, L. Magni, and E. F. Camacho. "Min-max model predictive control of nonlinear systems: a unifying overview on stability." In: *European J. Control* 15.1 (2009), pp. 5–21.

[118] J. B. Rawlings, D. Q. Mayne, and M. M. Diehl. *Model Predictive Control: Theory, Computation, and Design*. 3rd printing. Nob Hill Pub, 2020.

[119] J. B. Rawlings and K. R. Muske. "The stability of constrained receding horizon control." In: *IEEE Trans. Automat. Control* 38.10 (1993), pp. 1512–1516.

[120] B. Recht. "A tour of reinforcement learning: The view from continuous control." In: *Annual Review of Control, Robotics, and Autonomous Systems* 2 (2019), pp. 253–279.

[121] M. Rotulo, C. De Persis, and P. Tesi. "Online learning of data-driven controllers for unknown switched linear systems." In: *arXiv:2105.11523* (2021).

[122] J. G. Rueda-Escobedo and J. Schiffer. "Data-driven internal model control of second-order discrete Volterra systems." In: *Proc. 59th IEEE Conf. Decision and Control (CDC)*. 2020, pp. 4572–4579.

[123] J. R. Salvador, D. R. Ramirez, T. Alamo, D. M. de la Pena, and G. Garcia-Marin. "Data driven control: an offset free approach." In: *Proc. European Control Conf. (ECC)*. 2019, pp. 23–28.

[124] J. R. Salvador, D. M. de la Peña, T. Alamo, and A. Bemporad. "Data-based predictive control via direct weight optimization." In: *IFAC-PapersOnLine* 51.20 (2018), pp. 356–361.

[125] A. Sassella, V. Breschi, and S. Formentin. "Learning explicit predictive controllers: theory and applications." In: *arXiv:2108.08412* (2021).

[126] C. W. Scherer. "LPV control and full block multipliers." In: *Automatica* 37.3 (2001), pp. 361–375.

[127] B. Schuermann, N. Kochdumper, and M. Althoff. "Reachset model predictive control for disturbed nonlinear systems." In: *Proc. 57th IEEE Conf. Decision and Control (CDC)*. 2018, pp. 3463–3470.

[128] J. Sjöberg et al. "Nonlinear black-box modeling in system identification: a unified overview." In: *Automatica* 31.12 (1995), pp. 1691–1724.

[129] E. Sontag. *Mathematical Control Theory*. Springer-Verlag, New York, 1998.

[130] R. Sutton and A. Barto. *Reinforcement Learning: An Introduction*. Cambridge, MA, MIT Press, 1998.

[131] M. Tanaskovic, L. Fagiano, C. Novara, and M. Morari. "Data-driven control of nonlinear systems: An on-line direct approach." In: *Automatica* 75 (2017), pp. 1–10.

[132] M. Tanaskovic, L. Fagiano, R. Smith, and M. Morari. "Adaptive receding horizon control for constrained MIMO systems." In: *Automatica* 50.12 (2014), pp. 3019–3029.

[133] E. Terzi, L. Fagiano, M. Farina, and R. Scattolini. "Learning-based predictive control for linear systems: a unitary approach." In: *Automatica* 108 (2019), p. 108473.

[134] J. Umenberger. "Closed-loop data-enabled predictive control." In: *Proc. American Control Conf. (ACC)*. 2021, pp. 3349–3356.

[135] H. J. van Waarde. "Beyond persistent excitation: online experiment design for data-driven modeling and control." In: *IEEE Control Systems Lett.* 6 (2021), pp. 319–324.

[136] H. J. van Waarde and M. K. Camlibel. "A matrix Finsler's Lemma with applications to data-driven control." In: *Proc. 60th IEEE Conf. Decision and Control (CDC)*. 2021, pp. 5777–5782.

[137] H. J. van Waarde, M. K. Camlibel, and M. Mesbahi. "From noisy data to feedback controllers: non-conservative design via a matrix S-lemma." In: *IEEE Trans. Automat. Control* (2020). doi: 10.1109/TAC.2020.3047577.

[138] H. J. van Waarde, M. K. Camlibel, P. Rapisarda, and H. L. Trentelman. "Data-driven dissipativity analysis: application of the matrix S-lemma." In: *IEEE Control Systems Magazine* 42.3 (2022), pp. 140–149.

[139] H. J. van Waarde, C. De Persis, M. K. Camlibel, and P. Tesi. "Willems' Fundamental Lemma for state-space systems and its extension to multiple datasets." In: *IEEE Control Systems Lett.* 4.3 (2020), pp. 602–607.

[140] H. J. van Waarde, J. Eising, H. L. Trentelman, and M. K. Camlibel. "Data informativity: a new perspective on data-driven analysis and control." In: *IEEE Trans. Automat. Control* (2020). doi: 10.1109/TAC.2020.2966717.

[141] C. Verhoek, H. S. Abbas, R. Tóth, and S. Haesart. "Data-driven predictive control for linear parameter-varying systems." In: *IFAC-PapersOnLine* 54.8 (2021), pp. 101–108.

[142] C. Verhoek, R. Tóth, S. Haesart, and A. Koch. "Fundamental Lemma for data-driven analysis of linear parameter-varying systems." In: *Proc. 60th IEEE Conf. Decision and Control (CDC)*. 2021, pp. 5040–5046.

[143] J. C. Willems. "Paradigms and puzzles in the theory of dynamical systems." In: *IEEE Trans. Automat. Control* 36.3 (1991), pp. 259–294.

[144] J. C. Willems, P. Rapisarda, I. Markovsky, and B. De Moor. "A note on persistency of excitation." In: *Syst. Contr. Lett.* 54 (2005), pp. 325–329.

[145] T. M. Wolff, V. G. Lopez, and M. A. Müller. "Data-based moving horizon estimation for linear discrete-time systems." In: *arXiv:2111.04979* (2021).

[146] K. Worthmann, M. W. Mehrez, G. K. I. Mann, R. G. Gosine, and J. Pannek. "Interaction of open and closed loop control in MPC." In: *Automatica* 82 (2017), pp. 243–250.

[147] L. Xu, M. S. Turan, B. Guo, and G. Ferrari Trecate. "A data-driven convex programming approach to worst-case robust tracking controller design." In: *arXiv:2102.11918* (2021).

[148] L. Xu, M. S. Turan, B. Guo, and G. Ferrari Trecate. "Non-conservative design of robust tracking controllers based on input-output data." In: *Proc. 3rd Conference on Learning for Dynamics and Control*. Vol. 144. PMLR, 2021, pp. 138–149.

[149] A. Xue and N. Matni. "Data-driven system level synthesis." In: *Proc. 3rd Conference on Learning for Dynamics and Control*. Vol. 144. PMLR, 2021, pp. 189–200.

[150] H. Yang and S. Li. "A data-driven predictive controller design based on reduced hankel matrix." In: *Proc. Asian Control Conference*. 2015, pp. 1–7.

[151] M. Yin, A. Iannelli, and R. S. Smith. "Data-driven prediction with stochastic data: confidence regions and minimum mean-squared error estimates." In: *arXiv:2111.04789* (2021).

[152] M. Yin, A. Iannelli, and R. S. Smith. "Maximum likelihood estimation in data-driven modeling and control." In: *IEEE Trans. Automat. Control* (2021). doi: 10.1109/TAC.2021.3137788.

[153] M. Yin, A. Iannelli, and R. S. Smith. "Maximum likelihood signal matrix model for data-driven predictive conntrol." In: *Proc. 3rd Conference on Learning for Dynamics and Control*. Vol. 144. PMLR, 2021, pp. 1004–1014.

[154] S. Yu, M. Reble, H. Chen, and F. Allgöwer. "Inherent robustness properties of quasi-infinite horizon nonlinear model predictive control." In: *Automatica* 50.9 (2014), pp. 2269–2280.

[155] Y. Yu, S. Talebi, H. J. van Waarde, U. Topcu, M. Mesbahi, and B. Açıkmeşe. "On controllability and persistency of excitation in data-driven control: extensions of Willems' Fundamental Lemma." In: *Proc. 60th IEEE Conf. Decision and Control (CDC)*. 2021.

[156] Z. Yuan and J. Cortés. "Data-driven optimal control of bilinear systems." In: *arXiv:2112.15510* (2021).

[157] A. Zanelli, Q. Tran-Dinh, and M. Diehl. "A Lyapunov function for the combined system-optimizer dynamics in inexact model predictive control." In: *Automatica* 134 (2021), p. 109901.

[158] M. Zanon and S. Gros. "Safe reinforcement learning using robust MPC." In: *IEEE Trans. Automat. Control* 66.8 (2021), pp. 3638–3652.

[159] Y. Zheng, L. Furieri, M. Kamgarpour, and N. Li. "Sample complexity of linear quadratic gaussian (LQG) control for output feedback systems." In: *Proc. 3rd Conference on Learning for Dynamics and Control*. Vol. 144. PMLR, 2021, pp. 559–570.

[160] K. Zhou, J. C. Doyle, and K. Glover. *Robust and optimal control*. Prentice-Hall, Inc., Englewood Cliffs, N.J., 1996.

Publications of the Author

[JB1] M. Alsalti, J. Berberich, V. G. Lopez, F. Allgöwer, and M. A. Müller. "Data-based system analysis and control of flat nonlinear systems." In: *Proc. 60th IEEE Conf. Decision and Control (CDC)*. 2021, pp. 1484–1489.

[JB2] J. Berberich and F. Allgöwer. "A trajectory-based framework for data-driven system analysis and control." In: *Proc. European Control Conf. (ECC)*. 2020, pp. 1365–1370.

[JB3] J. Berberich, J. W. Dietrich, R. Hoermann, and M. A. Müller. "Mathematical modeling of the pituitary-thyroid feedback loop: role of a TSH-T_3-shunt and sensitivity analysis." In: *Frontiers in Endocrinology* 9 (2018), p. 91.

[JB4] J. Berberich, A. Koch, C. W. Scherer, and F. Allgöwer. "Robust data-driven state-feedback design." In: *Proc. American Control Conf. (ACC)*. 2020, pp. 1532–1538.

[JB5] J. Berberich, J. Köhler, F. Allgöwer, and M. A. Müller. "Dissipativity properties in constrained optimal control: a computational approach." In: *Automatica* 114 (2020), p. 108840.

[JB6] J. Berberich, J. Köhler, F. Allgöwer, and M. A. Müller. "Indefinite linear quadratic optimal control: Strict dissipativity and turnpike properties." In: *IEEE Control Systems Lett.* 2.3 (2018), pp. 399–404.

[JB7] J. Berberich, J. Köhler, M. A. Müller, and F. Allgöwer. "Data-driven model predictive control with stability and robustness guarantees." In: *IEEE Trans. Automat. Control* 66.4 (2021), pp. 1702–1717.

[JB8] J. Berberich, J. Köhler, M. A. Müller, and F. Allgöwer. "Data-driven model predictive control: closed-loop guarantees and experimental results." In: *at-Automatisierungstechnik* 69.7 (2021), pp. 608–618.

[JB9] J. Berberich, J. Köhler, M. A. Müller, and F. Allgöwer. "Data-driven tracking MPC for changing setpoints." In: *IFAC-PapersOnLine* 53.2 (2020), pp. 6923–6930.

[JB10] J. Berberich, J. Köhler, M. A. Müller, and F. Allgöwer. "Linear tracking MPC for nonlinear systems part I: the model-based case." In: *IEEE Trans. Automat. Control* (2022). doi: 10.1109/TAC.2022.3166872.

[JB11] J. Berberich, J. Köhler, M. A. Müller, and F. Allgöwer. "Linear tracking MPC for nonlinear systems part II: the data-driven case." In: *IEEE Trans. Automat. Control* (2022). doi: 10.1109/TAC.2022.3166851.

[JB12] J. Berberich, J. Köhler, M. A. Müller, and F. Allgöwer. "On the design of terminal ingredients for data-driven MPC." In: *IFAC-PapersOnLine* 54.6 (2021), pp. 257–263.

[JB13] J. Berberich, J. Köhler, M. A. Müller, and F. Allgöwer. "Robust constraint satisfaction in data-driven MPC." In: *Proc. 59th IEEE Conf. Decision and Control (CDC)*. 2020, pp. 1260–1267.

[JB14] J. Berberich, C. W. Scherer, and F. Allgöwer. "Combining prior knowledge and data for robust controller design." In: *arXiv:2009.05253* (2020).

[JB15] J. Berberich, M. Sznaier, and F. Allgöwer. "Signal estimation and system identification with nonlinear dynamic sensors." In: *Proc. Conference on Control Technology and Applications*. 2019, pp. 505–510.

[JB16] J. Berberich, S. Wildhagen, M. Hertneck, and F. Allgöwer. "Data-driven analysis and control of continuous-time systems under aperiodic sampling." In: *IFAC-PapersOnLine* 54.7 (2021), pp. 210–215.

[JB17] J. Bongard, J. Berberich, J. Köhler, and F. Allgöwer. "Robust stability analysis of a simple data-driven model predictive control approach." In: *IEEE Trans. Automat. Control* (2022). doi: 10.1109/TAC.2022.3163110.

[JB18] C. Klöppelt, J. Berberich, F. Allgöwer, and M. A. Müller. "A novel constraint tightening approach for robust data-driven predictive control." In: *arXiv:2203.07055* (2022).

[JB19] A. Koch, J. Berberich, and F. Allgöwer. "Provably robust verification of dissipativity properties from data." In: *IEEE Trans. Automat. Control* (2021). doi: 10.1109/TAC.2021.3116179.

[JB20] A. Koch, J. Berberich, and F. Allgöwer. "Verifying dissipativity properties from noise-corrupted input-state data." In: *Proc. 59th IEEE Conf. Decision and Control (CDC)*. 2020, pp. 616–621.

[JB21] A. Koch, J. Berberich, J. Köhler, and F. Allgöwer. "Determining optimal input-output properties: A data-driven approach." In: *Automatica* 134 (2021), p. 109906.

[JB22] J. Köhler, L. Schwenkel, A. Koch, J. Berberich, P. Pauli, and F. Allgöwer. "Robust and optimal predictive control of the COVID-19 outbreak." In: *Annual Reviews in Control* 51 (2021), pp. 525–539.

[JB23] M. Köhler, J. Berberich, M. A. Müller, and F. Allgöwer. "Data-driven distributed MPC of dynamically coupled linear systems." In: *Proc. 24th Int. Symp. Mathematical Theory of Networks and Systems (MTNS)*. to appear. 2022.

[JB24] P. Pauli, D. Gramlich, J. Berberich, and F. Allgöwer. "Linear systems with neural network nonlinearities: improved stability analysis via acausal Zames-Falb multipliers." In: *Proc. 60th IEEE Conf. Decision and Control (CDC)*. 2021, pp. 3611–3618.

[JB25] P. Pauli, A. Koch, J. Berberich, P. Kohler, and F. Allgöwer. "Training robust neural networks using Lipschitz bounds." In: *IEEE Control Systems Lett.* 6 (2022), pp. 121–126.

[JB26] P. Pauli, J. Köhler, J. Berberich, A. Koch, and F. Allgöwer. "Offset-free setpoint tracking using neural network controllers." In: *Proc. 3rd Conference on Learning for Dynamics and Control*. Vol. 144. PMLR, 2021, pp. 992–1003.

[JB27] A. Romer, J. Berberich, J. Köhler, and F. Allgöwer. "One-shot verification of dissipativity properties from input-output data." In: *IEEE Control Systems Lett.* 3.3 (2019), pp. 709–714.

[JB28] R. Strässer, J. Berberich, and F. Allgöwer. "Data-driven control of nonlinear systems: beyond polynomial dynamics." In: *Proc. 60th IEEE Conf. Decision and Control (CDC)*. 2021, pp. 4344–4351.

[JB29] J. Venkatasubramanian, J. Köhler, J. Berberich, and F. Allgöwer. "Robust dual control based on gain scheduling." In: *Proc. 59th IEEE Conf. Decision and Control (CDC)*. 2020, pp. 2270–2277.

[JB30] X. Wang, J. Berberich, J. Sun, G. Wang, F. Allgöwer, and J. Chen. "Data-driven control of event- and self-triggered discrete-time systems." In: *arXiv:2202.08019* (2022).

[JB31] X. Wang, J. Sun, J. Berberich, G. Wang, F. Allgöwer, and J. Chen. "Data-driven control of dynamic event-triggered systems with delays." In: *arXiv:2110.12768* (2021).

[JB32] N. Wieler, J. Berberich, A. Koch, and F. Allgöwer. "Data-driven controller design via finite-horizon dissipativity." In: *Proc. 3rd Conference on Learning for Dynamics and Control*. Vol. 144. PMLR, 2021, pp. 287–298.

[JB33] S. Wildhagen, J. Berberich, M. Hertneck, and F. Allgöwer. "Data-driven analysis and controller design for discrete-time systems under aperiodic sampling." In: *IEEE Trans. Automat. Control* (2022). doi: 10.1109/TAC.2022.3183969.

[JB34] S. Wildhagen, J. Berberich, M. Hirche, and F. Allgöwer. "Improved stability conditions for systems under aperiodic sampling: model- and data-based analysis." In: *Proc. 60th IEEE Conf. Decision and Control (CDC)*. 2021, pp. 5795–5801.